DATABASES AND INFORMATION SYSTEMS

Frontiers in Artificial Intelligence and Applications

FAIA covers all aspects of theoretical and applied artificial intelligence research in the form of monographs, doctoral dissertations, textbooks, handbooks and proceedings volumes. The FAIA series contains several sub-series, including "Information Modelling and Knowledge Bases" and "Knowledge-Based Intelligent Engineering Systems". It also includes the biannual ECAI, the European Conference on Artificial Intelligence, proceedings volumes, and other ECCAI – the European Coordinating Committee on Artificial Intelligence – sponsored publications. An editorial panel of internationally well-known scholars is appointed to provide a high quality selection.

Volume 118

Recently published in this series

Vol. 117. L. Castillo et al. (Eds.), Planning, Scheduling and Constraint Satisfaction: From Theory to Practice
Vol. 116. O. Corcho, A Layered Declarative Approach to Ontology Translation with Knowledge Preservation
Vol. 115. G. Phillips-Wren and L. Jain (Eds.), Intelligent Decision Support Systems in Agent-Based Environments
Vol. 114. A.C. Varzi and L. Vieu (Eds.), Formal Ontology in Information Systems – Proceedings of the Third International Conference (FOIS-2004)
Vol. 113. J. Vitrià et al. (Eds.), Recent Advances in Artificial Intelligence Research and Development
Vol. 112. W. Zhang and V. Sorge (Eds.), Distributed Constraint Problem Solving and Reasoning in Multi-Agent Systems
Vol. 111. H. Fujita and V. Gruhn (Eds.), New Trends in Software Methodologies, Tools and Techniques – Proceedings of the Third SoMeT_W04
Vol. 110. R. López de Mántaras and L. Saitta (Eds.), ECAI 2004 – Proceedings of the 16th European Conference on Artificial Intelligence
Vol. 109. E. Onaindia and S. Staab (Eds.), STAIRS 2004 – Proceedings of the Second Starting AI Researchers' Symposium
Vol. 108. V. Stefanuk and K. Kaijiri (Eds.), Knowledge-Based Software Engineering – Proceedings of the Sixth Joint Conference on Knowledge-Based Software Engineering
Vol. 107. F. Darses et al. (Eds.), Cooperative Systems Design – Scenario-Based Design of Collaborative Systems
Vol. 106. D. Bourcier (Ed.), Legal Knowledge and Information Systems – JURIX 2003: The Sixteenth Annual Conference
Vol. 105. Y. Kiyoki et al. (Eds.), Information Modelling and Knowledge Bases XV
Vol. 104. A. Abraham et al. (Eds.), Design and Application of Hybrid Intelligent Systems
Vol. 103. B. Tessem et al. (Eds.), Eighth Scandinavian Conference on Artificial Intelligence – SCAI'03
Vol. 102. C. Turchetti, Stochastic Models of Neural Networks
Vol. 101. G.L. Torres et al. (Eds.), Advances in Intelligent Systems and Robotics – LAPTEC'03
Vol. 100. I. Aguiló et al. (Eds.), Artificial Intelligence Research and Development
Vol. 99. D. Šuc, Machine Reconstruction of Human Control Strategies

ISSN 0922-6389

Databases and Information Systems

Selected Papers from the Sixth International Baltic Conference DB&IS'2004

Edited by

Janis Barzdins

Institute of Mathematics and Computer Science, University of Latvia, Riga, Latvia

and

Albertas Caplinskas

Institute of Mathematics and Informatics, Vilnius, Lithuania

IOS
Press

Amsterdam • Berlin • Oxford • Tokyo • Washington, DC

ISBN 1 58603 485 5
Library of Congress Control Number: 2004116333

Publisher
IOS Press
Nieuwe Hemweg 6B
1013 BG Amsterdam
The Netherlands
fax: +31 20 620 3419
e-mail: order@iospress.nl

Distributor in the UK and Ireland
IOS Press/Lavis Marketing
73 Lime Walk
Headington
Oxford OX3 7AD
England
fax: +44 1865 750079

Distributor in the USA and Canada
IOS Press, Inc.
4502 Rachael Manor Drive
Fairfax, VA 22032
USA
fax: +1 703 323 3668
e-mail: iosbooks@iospress.com

PRINTED IN THE NETHERLANDS

Conference Committee

Advisory Committee

Janis Bubenko, Sweden
Arne Solvberg, Norway

Programme Co-Chairs

Janis Barzdins, Latvia
Albertas Caplinskas, Lithuania
Rusins Freivalds, Latvia

Organising Co-Chairs

Juris Borzovs, Latvia
Inara Opmane, Latvia

Co-ordinators

Ahto Kalja, Estonia
Saulius Maskeliunas, Lithuania

Programme Committee

Witold Abramowicz, Poland
Mikhail Auguston, USA
Janis Bicevskis, Latvia
Johann Eder, Austria
Hans-Dieter Ehrich, Germany
Jorgen Fischer Nilsson, Denmark
Janis Grundspenkis, Latvia
Remigijus Gustas, Sweden
Hele-Mai Haav, Estonia
Igor Kabashkin, Latvia
Leonid Kalinichenko, Russia
Ahto Kalja, Estonia
Audris Kalnins, Latvia
Marite Kirikova, Latvia
Patrick Lambrix, Sweden
Jozef M. Zurada, USA
Saulius Maskeliunas, Lithuania
Mihhail Matskin, Norway
Boris Novikov, Russia
Monika Oit, Estonia
Algirdas Pakstas, UK
Jaan Penjam, Estonia
Jaroslav Pokorny, Czech Republic
Gunter Saake, Germany
Klaus-Dieter Schewe, New Zealand
Jaak Tepandi, Estonia
Bernhard Thalheim, Germany
Enn Tyugu, Estonia
Olegas Vasilecas, Lithuania
Benkt Wangler, Sweden
Mudasser Wyne, USA
Arkady Zaslavsky, Australia

Additional Referees

Peter Ahlbrecht, Germany
Arne Ansper, Estonia
Per Backlund, Sweden
Dmitry Barashev, Russia
Asa Dahlstedt, Sweden
Vahur Kotkas, Estonia
Algirdas Laukaitis, Lithuania
Marek Lehmann, Austria
Karl Neumann, Germany
Karlis Podnieks, Latvia

Rainer Eckstein, Germany
Silke Eckstein, Germany
Janis Eiduks, Latvia
Mohamed Medhat Gaber, Australia
Mika Hirvensalo, Finland
Hagen Hoepfner, Germany
Vaida Jakoniene, Sweden
Paulis Kikusts, Latvia
Markus Kirchberg, New Zealand
Christian Koncilia, Austria
Oleg Proskurnin, Russia
Eike Schallehn, Germany
Ingo Schmitt, Germany
Asko Seeba, Estonia
Andrey Simanovsky, Russia
Eva Soderstrom, Sweden
Mattias Strand, Sweden
Alexei Tretiakov, New Zealand
Viljar Tulit, Estonia
Juris Viksna, Latvia

Local Organising Committee

Janis Barzdins
Janis Benefelds
Ansis Ataols Berzins
Janis Bicevskis
Uldis Bojars
Juris Borzovs, co-chair
Edgars Celms
Andrejs Dubrovskis
Anita Ermusa
Rusins Freivalds
Martins Gills
Ija Haritonova
Maksims Kravcevs
Davis Kulis
Lelde Lace
Ilvars Mizniks
Laila Niedrite
Inara Opmane, co-chair
Martins Opmanis
Raimonds Praude
Oksana Scegulnaja-Dubrovska
Darja Smite
Agnis Stibe
Maris Treimanis
Valdis Vitolins
Aleksandrs Zelenkovs
Janis Zuters

Conference Secretary

Edgars Celms, Latvia

Preface

This volume contains revised versions of best papers presented at the 6th International Baltic Conference on Databases and Information Systems (Baltic DB&IS'2004). The papers present original results in database technologies, database applications, data mining, data warehousing, model transformations (MDA), specifications, ontologies, and information systems. The intended audience for this book are researchers and practitioners who are interested in advanced topics on databases and information systems.

The series of Baltic DB&IS biennial conferences began in 1994. The first conference was held in Trakai (1994), then followed conferences in Tallinn (1996), Riga (1998), Vilnius (2000), Tallinn (2002), and again in Riga .

During this period the Baltic DB&IS conferences have become real international forums that bring together academics and practitioners from many countries and provide an open forum for exchange of ideas in novel fields of databases and information systems. More information concerning these conferences you can find in the Opening Talk of Janis Bubenko, published in this volume. We are very thankful to Professor Janis Bubenko, who was one of the initiators of these conferences.

The 6th International Baltic Conference on Databases and Information Systems (Baltic DB&IS'2004) took place on June 6 – 9, 2004 in Riga, Latvia. The conference was organized by the Institute of Mathematics and Computer Science (University of Latvia), Department of Computer Science (University of Latvia), the Riga Information Technology Institute (Latvia) and the Institute of Mathematics and Informatics (Lithuania). The conference was approved by the IEEE Communication Society for Technical Co-Sponsorship.

The International Programme Committee representing 14 countries received 59 submissions from 17 countries for the main conference, and 27 papers from 5 countries for the Doctoral Consortium. Each paper was reviewed by 3 referees from different countries. As a result, 35 papers were accepted for the main conference and included in the local conference proceedings. After the conference the Programme Committee selected the best 22 papers to be published in this book.

Many people contributed to the organization of the Baltic DB&IS'2004. We express our warmest thanks to all authors who contributed to the conference. We are very grateful to the members of Programmee Committee and the additional referees for carefully reviewing the submissions. We thank all the organizing team led by Juris Borzovs, the Riga Information Technology Institute, and Inara Opmane, the Institute of Mathematics and Computer Science, University of Latvia. We are obliged to all the supporting institutions and sponsors, namely, the University of Latvia, VLDB Baltic Fund, Unibanka, DATI, Exigen, IT Alise, and Lattelekom.

We express our special thanks to Edgars Celms for all his technical assistance during the preparation of the manuscript.

September 2004

Janis Barzdins
Albertas Caplinskas

Opening talk at the 6th Baltic Conference on Databases and Information Systems

Janis Bubenko jr

Dear colleagues,

It is a great honour and pleasure for me to give the opening address at the 6th Baltic conference on Databases and Information Systems in Riga.

When the conference was initiated in 1994 in Trakai, Lithuania, we hoped it would survive a number of years and alternate each second year between the three Baltic countries. It is now the sixth time the conference is arranged, and the second time it is held in Riga, Latvia. The quality and content of the conference has indeed grown over our initial expectations.

Let me first note some changes the conference has experienced since its start. Firstly, the name and scope has changed. In 1994 the conference was called *The 1994 Baltic Workshop on National Infrastructure Databases – problems, methods, and experiences*. The conference addressed researchers as well as practitioners in the field of databases for public administration in the Baltic states. Such databases are operating as vital components of national infrastructures. The workshop in 1994 gathered professionals from the Baltic states as well as experts from the rest of the world to discuss their findings and to establish ties, relationships, and cooperation in the future.

A particular, additional objective of the workshop was to stimulate contacts and cooperation between professionals in the Baltic states themselves. It was believed that the topic area of national databases for public administration were essential building blocks in the infrastructures of these new democracies.

Today the name has changed to *Baltic Conference on Databases and Information Systems*. The scope of the conference has widened considerably. It now includes many issues and topics of the academic field of "Informatics" (or Computer Science). Also, as I can see from the programme, it includes a mixture of theoretical as well as practical, engineering type and empirical contributions. In this way the conference has become a forum for a larger community of Baltic professionals and scholars in the area of Information and Communication Technologies (ICT). Also, the fraction of non-Baltic, international contributors has been growing steadily.

Of a particular interest is, I feel, the specific consideration of our young doctoral candidates in the midst of their educational career. I am, of course, thinking of the *doctoral consortium* arranged in connection with the conference. A consortium of this kind gives young researchers excellent possibilities to meet doctoral candidates and professors from other countries and to stimulate each other's work with constructive critique and discussions. I see this initiative as a very promising development of the Baltic Conference on Databases and Information Systems of the future.

In 1994 the conference was attended largely by invitation. About one third of the speakers were invited international experts, one third were invited Baltic professionals, and one third were selected papers contributed by primarily Baltic professionals. This pattern has radically changed. Today practically all papers to be presented are selected from a set of submitted papers by a thorough reviewing process. The programme committee includes well known researchers from a number of countries. This makes the Baltic Databases and Information Systems a truly international conference, but in a Baltic setting. From the programme I can see that the conference has attracted submissions from a large number of countries. From a Baltic point of view this is, of course, very encouraging indeed. It is what we hoped for when the conference series was started in 1994.

One important objective set in 1994 was to increase the cooperation between researchers in databases and in information systems in the Baltic states. It is perhaps not easy to measure to what extent this has actually happened, but my firm opinion, after having observed the conference five times, is that it is well under way. Many new contacts have been made and old ones have deepened during the last 10 years. What I could wish for in the future is more direct collaboration in day-to-day research work, and in research projects. I would also love to see more international projects which include members of all three Baltic nations, as well as members from other European nations.

The seed we planted 10 years ago has grown into a small tree. The soil is good and the tree is growing well, but as for all plants it needs constant attendance and continued care. I am fully confident the Baltic conference organisers, with the enthusiastic contribution of new, young Baltic researchers in databases and in information systems, will continue to cultivate this important Baltic entity for many years to come.

There is another important distinction between the conference in 1994 and in 2004. In 1994 most of us needed a visa to enter a Baltic country. Today we all are members of the European union. What a fantastic progress and what a wealth of possibilities this opens up! Who could believe this development 10 years ago?

An additional interesting development is that the sixteenth international *Conference on Advanced information Systems Engineering*, CAiSE, has this year come to Riga, Latvia. CAiSE was started in 1989 in Stockholm, Sweden. It has since then been held in most west-European countries. This is the first time CAiSE is being arranged in an East European country.

The CAiSE event, together with the Baltic Databases and Information Systems conference, I am convinced, really puts the Baltic states on the map in the discipline of databases and information systems. It is, therefore, a great pleasure and satisfaction for me to wish you all a very interesting and rewarding conference 2004 in Riga, Latvia.

Contents

Specifications

Modeling and Model Transformations

Ontologies

Information Systems

Database Integration

Databases and Information Systems
J. Barzdins and A. Caplinskas (Eds.)
IOS Press, 2005

Generating and Optimising Views from Both as View Data Integration Rules

Edgar JASPER[1], Nerissa TONG[2], Peter McBRIEN[2], and Alexandra POULOVASSILIS[1]
[1] *School of Computer Science and Information Systems, Birkbeck College, Univ. of London,* {edgar,ap}@dcs.bbk.ac.uk
[2] *Dept. of Computing, Imperial College,* {nnyt98,pjm}@doc.ic.ac.uk

Abstract

This paper describes the generation and logical optimisation of views in the AutoMed heterogeneous data integration framework, which is based on the use of reversible schema transformation sequences called both as view (BAV) rules. We show how views can be generated from such sequences, for global as view (GAV), local as view (LAV) and GLAV query processing. We also present techniques for optimising these generated views, firstly by optimising the transformation sequences, and secondly by optimising the view definitions generated from them.

1 Introduction

Data integration is a process by which several databases, with associated **local schemas**, are integrated to form a single virtual database with an associated **global schema**. The two most common data integration approaches are **global as view (GAV)** (used in TSIMMIS [4], InterViso [19] and Garlic [18]), and **local as view (LAV)** (used in IM [9] and Agora [11]). In GAV, the constructs of a global schema are described as views over the local schemas. These view definitions are used to rewrite queries over a global schema into distributed queries over the local databases. In LAV, the constructs of the local schemas are defined as views over the global schema, and processing queries over the global schema involves rewriting queries using views [8].

Both LAV and GAV lack a certain degree of expressiveness. GAV is unable to fully capture data integration semantics where a source schema construct can be defined by a non-reversible function over global schema constructs. For example, if source schema attribute money is the sum of global schema attributes coins and notes, neither coins nor notes in the global schema can be defined by views over the source schema. Thus a query on the global schema asking for the sum of coins and notes cannot be answered even though the answer (money) is present in the source schema. In LAV, the attribute money can be defined by a view as the sum of global schema attributes coins and notes. Reversing the presence of the attributes, so that coins and notes are in the local schema and money in the global schema, leads to a situation which GAV can express but LAV cannot.

GLAV [5] is a variation of LAV that allows the head of the view definition rules to contain conjunctions of relations from a source schema as a natural join, and is thus able to capture situations where a non-reversible function is a natural-join between attributes. In [10] GLAV was extended to allow any source schema query in the head of the rule, and hence is able to express the case where a single source schema is used to define the global schema constructs referenced in the body of the rule.

We have developed a richer integration framework which is based on the use of reversible sequences of primitive schema transformations, called transformation **pathways**. In [15] we showed how these pathways incorporate the semantics of GAV rule definitions and LAV rule definitions, and hence termed our approach **both as view** (**BAV**). The BAV data integration approach has been implemented within the AutoMed system (the software and documentation being available at `http://www.doc.ic.uk/automed`).

Since BAV integration is based on sequences of primitive schema transformations, it could be argued that the pathways resulting from BAV are likely to be more costly to reason with and process (*e.g.* for global query processing) than the corresponding LAV, GAV or GLAV view definitions would be. However, in Section 5 of this paper we show how BAV pathways are amenable to considerable simplification. Moreover, standard query optimisation techniques can be applied to the view definitions derived from BAV pathways.

The outline of this paper is as follows. Section 2 gives a review and examples of the BAV integration approach, and compares it with the GAV, LAV and GLAV approaches. Section 3 shows how view definitions can be generated from BAV pathways for GAV, LAV or GLAV query processing. Section 4 presents techniques for optimising these generated views, and Section 5 gives techniques for optimising the BAV pathways themselves. Section 6 gives our concluding remarks and directions of further work.

2 The BAV Integration Approach

In previous work (see `http://www.doc.ic.uk/automed` for publications to download) we have developed a framework to support schema transformation and integration in heterogeneous database architectures. The framework consists of a low-level **hypergraph-based data model (HDM)** and a set of primitive schema transformations defined for this model. Higher-level data models and primitive schema transformations for them are defined in terms of this lower-level common data model.

In BAV, schemas are incrementally transformed by applying a sequence of primitive transformations $t_1, \ldots, t_r$, where each t_i adds, deletes or renames just one schema construct. Each add or delete transformation is accompanied by a query, expressed in the **intermediate query language** (**IQL**), specifying the extent of the new or deleted construct in terms of the rest of the constructs in the schema. All primitive transformations have an optional additional argument which specifies a constraint (also expressed in the IQL) on the current schema extension that must hold if the transformation is to be applied.

A **composite transformation** is a sequence of primitive transformations. We term the composite transformation defined for transforming schema S_1 to schema S_2 a transformation **pathway** $S_1 \rightarrow S_2$. All source schemas, intermediate schemas and global schemas, and the pathways between them are stored in AutoMed's metadata repository [1].

AutoMed supports a variety of methodologies for performing data integration and hence forming a **network** of pathways joining schemas together. For example, Figure 1 illustrates the integration of n local schemas, $LS_1, \ldots, LS_n$, into a global schema GS. In order to integrate these n local schemas, each LS_i is first transformed into a "union" schema US_i. These n union schemas are syntactically identical, and this is asserted by creating a sequence of id transformation steps between each pair US_i and US_{i+1}, of the form id $(US_i : c, US_{i+1} : c)$ for each schema construct.

id is an additional type of primitive transformation, and the notation $US_i : c$ is used to denote construct c appearing in schema US_i. These id transformations are generated auto-

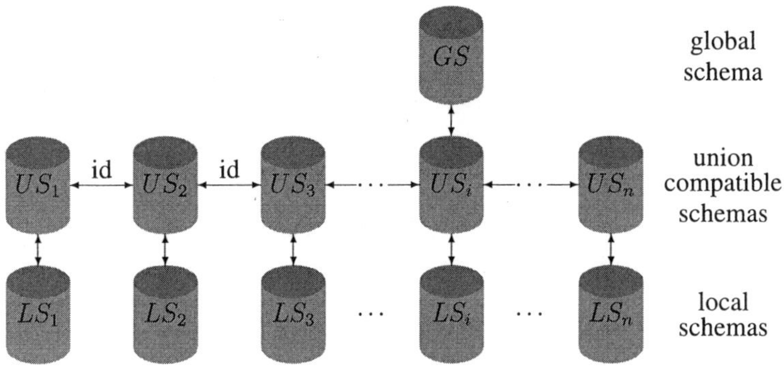

Figure 1: A general AutoMed Integration

matically by the AutoMed software. An arbitrary one of the US_i can then be selected for further transformation into a global schema GS. This is where constructs sourced from different local schemas can be combined together by unions, joins, outer-joins *etc*.

There may be information within a US_i which is not semantically derivable from the corresponding LS_i. This is asserted by means of extend transformation steps within the pathway $LS_i \rightarrow US_i$. Conversely, not all of the information within a local schema LS_i need be transferred into US_i, and this is asserted by means of contract transformation steps within $LS_i \rightarrow US_i$. These extend and contract transformations behave in the same way as add and delete, respectively, except that they indicate that only partial information can be derived about the new or deleted construct. Rather than a single query, they take a pair of queries which specify a lower and upper bound on the extent of the new or deleted construct. The lower bound query may be the constant Void if no lower bound can be specified, and the upper bound query may be the constant Any if no upper bound can be specified.

Each primitive transformation t has an **automatically derivable** reverse transformation $\bar{t}$. In particular, each add or extend transformation is reversed by a delete or contract transformation with the same arguments, and vice versa, while each rename or id transformation is reversed by another rename or id transformation with the two arguments swapped. This holds for the primitive transformations of **any** modelling language defined in AutoMed. In [12] we show how this reversibility of schema transformations allows automatic data query translation between schemas.

In [13] we described how our framework can be applied to different high-level modelling languages such as relational, ER and UML, and more recently we have extended AutoMed to also support semi-structured data models (flat file, XML, RDF). For our examples in this paper we will use a simplified relational data model. However, we stress that the techniques that we describe here are equally applicable to any data modelling language supported by AutoMed.

In our simple relational model, there are two kinds of schema construct: Rel and Att (see [13] for an encoding of a richer relational data model, including the modelling of constraints).

The extent of a Rel construct $\langle\langle R \rangle\rangle$ is the projection of the relation R onto its primary key attributes $k_1, ..., k_n$. The extent of each Att construct $\langle\langle R, a \rangle\rangle$ where a is an attribute (key or non-key) is the projection of relation R onto $k_1, ..., k_n, a$. For example, a relation student(id,sex,dname) would be modelled by a Rel construct $\langle\langle \mathsf{student} \rangle\rangle$, and three Att constructs $\langle\langle \mathsf{student, id} \rangle\rangle$, $\langle\langle \mathsf{student, sex} \rangle\rangle$ and $\langle\langle \mathsf{student, dname} \rangle\rangle$.

Once the constructs of modelling language $\mathcal{M}$ have been defined in terms of the HDM (via the API of AutoMed's metadata repository [1]), a set of primitive schema transformations

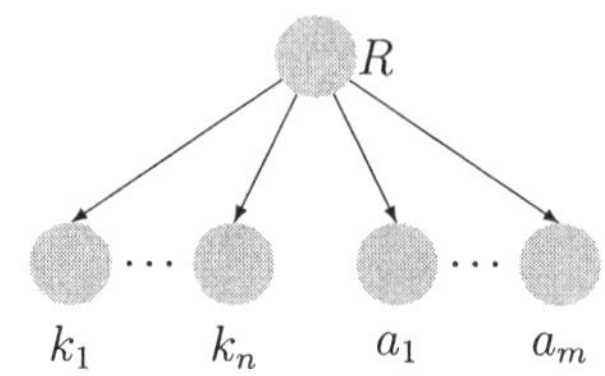

Figure 2: A simple relational data model

for $\mathcal{M}$ are automatically available. For the simple relational model above, these would be as follows:

- addRel($\langle\langle R\rangle\rangle, q$) adds to the schema a new relation R. The query q specifies the set of primary key values in the extent of R in terms of the already existing schema constructs.
- addAtt($\langle\langle R, a\rangle\rangle, q$) adds to the schema an attribute a (key or non-key) for relation R. The query q specifies the extent of the binary relationship between the primary key attribute(s) of R and this new attribute a in terms of the already existing schema constructs.
- deleteRel($\langle\langle R\rangle\rangle, q$) deletes from the schema the relation R (provided all its attributes have first been deleted). The query q specifies how the extent of R can be restored from the remaining schema constructs.
- deleteAtt($\langle\langle R, a\rangle\rangle, q$) deletes from the schema attribute a of relation R. The query q specifies how the extent of the binary relationship between the primary key attribute(s) of R and a can be restored from the remaining schema constructs.
- renameRel($\langle\langle R\rangle\rangle, \langle\langle R'\rangle\rangle$) renames the relation R to R' in the schema.
- renameAtt($\langle\langle R, a\rangle\rangle, \langle\langle R, a'\rangle\rangle$) renames the attribute a of R to a'.

There is also a set of extendRel, extendAtt, contractRel and contractAtt primitive transformations.

2.1 *An Example Integration*

Figure 3 gives some specific schemas to illustrate the integration approach of Figure 1. Primary key attributes are underlined, foreign key attributes are in italics and nullable attributes are suffixed by a question mark.

In Example 1, transformations t_1–t_5 use a composite transformation extendTable to state that the tables student, enrolled, university, campus and degree in US_1 cannot be derived from LS_1. The definition of extendTable is:

$$\text{extendTable}(\langle\langle R, a_1, \ldots, a_n\rangle\rangle) = \text{extendRel}(\langle\langle R\rangle\rangle, \text{Void}, \text{Any});$$
$$\text{extendAtt}(\langle\langle R, a_1\rangle\rangle, \text{Void}, \text{Any}); \ldots; \text{extendAtt}(\langle\langle R, a_n\rangle\rangle, \text{Void}, \text{Any})$$

Then transformations t_6–t_9 use the dname attribute of staff to derive the dept table in US_1, and use extend transformations for the two attributes street and uname that cannot be derived from LS_1. Finally, in t_{10}–t_{14} the male and female relations of LS_1 are restructured into the single sex attribute of staff.

The queries accompanying the add and delete transformations are expressed in our IQL intermediate query language. In IQL, ++ is the bag union operator and the construct $[e \mid Q_1; \ldots Q_n]$ is a **comprehension** [2]. The expressions Q_1 to Q_n are termed **qualifiers**, each

LS_1	staff(<u>id</u>,name,dname) male(<u>*id*</u>) female(<u>*id*</u>)
LS_2	university(uname) campus(<u>cmname</u>,*uname*) dept(<u>deptname</u>,*cmname*) degree(<u>dcode</u>,title,dtype,*deptname*)
LS_3	student(<u>id</u>,name,sex) enrolled(<u>*id*,from</u>,to,*dcode*) degree(<u>dcode</u>)
LS_4	university(uname) college(<u>cname</u>,*uname*) dept(<u>dname</u>,street,*cname*)
US_i	university(<u>uname</u>) campus(<u>cmname</u>,*uname*) dept(<u>dname</u>,street,*cmname*) degree(<u>dcode</u>,title,dtype,*dname*) staff(<u>id</u>,name,sex,*dname*) student(<u>id</u>,name,sex) enrolled(<u>*id*,from</u>,to,*dcode*)
GS	university(<u>uname</u>) campus(<u>cmname</u>,*uname*) dept(<u>dname</u>,street,*cmname*) degree(<u>dcode</u>,title,dtype,*dname*) person(<u>id</u>,name,sex,*dname?*) enrolled(<u>*id*,from</u>,to,*dcode*)

Figure 3: Example schemas

qualifier being either a **filter** or a **generator**. A filter is a boolean-valued expression. A generator has syntax $p \leftarrow c$ where p is a **pattern** and c is a bag-valued expression. In IQL, the patterns p are restricted to be single variables or tuples of variables.

Example 1 Pathway $LS_1 \rightarrow US_1$
t_1 extendTable(⟨⟨student, id, name, sex⟩⟩)
t_2 extendTable(⟨⟨university, uname⟩⟩)
t_3 extendTable(⟨⟨campus, cmname, uname⟩⟩)
t_4 extendTable(⟨⟨degree, dcode, title, dtype, dname⟩⟩)
t_5 extendTable(⟨⟨enrolled, id, from, to, dcode⟩⟩)
t_6 addRel(⟨⟨dept⟩⟩, [x | (y, x) ← ⟨⟨staff, dname⟩⟩])
t_7 addAtt(⟨⟨dept, dname⟩⟩, [(x, x) | x ← ⟨⟨dept⟩⟩])
t_8 extendAtt(⟨⟨dept, street⟩⟩, Void, Any)
t_9 extendAtt(⟨⟨dept, uname⟩⟩, Void, Any)
t_{10} addAtt(⟨⟨staff, sex⟩⟩, [(x, 'M') | x ← ⟨⟨male⟩⟩] ++ [(x, 'F') | x ← ⟨⟨female⟩⟩])
t_{11} deleteAtt(⟨⟨male, id⟩⟩, [(x, x) | x ← ⟨⟨male⟩⟩])
t_{12} deleteRel(⟨⟨male⟩⟩, [x | (x, 'M') ← ⟨⟨staff, sex⟩⟩])
t_{13} deleteAtt(⟨⟨female, id⟩⟩, [(x, x) | x ← ⟨⟨female⟩⟩])
t_{14} deleteRel(⟨⟨female⟩⟩, [x | (x, 'F') ← ⟨⟨staff, sex⟩⟩])

The pathway $LS_2 \rightarrow US_2$ contains extend steps to add the missing staff, student, and enrolled tables. It then renames deptname, and adds the missing attributes of dept:

Example 2 Pathway $LS_2 \rightarrow US_2$
t_{15} extendTable(⟨⟨student, id, name, sex⟩⟩)
t_{16} extendTable(⟨⟨staff, id, name, sex, dname⟩⟩)
t_{17} extendTable(⟨⟨enrolled, id, from, to, dcode⟩⟩)
t_{18} renameAtt(⟨⟨dept, deptname⟩⟩, ⟨⟨dept, dname⟩⟩)
t_{19} renameAtt(⟨⟨degree, deptname⟩⟩, ⟨⟨degree, dname⟩⟩)
t_{20} extendAtt(⟨⟨dept, street⟩⟩, Void, Any)
t_{21} extendAtt(⟨⟨dept, uname⟩⟩, Void, Any)

The pathway $\mathsf{LS}_3 \rightarrow \mathsf{US}_3$ contains a sequence of extend steps for its missing information. The pathway $\mathsf{LS}_4 \rightarrow \mathsf{US}_4$ creates in t_{22} a new attribute ⟨⟨dept, uname⟩⟩ by joining the dept and college relations, and then deletes in t_{23}–t_{25} the college table that can be recovered from the remaining ⟨⟨dept, cname⟩⟩ attribute. Transformation t_{26} is unable to put any restriction on the values of ⟨⟨dept, cname⟩⟩, since that association cannot be recovered from the global schema. Transformations t_{27}–t_{31} then perform the logical inverse of t_{22}–t_{26} to partially extract the campus table from the direct association between departments and universities represented by ⟨⟨dept, uname⟩⟩.

Example 3 Pathway $\mathsf{LS}_4 \rightarrow \mathsf{US}_4$

t_{22} addAtt(⟨⟨dept, uname⟩⟩,
  [(x, y) | (x, z) ← ⟨⟨dept, cname⟩⟩; (z, y) ← ⟨⟨college, uname⟩⟩])
t_{23} deleteAtt(⟨⟨college, uname⟩⟩,
  [(x, y) | (z, x) ← ⟨⟨dept, cname⟩⟩; (z, y) ← ⟨⟨dept, uname⟩⟩])
t_{24} deleteAtt(⟨⟨college, cname⟩⟩, [(x, x) | x ← ⟨⟨college⟩⟩])
t_{25} deleteRel(⟨⟨college⟩⟩, [y | (x, y) ← ⟨⟨dept, cname⟩⟩])
t_{26} contractAtt(⟨⟨dept, cname⟩⟩, Void, Any)
t_{27} extendAtt(⟨⟨dept, cmname⟩⟩, Void, Any)
t_{28} addRel(⟨⟨campus⟩⟩, [y | (x, y) ← ⟨⟨dept, cmname⟩⟩])
t_{29} addAtt(⟨⟨campus, cmname⟩⟩, [(x, x) | x ← ⟨⟨campus⟩⟩])
t_{30} addAtt(⟨⟨campus, uname⟩⟩,
  [(x, y) | (z, x) ← ⟨⟨dept, cmname⟩⟩; (z, y) ← ⟨⟨dept, uname⟩⟩])
t_{31} delAtt(⟨⟨dept, uname⟩⟩,
  [(x, y) | (x, z) ← ⟨⟨dept, cmname⟩⟩; (z, y) ← ⟨⟨campus, uname⟩⟩])
t_{32} extendTable(⟨⟨student, id, name, sex⟩⟩)
t_{33} extendTable(⟨⟨staff, id, name, sex, dname⟩⟩)
t_{34} extendTable(⟨⟨enrolled, id, from, to, dcode⟩⟩)

Finally, we list in Example 4 the pathway from the union schema US_1 to the global schema GS. The pathway from US_2, US_3 or US_4 would be identical.

Example 4 Pathway $\mathsf{US}_1 \rightarrow \mathsf{GS}$

t_{35} addRel(⟨⟨person⟩⟩, ⟨⟨staff⟩⟩ ++ [x | x ← ⟨⟨student⟩⟩; not (member x ⟨⟨staff⟩⟩)])
t_{36} addAtt(⟨⟨person, id⟩⟩, ⟨⟨staff, id⟩⟩ ++
  [(x, y) | (x, y) ← ⟨⟨student, id⟩⟩; not (member x ⟨⟨staff⟩⟩)])
t_{37} addAtt(⟨⟨person, name⟩⟩, ⟨⟨staff, name⟩⟩ ++
  [(x, y) | (x, y) ← ⟨⟨student, name⟩⟩; not (member x ⟨⟨staff⟩⟩)])
t_{38} addAtt(⟨⟨person, sex⟩⟩, ⟨⟨staff, sex⟩⟩++
  [(x, y) | (x, y) ← ⟨⟨student, sex⟩⟩; not (member x ⟨⟨staff⟩⟩)])
t_{39} addAtt(⟨⟨person, dname⟩⟩, ⟨⟨staff, dname⟩⟩)
t_{40} contractAtt(⟨⟨student, id⟩⟩, [(x, y) | (x, y) ← ⟨⟨person, id⟩⟩;
  not (member x ⟨⟨staff⟩⟩)], [(x, y) | (x, y) ← ⟨⟨person, id⟩⟩])
t_{41} contractAtt(⟨⟨student, name⟩⟩, [(x, y) | (x, y) ← ⟨⟨person, name⟩⟩;
  not (member x ⟨⟨staff⟩⟩)], [(x, y) | (x, y) ← ⟨⟨person, name⟩⟩])
t_{42} contractAtt(⟨⟨student, sex⟩⟩, [(x, y) | (x, y) ← ⟨⟨person, sex⟩⟩;
  not (member x ⟨⟨staff⟩⟩)], [(x, y) | (x, y) ← ⟨⟨person, sex⟩⟩])

t_{43} contractRel(⟨⟨student⟩⟩), [x | x ← ⟨⟨person⟩⟩; not (member x ⟨⟨staff⟩⟩)],
[x | x ← ⟨⟨person⟩⟩])
t_{44} deleteAtt(⟨⟨staff, id⟩⟩, [(x, y) | (x, y) ← ⟨⟨person, id⟩⟩; member x ⟨⟨staff⟩⟩])
t_{45} deleteAtt(⟨⟨staff, name⟩⟩, [(x, y) | (x, y) ← ⟨⟨person, name⟩⟩; member x ⟨⟨staff⟩⟩])
t_{46} deleteAtt(⟨⟨staff, sex⟩⟩, [(x, y) | (x, y) ← ⟨⟨person, sex⟩⟩; member x ⟨⟨staff⟩⟩])
t_{47} deleteAtt(⟨⟨staff, dname⟩⟩, ⟨⟨person, dname⟩⟩)
t_{48} deleteRel(⟨⟨staff⟩⟩, [x | (x, y) ← ⟨⟨person, dname⟩⟩])

We assume in this example integration that a person may be both a member of staff and a student. For such people, their information in the staff table is preferred for propagation to the global person table in steps t_{35}–t_{38} above. Thus, there is not sufficient information in the global schema to totally derive the student table, and only contract statements can be given in steps t_{40}–t_{43}, where as a lower bound we know all persons not in the staff table are students, but as an upper bound know that all persons could be in student (if it were the case that all staff members were former students). Conversely, there is sufficient information to totally derive the staff table.

2.2 *Comparison of BAV with GAV, LAV and GLAV*

We see from the above example that the add and extend steps in the transformation pathways from the local schemas to the global schema correspond to GAV, since it is these steps that are incrementally defining global constructs in terms of local ones. Similarly, it is the delete and contract steps in the transformation pathways from the local schemas to the global schema that correspond to LAV, since it is these steps that are incrementally defining local constructs in terms of global ones. We will see in Section 3 how these pathways can be traversed to derive GAV and LAV views.

If a GAV view is derived from solely add steps it will be *exact* in the terminology of [7]. If, in addition, it is derived from one or more extend steps using their lower-bound (upper-bound) queries, then the GAV view will be *sound* (*complete*) in the terminology of [7]. Similarly, if a LAV view is derived solely from delete steps it will be exact. If, in addition, it is derived from one or more contract steps using their lower-bound (upper-bound) queries, then the LAV view will be *complete* (*sound*) in the terminology of [7]. For example, in pathway $\mathsf{US}_1 \rightarrow \mathsf{GS}$ above, we could enhance t_{43} above to:
contractRel(⟨⟨student⟩⟩, [x | x ← ⟨⟨person⟩⟩; not (member x ⟨⟨staff⟩⟩)], ⟨⟨person⟩⟩])
asserting that ⟨⟨student⟩⟩ contains the set of people who are not staff (completeness) and is contained by the whole set of people (soundness).

As we discussed in the introduction, BAV is a more expressive data integration language than LAV, GAV or GLAV, since it allows for the expression of mappings in both directions, and since it is not limited on how many source schemas are associated by a mapping. Indeed, in the context of peer-to-peer integration, [3] has suggested using GLAV rules in both directions in a similar manner to BAV, in order to overcome weaknesses of using GLAV alone.

As discussed in [14, 15], a further advantage of BAV over GAV and LAV is that it readily supports the evolution of both global and local schemas, by allowing pathways and schemas to be incrementally modified as opposed to having to be regenerated.

A further difference between BAV and GAV, LAV or GLAV (including the approach of using GLAV in each direction of [3]) is that statements about the relationships between global and local schemas are made at a finer level of detail, *i.e.* at the level of individual attributes as opposed to entire tables. So we can assert exact knowledge about some attributes of a table, and sound or complete knowledge about other attributes. We are also able to introduce intermediate constructs in the mapping, such as in $\mathsf{LS}_4 \rightarrow \mathsf{US}_4$.

3 Generating Views

We now present a general technique for generating GAV, LAV and GLAV view definitions from a BAV pathway. This ability to generate any of these kinds of view definitions from a single BAV pathway means that we can select a query processing technique that can vary between queries as appropriate.

To define a construct c in S_x in terms of the constructs in schema S_y, we consider in turn the transformations of $S_x \rightarrow S_y$. The only transformations that are significant are those that delete, contract or rename a construct[1]. These transformations are significant because the current view definitions may query constructs that no longer exist after such a transformation. Each of these types of transformation is handled as follows if it is encountered during the traversal of $S_x \rightarrow S_y$:

- delete: This has an associated query which shows how to reconstruct the extent of the construct being deleted. Any occurrence of the deleted construct within the current view definitions is replaced by this query.
- contract: Any occurrence of the contracted construct within the current view definitions is replaced by either the lower-bound or the upper-bound query accompanying this transformation step, depending on whether sound or complete views are required.
- rename: All references to the old construct in the current view definitions are replaced by references to the new construct.

3.1 *Generating GAV Views*

To generate the set of GAV views for a global schema, the pathways from it to each local schema are retrieved from AutoMed's metadata repository. For some part of the start of their length these pathways may be the same, as may be seen from the tree structure of Figure 1. Each node of this transformations tree is a schema (global, intermediate or local) linked to its neighbours by a single transformation step. View definitions for each global schema construct are derived by traversing the tree from top to bottom. Initially, each construct's view definition is just the construct itself. Each node in the tree is then visited in a downwards direction, and delete, contract and rename transformations are handled as described above. In particular, if a contract transformation step is encountered, any occurrence of the contracted construct within the current GAV view definitions is replaced by the lower-bound query accompanying this transformation step (so that sound GAV views will be generated).

At some points the tree may branch. When this happens, constructs of the parent schema are semantically identical to constructs that have the same scheme within the child schemas. The possibility of using all paths is retained within the view definitions by replacing each construct of the parent schema by a disjunction (OR) of the corresponding constructs in the child schemas.

The tree is traversed in this fashion from the root to the leaves until all the nodes are visited. The resulting view definitions are the GAV definitions for the global schema constructs over the local schemas. Referring again to the example of Section 2.1, consider the construct GS : ⟨⟨person, sex⟩⟩ in the global schema. The pathway GS $\rightarrow$ US_1 would be processed first (*i.e.* the reverse of the pathway US_1 $\rightarrow$ GS listed in Section 2.1). The only significant transformation is $\overline{t_{38}}$ that deletes ⟨⟨person, sex⟩⟩, resulting in an intermediate view definition:

[1]Note that this is equivalent to considering the add, extend and rename steps in the reverse $S_y \rightarrow S_x$

GS : ⟨⟨person, sex⟩⟩ :- US_1 : ⟨⟨staff, sex⟩⟩ ++
[(x, y) | (x, y) ← US_1 : ⟨⟨student, sex⟩⟩; not (member x US_1 : ⟨⟨staff⟩⟩)]

at one copy, US_1, of the four union schemas. Traversing the pathways $US_1 \rightarrow LS_1$ and $US_1 \rightarrow US_2$, we replace the body of the view definition with:

([(x, 'M') | x ← LS_1 : ⟨⟨male⟩⟩] ++ [(x, 'F') | x ← LS_1 : ⟨⟨female⟩⟩] OR
US_2 : ⟨⟨staff, sex⟩⟩)
++ ([(x, y) | (x, y) ← Void OR US_2 : ⟨⟨student, sex⟩⟩;
not (member x (LS_1 : ⟨⟨staff⟩⟩ OR US_2 : ⟨⟨staff⟩⟩))])

Traversing next $US_2 \rightarrow LS_2$ and $US_2 \rightarrow US_3$, we get:

([(x, 'M') | x ← LS_1 : ⟨⟨male⟩⟩] ++ [(x, 'F') | x ← LS_1 : ⟨⟨female⟩⟩]) OR
Void OR US_3 : ⟨⟨staff, sex⟩⟩)
++ ([(x, y) | (x, y) ← Void OR Void OR US_3 : ⟨⟨student, sex⟩⟩;
not (member x (LS_1 : ⟨⟨staff⟩⟩ OR Void OR US_3 : ⟨⟨staff⟩⟩))])

Continuing with $US_3 \rightarrow LS_3$, $US_3 \rightarrow US_4$ and finally $US_4 \rightarrow LS_4$, we obtain the view definition:

GS : ⟨⟨person, sex⟩⟩ :-
([(x, 'M') | x ← LS_1 : ⟨⟨male⟩⟩] ++ [(x, 'F') | x ← LS_1 : ⟨⟨female⟩⟩])OR
Void OR Void OR Void)
++ ([(x, y) | (x, y) ← Void OR Void OR LS_3 : ⟨⟨student, sex⟩⟩ OR Void;
not (member x (LS_1 : ⟨⟨staff⟩⟩ OR Void OR Void OR Void))])

Such view derivations can be substituted into any query posed on a global schema in order to obtain an equivalent query distributed over the local schemas — this is the GAV approach to global query processing, which is what the AutoMed implementation currently supports. Section 4 will justify how this view definition can be simplified further.

3.2 *Generating LAV Views*

LAV views are derived similarly: the pathway from a local schema to the global schema is again retrieved from the metadata repository and is processed as above to derive the view definitions, except that it is the local schema end of the pathway that is now taken as the root of the tree. The derivation of LAV views is simpler because there is now only a single pathway being processed, with no branching. Also, if a contract transformation step is encountered, any occurrence of the contracted construct within the current LAV view definitions is replaced by the upper-bound query accompanying this transformation step (so that sound LAV views will be generated).

For example, to generate a LAV definition of LS_1 : ⟨⟨male⟩⟩, we inspect the pathway t_1,..,t_{14},t_{35},..,t_{48}. The transformation t_{12} deletes ⟨⟨male⟩⟩, and therefore we have an intermediate view definition on US_1:

LS_1 : ⟨⟨male⟩⟩ :- [x | (x, 'M') ← US_1 : ⟨⟨staff, sex⟩⟩]

Then ⟨⟨staff, sex⟩⟩ construct is deleted by t_{46}, which substitutes(x, y) ← ⟨⟨staff, sex⟩⟩ with (x, y) ← ⟨⟨person, sex⟩⟩; member x ⟨⟨staff⟩⟩, and the ⟨⟨staff⟩⟩ construct in this query is deleted by t_{40} giving a final LAV rule:

LS_1 : ⟨⟨male⟩⟩ :-
[x | (x, 'M') ← GS : ⟨⟨person, sex⟩⟩; member x [x | (x, y) ← GS : ⟨⟨person, dname⟩⟩]]

3.3 *Generating GLAV Views*

First, it should be noted that GLAV view definitions will include all the LAV view definitions, and all the GAV view definitions where the body of the rule is a query that matches the conditions required for the GLAV query processing system in use (which in [10] would be queries over a single source). In addition, we inspect now all the add and extend transformations of the pathway that would be ignored by LAV view generation, and for each one use the query to form the head of a new GLAV rule.

For example, in $\mathsf{LS}_4 \rightarrow \mathsf{US}_4$, the query in t_{22} gives a new view rule head: $[(\mathsf{x},\mathsf{y}) \mid (\mathsf{x},\mathsf{z}) \leftarrow \mathsf{LS}_4 : \langle\langle \mathsf{dept}, \mathsf{cname} \rangle\rangle; (\mathsf{z},\mathsf{y}) \leftarrow \mathsf{LS}_4 : \langle\langle \mathsf{college}, \mathsf{uname} \rangle\rangle]$ which is defined by $\langle\langle \mathsf{dept}, \mathsf{uname} \rangle\rangle$ at this stage. We then use our standard algorithm on construct $\langle\langle \mathsf{dept}, \mathsf{uname} \rangle\rangle$, detect that it is deleted in t_{31}, and hence replace it with the query from t_{31} to result in the GLAV rule:

$[(\mathsf{x},\mathsf{y}) \mid (\mathsf{x},\mathsf{z}) \leftarrow \mathsf{LS}_4 : \langle\langle \mathsf{dept}, \mathsf{cname} \rangle\rangle; (\mathsf{z},\mathsf{y}) \leftarrow \mathsf{LS}_4 : \langle\langle \mathsf{college}, \mathsf{uname} \rangle\rangle]$:-
  $[(\mathsf{x},\mathsf{z}) \mid \mathsf{GS} : \langle\langle \mathsf{dept}, \mathsf{cmname} \rangle\rangle; (\mathsf{z},\mathsf{y}) \leftarrow \mathsf{GS} : \langle\langle \mathsf{campus}, \mathsf{uname} \rangle\rangle]$

Note however that the BAV integration would still hold if LS_4 were fragmented, with campus and departments held on separate sources, whereas GLAV would cease to be valid in this situation.

4 Logical Optimisation of the Generated Views

The view definitions generated by the process described above can be simplified by a process of logical optimisation, where redundant parts of the query are removed. This saves later work for the query optimiser, when these definitions are substituted into specific global queries for query processing. It also generates views that are similar to the views that would have been specified directly in a GAV, LAV or GLAV framework.

4.1 *The* OR *Operator and Void*

The OR operator has the semantics that each of its arguments has the same real-world semantics. The Void value represents a construct that is unobtainable from a data source. Hence we define e OR Void = Void OR e = e for any IQL expression e. Also, any IQL expression that contains Void evaluates to Void. Hence the expression $\langle\langle \mathsf{staff}, \mathsf{name} \rangle\rangle$ OR $[\langle \mathsf{x},\mathsf{y} \rangle \mid \langle \mathsf{x},\mathsf{y} \rangle \leftarrow \mathsf{Void}]$ first simplifies to $\langle\langle \mathsf{staff}, \mathsf{name} \rangle\rangle$ OR Void, and then simplifies to $\langle\langle \mathsf{staff}, \mathsf{name} \rangle\rangle$. Applying this simplification method to the GAV view definition derived in Section 3.1 results in:

GS : $\langle\langle$person, sex$\rangle\rangle$:-
  ($[(\mathsf{x}, \text{'M'}) \mid \mathsf{x} \leftarrow \mathsf{LS}_1 : \langle\langle \mathsf{male} \rangle\rangle]$ ++ $[(\mathsf{x}, \text{'F'}) \mid \mathsf{x} \leftarrow \mathsf{LS}_1 : \langle\langle \mathsf{female} \rangle\rangle]$))
++ ($[(\mathsf{x},\mathsf{y}) \mid (\mathsf{x},\mathsf{y}) \leftarrow \mathsf{LS}_3 : \langle\langle \mathsf{student}, \mathsf{sex} \rangle\rangle; \mathsf{not}\ (\mathsf{member}\ \mathsf{x}\ (\mathsf{LS}_1 : \langle\langle \mathsf{staff} \rangle\rangle))]$)

It may be the case that two data sources supply information for a single schema construct. For example, the global schema attribute $\langle\langle \mathsf{university}, \mathsf{uname} \rangle\rangle$ has the GAV view definition:
$\mathsf{GS} : \langle\langle \mathsf{university}, \mathsf{uname} \rangle\rangle$:-$\mathsf{LS}_2 : \langle\langle \mathsf{university}, \mathsf{uname} \rangle\rangle$ OR $\mathsf{LS}_4 : \langle\langle \mathsf{university}, \mathsf{uname} \rangle\rangle$
which expresses the fact that either LS_2 or LS_4 can be used to extract information about university names. This leads to several possibilities for operational semantics that may be used for the OR operator:

1. **ident** semantics would choose one of the expressions to evaluate, since the integration rules specify that they are the same. This may be defined by the equivalence rule e1 OR e2 = e1 = e2, *i.e.* that we has replace e1 OR e2 by either e1 or e2.

2. **intersect** semantics would determine that a value should be returned only if it is present in all data sources, defined by e1 OR e2 = intersect e1 e2.

3. **append** semantics would determine that all values in all data sources should be returned, defined by e1 OR e2 = e1 ++ e2.

4. **union** semantics would determine that one copy of a value should be returned if present in any data source, defined by e1 OR e2 = distinct (e1 ++ e2).

Option (1) is that which should be used if it is known that the data sources obey the semantics specified by the data integration rules *i.e.* that their extents are identical and there are no distributed data integrity violations. In this circumstance, the OR operator may also be used during distributed data integrity checking, where both expressions are evaluated, and the results compared to determine if the data sources contain consistent data.

Options (2)–(4) provide different mechanisms for handling situations where the data sources are possibly inconsistent, and thus may not share information that they should share. Option (3) provides a result that may be used to derive Options (2) and (4), and therefore is the default semantics provided by the AutoMed's view generation algorithm. Note also that Option (4) gives the same result as Option (1) if the data sources are identical.

4.2 *Other IQL Operators*

The AutoMed intermediate query language IQL supports several primitive operators for manipulating lists. The list **append** operator, ++, concatenates two lists together. The **distinct** operator removes duplicates from a list. The **monus** operator −− subtracts each instance of the second list from the first. For example, [1, 2, 3, 2, 4] −− [4, 4, 2, 1] = [3, 2]. The **fold** operator applies a given function f to each element of a list and then 'folds' a binary operator op into the resulting values, and is defined as follows:

fold f op e [] = e
fold f op e [x] = f x
fold f op e (b1 ++ b2) = (fold f op e b1) op (fold f op e b2)

Other IQL list manipulation operators may be defined using fold together with the usual set of built-in operators and also the support of lambda abstractions. For example, the IQL functions sum and count are equivalent to the SQL SUM and COUNT aggregation functions and are defined respectively as sum = fold (id) (+) 0 and count = fold (lambda x.1) (+) 0.

The function flatmap applies a list-valued function f to each member of a list b and is defined as flatmap f b = fold f (++) [] b. flatmap can in turn be used to define selection, projection, join and, more generally, the comprehension syntax used in the view definitions of the previous section. For example, the list comprehension

[x | x ← ⟨⟨student⟩⟩; not (member x ⟨⟨staff⟩⟩)]

translates to:

flatmap (lambda x.if (not (member x ⟨⟨staff⟩⟩)) then [x] else []) ⟨⟨student⟩⟩

Optimisations for fold apply to all the operators defined in terms of it. Regarding the view definitions generated from BAV pathways there are two particular optimisations that can be applied to them. First, any instances of fold applied to Void can be simplied by treating Void as identical to the empty bag, so that fold f op e Void = e for any f, op, e. Second, due to the step-wise specification of our schema transformations, **loop fusion** may be applicable. This replaces two successive iterations over a collection by one iteration provided the operators in question satisfy certain algebraic properties. A simple instance of loop fusion is the standard relational query optimisation $\pi_A(\pi_B(R)) = \pi_{A,B}(R)$. Loop fusion does not arise in the schema integration example of Section 2.1 but consider the following fragment of an AutoMed pathway. This first joins two schemes $\langle\langle R, a\rangle\rangle$ and $\langle\langle R, b\rangle\rangle$, creating an intermediate relation $\langle\langle I\rangle\rangle$, and then projects onto the a and b attributes, creating a relation $\langle\langle V\rangle\rangle$, and finally deletes $\langle\langle I\rangle\rangle$:

$\text{addRel}(\langle\langle I\rangle\rangle, [(x, y, z) \mid (x, y) \leftarrow \langle\langle R, a\rangle\rangle; (x, z) \leftarrow \langle\langle R, b\rangle\rangle])$
$\text{addRel}(\langle\langle V\rangle\rangle, [(y, z) \mid (x, y, z) \leftarrow \langle\langle I\rangle\rangle])$
$\text{deleteRel}(\langle\langle I\rangle\rangle, [(x, y, z) \mid (x, y) \leftarrow \langle\langle R, a\rangle\rangle; (x, z) \leftarrow \langle\langle R, b\rangle\rangle])$

The view definition generated for $\langle\langle V\rangle\rangle$ would be

$$[(y, z) \mid (x, y, z) \leftarrow [(x, y, z) \mid (x, y) \leftarrow \langle\langle R, a\rangle\rangle; (x, z) \leftarrow \langle\langle R, b\rangle\rangle]])$$

and the generator $(x, y, z) \leftarrow$ in the outer comprehension can be fused with the head expression of the inner comprehension, giving:

$$[(y, z) \mid (x, y) \leftarrow \langle\langle R, a\rangle\rangle; (x, z) \leftarrow \langle\langle R, b\rangle\rangle]$$

There are a range of other standard algebraic optimisations that could be performed on the view definitions *e.g.* pushing down selections and projections. However, these kinds of optimisations will also be applied later, when a specific global query is reformulated by substituting into it the view definitions. Further optimisations and rewrites will be applied at this stage *e.g.* to bring constructs from the same local schemas together into sub-queries which can be posed entirely on one local schema and it is these sub-queries (appropriately translated) that will be sent to local data sources for evaluation.

We finally note that, although IQL is list-based, if the ordering of elements within lists is ignored then its operators are faithful to the expected bag semantics. Moreover, use of the distinct operator can be used to obtain set semantics as needed. We refer the reader to [17, 6] for more details of IQL and for references to work on fold-based functional query languages and optimisation techniques for such languages.

5 Validating and Optimising Pathways

One important feature of the AutoMed approach is that once a set of schemas have been joined in a network of pathways, data and queries may be translated or migrated between any pair of schemas in the network. Such networks may be complex to analyse, so we need to support automated validation that a network is well-formed. We also need to support automated optimisation of the pathways between schemas, since they may contain redundant transformations.

To support such validation and optimisation of pathways, we have developed the **Transformation Manipulation Language (TML)** [20, 21], which represents each transformation in a form suited to analysis of the schema constructs that are created, deleted or are required to be present for the transformation to be correct. Our definitions below require two functions sc and rc. Given a query q on schema S containing n number of constructs, sc determines all schema constructs that must exist in S if the query is valid, rc determines all schema constructs in S referencing the constructs in q. For the IQL language constructs used in our earlier examples, sc and rc are defined as:

$$\begin{aligned}
sc(\langle\langle \mathsf{r}\rangle\rangle) &= \langle\langle \mathsf{r}\rangle\rangle\\
sc(\langle\langle \mathsf{r}, \mathsf{a}\rangle\rangle) &= \{\langle\langle \mathsf{r}\rangle\rangle, \langle\langle \mathsf{r}, \mathsf{a}\rangle\rangle\}\\
sc([q_1, \ldots, q_n]) &= sc(q_1) \cup \ldots \cup sc(q_n)\\
sc(q_1 \mathbin{++} q_2) &= sc(q_1) \cup sc(q_2)\\
sc([q \mid q_1, \ldots, q_n]) &= sc(q) \cup sc(q_1) \cup \ldots \cup sc(q_n)\\
rc(\langle\langle \mathsf{r}\rangle\rangle) &= \bigcup_{1 \le i \le n} c_i (c_i \in S \wedge \langle\langle \mathsf{r}\rangle\rangle \in sc(c_i))
\end{aligned}$$

Note that as a shorthand, we will write the pair of queries q_l, q_u in extend or contract as just q, with the semantics in such cases that $sc(q_l, q_u) = sc(q_l) \cup sc(q_u)$. The TML formalises each transformation t_i of schema S_i into schema S_{i+1} as having four **conditions** $a_i^+, b_i^-, c_i^+, d_i^-$:

- The positive precondition a_i^+ is the set of constructs that t_i implies must be present in S_i. It comprises those constructs that are present in the query of the transformation (given by $sc(q)$) together with any constructs implied as being present by the construct c:

  $t_i \in \{\mathsf{add}(c,q), \mathsf{extend}(c,q)\} \rightarrow a_i^+ = (sc(c) - c) \cup sc(q)$

  $t_i \in \{\mathsf{delete}(c,q), \mathsf{contract}(c,q), \mathsf{id}(c,c')\} \rightarrow a_i^+ = sc(c) \cup sc(q)$

  $t_i = \mathsf{rename}(c,c') \rightarrow a_i^+ = rc(c)$

- The negative precondition b_i^- is the set of constructs that t_i implies must not be present in S_i. It comprises those constructs which the transformation will add to the schema, and thus must not already be present:

  $t_i \in \{\mathsf{add}(c,q), \mathsf{extend}(c,q), \mathsf{id}(c',c)\} \rightarrow b_i^- = c$

  $t_i \in \{\mathsf{delete}(c,q), \mathsf{contract}(c,q)\} \rightarrow b_i^- = \emptyset$

  $t_i = \mathsf{rename}(c,c') \rightarrow b_i^- = \{c/c'\}rc(c)$

- The positive postcondition c_i^+ is the set of constructs that t_i implies must be present in S_{i+1}, and is derived in the same way as $\overline{a_i^+}$ (*i.e.* the positive precondition of the $\overline{t_i}$):

  $t_i \in \{\mathsf{add}(c,q), \mathsf{extend}(c,q), \mathsf{id}(c',c)\} \rightarrow c_i^+ = sc(c) \cup sc(q)$

  $t_i \in \{\mathsf{delete}(c,q), \mathsf{contract}(c,q)\} \rightarrow c_i^+ = (sc(c) - c) \cup sc(q)$

  $t_i = \mathsf{rename}(c,c') \rightarrow c_i^+ = \{c/c'\}rc(c)$

- The negative postcondition d_i^- is the set of constructs that t_i implies must not be present in S_{i+1}, and is derived in the same way as $\overline{b_i^-}$:

  $t_i \in \{\mathsf{delete}(c,q), \mathsf{contract}(c,q), \mathsf{id}(c,c')\} \rightarrow d_i^- = c,$

  $t_i \in \{\mathsf{add}(c,q), \mathsf{extend}(c,q)\} \rightarrow d_i^- = \emptyset$

  $t_i = \mathsf{rename}(c,c') \rightarrow d_i^- = rc(c)$

Below we show how the compounded transformation t_1 and the primitive transformation t_6 are represented in the TML.

$t_{1.1} : [\emptyset, \{\langle\langle \mathsf{student}\rangle\rangle\}, \{\langle\langle \mathsf{student}\rangle\rangle\}, \emptyset]$
$t_{1.2} : [\emptyset, \{\langle\langle \mathsf{student, id}\rangle\rangle\}, \{\langle\langle \mathsf{student}\rangle\rangle, \langle\langle \mathsf{student, id}\rangle\rangle\}, \emptyset]$
$t_{1.3} : [\emptyset, \{\langle\langle \mathsf{student, id}\rangle\rangle\}, \{\langle\langle \mathsf{student}\rangle\rangle, \langle\langle \mathsf{student, sex}\rangle\rangle\}, \emptyset]$
$t_{1.4} : [\emptyset, \{\langle\langle \mathsf{student, id}\rangle\rangle\}, \{\langle\langle \mathsf{student}\rangle\rangle, \langle\langle \mathsf{student, dname}\rangle\rangle\}, \emptyset]$
$t_6 : [\{\langle\langle \mathsf{staff}\rangle\rangle, \langle\langle \mathsf{staff, dname}\rangle\rangle\}, \{\langle\langle \mathsf{dept}\rangle\rangle\}, \{\langle\langle \mathsf{dept}\rangle\rangle, \langle\langle \mathsf{staff}\rangle\rangle, \langle\langle \mathsf{staff, dname}\rangle\rangle\}, \emptyset]$

5.1 *Well-formed Transformation Pathways*

A pathway T from schema S_m to S_n is said to be **well-formed** if for each transformation step $t_i : S_i \rightarrow S_{i+1}$ within it:

- The only difference between the schema constructs in S_{i+1} and S_i is those constructs specifically changed by transformation t_i, implying that $S_{i+1} = (S_i \cup c_i^+) - d_i^-$ and $S_i = (S_{i+1} \cup a_i^+) - b_i^-$
- The constructs required by t_i are in the schemas, implying that $a_i^+ \subseteq S_i$, $b_i^- \cap S_i = \emptyset$, $c_i^+ \subseteq S_{i+1}$ and $d_i^- \cap S_{i+1} = \emptyset$

The above definition leads to the recursive definition of a well-formed pathway, wf, given below. The first rule applies each transformation step in turn, and the second rule ensures that the schema that results from applying all the transformation steps is equal to the schema at the end of the pathway (equal both in terms of the schema constructs found in each schema and the extent of the schemas). Note that any implementation may use these rules in two ways. Firstly, given a schema S_m representing a data source, and pathway P, a new data source schema S_n and its extent can be derived. Secondly, if S_n exists as a data source already, a check can be made to verify that P correctly derives its schema and extent from that of S_m.

$$\begin{array}{rcl} wf(S_m, S_n, [t_m, t_{m+1}, \ldots, t_{n-1}]) & \leftarrow & a_m^+ \subseteq S_m \wedge b_m^- \cap S_m = \emptyset \wedge \\ & & wf((S_m \cup c_m^+) - d_m^-, S_n, [t_{m+1}, \ldots, t_{n-1}]) \\ wf(S_m, S_n, []) & \leftarrow & S_m = S_n \wedge Ext(S_m) = Ext(S_n) \end{array}$$

5.2 *Reordering of Transformations*

Certain transformations may be performed in any order, whilst others must be performed in a specific order. For example, in $\mathsf{LS}_1 \rightarrow \mathsf{US}_1$, t_{11} must be performed before t_{12}, since the attribute ⟨⟨male, id⟩⟩ must be deleted before the ⟨⟨male⟩⟩ relation is deleted. However the sub-pathway t_{11},t_{12} could be performed before or after the sub-pathway t_{13},t_{14} since it does not matter which of the ⟨⟨male⟩⟩ or ⟨⟨female⟩⟩ relations is deleted first.

In the TML, this intuition is expressed by stating that transformations may be swapped provided the pathway remains well-formed. This may be verified by inspecting the conditions of each transformation. In particular, a pair of transformations t_i,t_{i+1} may be reordered to t_{i+1},t_i provided:

1. t_i does not add a construct required by t_{i+1}, and $\overline{t_{i+1}}$ does not add a construct required by $\overline{t_i}$, *i.e.* $(c_i^+ - a_i^+) \cap a_{i+1}^+ = \emptyset$ and $(a_{i+1}^+ - c_{i+1}^+) \cap c_i^+ = \emptyset$

2. t_i does not delete a construct required not to be present by t_{i+1}, and $\overline{t_{i+1}}$ does not delete a construct required not to be present by $\overline{t_i}$, *i.e.* $d_i^+ \cap b_{i+1}^+ = \emptyset$

3. if t_i is preceded by t_{i-1}, the preconditions of t_{i+1} do not conflict with the postconditions of t_{i-1}, *i.e.* $c_{i-1}^+ \cap b_{i+1}^- = \emptyset$ and $d_{i-1}^- \cap a_{i+1}^+ = \emptyset$

4. if t_{i+1} is followd by t_{i+2}, the preconditions of t_{i+2} do not conflict with the postconditions of t_i, *i.e.* $c_i^+ \cap b_{i+2}^- = \emptyset$ and $d_i^- \cap a_{i+2}^+ = \emptyset$

We can now formalise the two examples given above from $\mathsf{LS}_1 \rightarrow \mathsf{US}_1$. For t_{11},t_{12}, (1) is broken, and hence they may not be swapped. The changing of t_{11},t_{12},t_{13},t_{14} to t_{13},t_{14},t_{11},t_{12} may be performed by iteratively swapping pairs of transformations. Considering first t_{12},t_{13}, we find neither rule is broken, and they may be reordered to t_{13},t_{12}. Then t_{12},t_{14} breaks neither rule, and may be reordered to t_{14},t_{12}. This leaves a sub-pathway t_{11},t_{13},t_{14},t_{12}, and a similar argument allows t_{11} swap with t_{13} and then t_{14}, to give the sub-pathway t_{13},t_{14},t_{11},t_{12}.

5.3 *Redundant and Partially Redundant Transformations*

Two transformations t_x and t_y in a well-formed pathway T are **redundant** if T may be reordered such that t_x and t_y become consecutive within it, and T remains well-formed if they are then removed. Such redundant transformations will occur if a source schema evolves to model information in the same way as the global schema when previously it modelled the information in a different way. For example, suppose LS_1 is evolved by transformations t_{49},t_{50},t_{51},t_{52},t_{53}, textually identical to transformations t_{10},t_{11},t_{12},t_{13},t_{14}, to model the gender of staff as a single sex attribute in a new version of the schema LS_1'. By reversing these transformation steps we can derive the pathway from the new to the old schema $\mathsf{LS}_1' \rightarrow \mathsf{LS}_1$:

Example 5 Pathway $\mathsf{LS}_1' \rightarrow \mathsf{LS}_1$
$\overline{t_{53}}$ addRel(⟨⟨female⟩⟩, [x | (x, 'F') ← ⟨⟨staff, sex⟩⟩])
$\overline{t_{52}}$ addAtt(⟨⟨female, id⟩⟩, [(x, x) | x ← ⟨⟨female⟩⟩])
$\overline{t_{51}}$ addRel(⟨⟨male⟩⟩, [x | (x, 'M') ← ⟨⟨staff, sex⟩⟩])
$\overline{t_{50}}$ addAtt(⟨⟨male, id⟩⟩, [(x, x) | x ← ⟨⟨male⟩⟩])
$\overline{t_{49}}$ deleteAtt(⟨⟨staff, sex⟩⟩, [(x, 'M') | x ← ⟨⟨male⟩⟩] ++ [(x, 'F') | x ← ⟨⟨female⟩⟩])

If we inspect the entire path $\mathsf{LS}_1' \rightarrow \mathsf{US}_1$, consisting of $\mathsf{LS}_1' \rightarrow \mathsf{LS}_1$ followed by $\mathsf{LS}_1 \rightarrow \mathsf{US}_1$, it may be reordered to contain the sub-pathway:
$\overline{t_{51}}$ addRel(⟨⟨male⟩⟩, [x | (x, 'M') ← ⟨⟨staff, sex⟩⟩])
$\overline{t_{50}}$ addAtt(⟨⟨male, id⟩⟩, [(x, x) | x ← ⟨⟨male⟩⟩])
$\overline{t_{49}}$ deleteAtt(⟨⟨staff, sex⟩⟩, [(x, 'M') | x ← ⟨⟨male⟩⟩] ++ [(x, 'F') | x ← ⟨⟨female⟩⟩])
t_{10} addAtt(⟨⟨staff, sex⟩⟩, [(x, 'M') | x ← ⟨⟨male⟩⟩] ++ [(x, 'F') | x ← ⟨⟨female⟩⟩])
t_{11} deleteAtt(⟨⟨male, id⟩⟩, [(x, x) | x ← ⟨⟨male⟩⟩])
t_{12} deleteRel(⟨⟨male⟩⟩, [x | (x, 'M') ← ⟨⟨staff, sex⟩⟩])

Clearly $\overline{t_{49}}$,t_{10} forms a redundant pair, because we are adding and deleting the same construct *with the same extent* since the query is the same. Once this has been performed $\overline{t_{50}}$,t_{11} may be removed for the same reason, and then $\overline{t_{51}}$,t_{12}. Once all other redundant pairs have been removed, $\mathsf{LS}_1' \rightarrow \mathsf{US}_1$ would comprise of just t_1–t_9.

Using the TML, we can identify redundant transformations as satisfying:
$(a_x^+ = c_y^+) \wedge (b_x^- = d_y^-) \wedge (c_x^+ = a_y^+) \wedge (d_x^- = b_y^-) \wedge \ Ext(c_x^+ \oplus a_x^+) = Ext(c_y^+ \oplus a_y^+)$
where $(x \oplus y) = (x - y) \cup (y - x)$, and thus serves to find all the constructs being added or deleted by the pair of transformations. In practice, this rule means that any pair of transformations which add/extend and then delete/contract (in either order) the same construct are redundant, providing the query can be demonstrated to result in the same extent.

Two transformations t_x and t_y in a well-formed pathway T are **partially redundant** if T may be reordered to make t_x and t_y consecutive, and T remains well-formed if they are then replaced by a single transformation t_{xy}.

The pathway $\mathsf{LS}_1 \rightarrow \mathsf{LS}_2$ has a pair of such partially redundant transformations, since it can be reordered to obtain the sub-pathway:
t_7 addAtt(⟨⟨dept, dname⟩⟩, [(x, x) | x ← ⟨⟨dept⟩⟩])
$\overline{t_{18}}$ renameAtt(⟨⟨dept, dname⟩⟩, ⟨⟨dept, deptname⟩⟩)

This may be replaced by the new transformation given below, which leaves a fully optimised pathway $\mathsf{LS}_1 \rightarrow \mathsf{LS}_2$.
t_{54} addAtt(⟨⟨dept, deptname⟩⟩, [(x, x) | x ← ⟨⟨dept⟩⟩])

Using the TML, we can identify partially redundant transformations as satisfying the following rules, where $\oplus$ indicates the exclusive-or operator:
$((a_x^+ = c_y^+) \wedge a_x^+ \neq \emptyset \oplus (b_x^- = d_y^-) \wedge b_x^- \neq \emptyset \wedge d_x^- \cap b_y^- = \emptyset \wedge d_x^- \neq \emptyset \wedge b_y^- \neq \emptyset$

The simplifications for removing partially redundant and fully redundant transformations are summarised in the table below. The table shows what simplifications may be applied where a pair of transformations is found to operate on the same construct c. NWF denotes 'not well-founded' and [] denotes the removal of the pair. The table would remain correct if extend were to replace add, contract replace delete, and id replace rename. Further details of redundant and partially redundant transformations may be found in [20, 21].

		t_y	
	add(c,q)	delete(c,q)	rename(c,c')
add(c,q)	NWF	[]	add(c',q)
t_x delete(c,q)	[]	NWF	NWF
rename(c',c)	NWF	delete(c',q)	[]
rename(c",c)	NWF	delete(c",q)	rename(c",c')

6 Concluding Remarks

In this paper we have described view generation and view optimisation in the AutoMed heterogeneous database integration framework. We have shown how the AutoMed schema pathways and views generated from them are amenable to considerable simplification, resulting in view definitions that look much like the views that would have been specified directly in a GAV, LAV or GLAV framework.

Since BAV integration is based on sequences of primitive schema transformations, it could be argued that data integration using it is more complex than with GAV, LAV or GLAV. However, the integration process can be greatly simplified by specifying well-known schema equivalences as higher-level composite transformations. We gave such an example, extendTable, in Section 2.1 above, and further examples are given in [15]. Moreover, we are working on techniques for semi-automatically generating transformation pathways to convert a source schema expressed in one modelling language into an equivalent target schema expressed in another modelling language, based on well known schema equivalences. We are also investigating schema matching techniques to automatically or semi-automatically integrate two specific schemas.

Finally, it should be noted that BAV is well-suited to peer-to-peer data integration (see [16]) since it lacks the directionality inherent in LAV, GAV and GLAV, all of which are tied to the concept of there being a global schema which may not always be the case in peer-to-peer environments.

References

[1] M. Boyd, S. Kittivoravitkul, C. Lazanitis, P.J. McBrien, and N. Rizopoulos. AutoMed: A BAV data integration system for heterogeneous data sources. In *Proc. CAiSE2004*, 2004.

[2] P. Buneman *et al.* Comprehension syntax. *SIGMOD Record*, 23(1):87–96, 1994.

[3] D. Calvanese, E. Damagio, G. De Giacomo, M. Lenzerini, and R. Rosati. Semantic data integration in P2P systems. In *Proc. DBISP2P*, Berlin, Germany, 2003.

[4] S.S. Chawathe *et al.* The TSIMMIS project: Integration of heterogeneous information sources. In *Proc. 10th Meeting of the Information Processing Society of Japan*, pages 7–18, October 1994.

[5] M. Friedman, A. Levy, and T. Millstein. Navigational plans for data integration. In *Proc.16th National Conf. on AI*, pages 67–73. AAAI Press, 1999.

[6] E. Jasper, A. Poulovassilis, and L. Zamboulis. Processing IQL Queries and Migrating Data in the AutoMed toolkit. Technical Report No. 20, AutoMed, 2003.

[7] M. Lenzerini. Data integration: A theorectical perspective. In *Proc. PODS02*, pages 247–258, 2002.

[8] A.Y. Levy, A.O. Mendelzon, Y. Sagiv, and D. Srivastava. Answering queries using views. In *Proc. PODS'95*, pages 95–104. ACM Press, May 1995.

[9] A.Y. Levy, A. Rajamaran, and J.Ordille. Querying heterogeneous information sources using source description. In *Proc. VLDB'96*, pages 252–262, 1996.

[10] J. Madhavan and A.Y. Halevy. Composing mappings among data sources. In *Proc. 29th Conference on VLDB*, pages 572–583, 2003.

[11] I. Manolescu, D. Florescu, and D. Kossmann. Answering XML queries on heterogeneous data sources. In *Proc. VLDB'01*, pages 241–250, 2001.

[12] P.J. McBrien and A. Poulovassilis. Automatic migration and wrapping of database applications — a schema transformation approach. In *Proc. ER'99, LNCS 1728*, pages 96–113, 1999.

[13] P.J. McBrien and A. Poulovassilis. A uniform approach to inter-model transformations. In *Proc. CAiSE'99, LNCS 1626*, pages 333–348, 1999.

[14] P.J. McBrien and A. Poulovassilis. Schema evolution in heterogeneous database architectures, a schema transformation approach. In *Proc. CAiSE'02, LNCS 2348*, pages 484–499, 2002.

[15] P.J. McBrien and A. Poulovassilis. Data integration by bi-directional schema transformation rules. In *Proc. ICDE'03*, 2003.

[16] P.J. McBrien and A. Poulovassilis. Defining peer-to-peer data integration using both as view rules. In *Proc. DBISP2P*, Berlin, Germany, 2003.

[17] A. Poulovassilis. The AutoMed Intermediate Query Language. Technical report, AutoMed Project, 2001.

[18] M.T. Roth and P. Schwarz. Don't scrap it, wrap it! A wrapper architecture for data sources. In *Proc. VLDB'97*, pages 266–275, Athens, Greece, 1997.

[19] M. Templeton, H.Henley, E.Maros, and D.J. Van Buer. InterViso: Dealing with the complexity of federated database access. *VLDB Journal*, 4(2):287–317, 1995.

[20] N. Tong. Database schema transformation optimisation techniques for the AutoMed system. Technical report, AutoMed Project, 2002.

[21] N. Tong. Database schema transformation optimisation techniques for the AutoMed system. In *Proc. BNCOD'03*, volume 2712 of *LNCS*, pages 157–171. Springer, 2003.

Databases and Information Systems
J. Barzdins and A. Caplinskas (Eds.)
IOS Press, 2005

A Heuristic Approach to Horizontal Fragmentation in Object Oriented Databases

Hui MA, Klaus-Dieter SCHEWE

Massey University, Department of Information Systems & Information Science Research Centre
Private Bag 11 222, Palmerston North, New Zealand
[h.ma|k.d.schewe]@massey.ac.nz

Abstract. Distribution design for databases involves fragmentation, allocation and replication. This paper addresses fragmentation in the context of a structurally rich object oriented data model. Horizontal and vertical fragmentation operations are extended from the relational data model to the object oriented data model, and a third fragmentation operation called splitting is introduced. The goodness of a distribution design should be assessed by the performance of a system. For this we present a query processing cost model to evaluate the performance of a system. The core of the paper is a heuristic approach for horizontal fragmentation, which uses the cost model and is targeted at globally minimising these costs.

Keywords. fragmentation, distribution design, cost model, heuristic procedure

1 Introduction

It is commonly accepted that the design of distributed databases extends the design of centralised databases by three additional activities: fragmentation, allocation and replication [4, 13]. Fragmentation replaces relation schemata or classes in a given database schema by several new relation schemata or classes without losing or adding information. Allocation associates with each fragment a node in the computer network. Thus, fragmentation and allocation together replace a global database schema by local database schemata for each node of the network. In addition, replication allows a fragment to be stored not just on one node, but on several ones, which impacts on the way queries and updates using replicated fragments have to be realised.

The main objective of distribution design is to improve the overall performance and reliability of a distributed database system. Improved performance can be achieved by avoiding or at least reducing unnecessary data transport between the nodes of the network.

Fragmentation and allocation have mainly been investigated in the context of the relational data model. Horizontal fragmentation [8, 14, 19] exploits relations being sets of tuples. These sets are partitioned using selection predicates. Depending on whether these selection predicates are based on the relation schema at hand or on another relation we distinguish between primary and derived horizontal fragmentation. Vertical fragmentation [12, 13] starts from relation schemata as sets of attributes. Fragments are defined by projections onto subsets and the original relation is recovered by a natural join. This join must be lossless, which is usually achieved by letting a selected minimal be part of all fragment relation schemata. Alternatively, an additional virtual attribute can be used. Allocation [2] assigns the fragments to the nodes in a network. More generally, we may extend allocation to associate nodes also to all intermediate query results, but to our knowledge this generalised view has not yet been approached. In most work in the literature fragmentation and allocation are studied separately, but in some papers these activities are treated jointly [18].

Search at the DBLP digital bibliography service (see `http://dblp.uni-trier.de`) shows that distribution design has been a hot research topic in the early years of last decade, but the interest in the topic has decreased despite the fact that the problems still need further investigation. Especially, with the advent of coupling databases and web-based systems using XML, we obviously create distributed databases, but the way we do it is more ad-hoc than planned and systematically. This current trend also indicates the need for understanding database distribution on the basis of more sophisticated data models than the relational one, in particular object oriented data models and similarly semi-structured data and XML. However, most of the research on distribution design is still based on the relational model, and very little attention is paid to other data models, especially object oriented models.

The work in [3, 6, 9, 16] generalises horizontal fragmentation to the object oriented case. The work in [5, 7, 11, 16] does the analogue for vertical fragmentation. The work in [10] contains first steps to carry fragmentation over to XML. In a nutshell, common horizontal fragmentation techniques start from simple selection predicates found in the most frequent queries and their usage in these queries. The fragments are defined by clustering these simple predicates using either graphical techniques [14] or variants of the Bond Energy Algorithm (BEA) [19], which has originally been used in the context of vertical fragmentation. Vertical fragmentation is based on the clustering of attributes using the affinity between them. Again, approaches are graph-based [12] or use BEA [13]. Most of the work on horizontal and vertical fragmentation follows these basic ideas.

The existing approaches to fragmentation in the context of object oriented databases suffers from three major drawbacks:

- The assumed data model is usually extremely simplified in comparison to more advanced models such as the OODM [17] or IQL [1]. In particular, they miss out on deeply nested structures and consequently on splitting as a primitive for fragmentation [16].
- Despite the fact that distribution design is largely concerned with performance optimisation, little reference is made to the actual query (and transaction) computing costs. Intermediate results of the queries, which have to be stored somewhere in the network and transported to other nodes should be taken into account [9].
- Approaches rely on dependencies between simple queries, though these can hardly be determined. In the end the major use of these dependencies is to determine the satisfiability of a conjunction of simple predicates which is undecidable, and if satisfiable to simplify them.

Our work here is a first attempt to overcome these drawbacks. The remainder of the paper is organised as follows. The fundamentals of the object oriented data model will be reviewed in Section 2 as the preliminaries of the following discussion. In Section 3 we will review fragmentation techniques for the OODM in [16] which leads us to distinguish not only horizontal and vertical fragmentation, but also splitting. Then in Section 4 we present a general query processing cost model and use it as the basis for a new heuristic method for cost optimisation based on reducing the number of relevant simple predicates and thus fragments. The assumption underlying the heuristics is that building fragments with respect to the most frequent simple predicates will first reduce costs, but using too many of them will increase costs again. The approach will determine a suitable cut-off point. Finally Section 5 presents a brief conclusion.

It has been argued that object oriented databases are out of date and after all have not fulfilled the expectations. Whether this is true or not, the object oriented model we use captures the gist of XML, and thus the approach to fragmentation can be carried over. In [16] it has already been shown how to carry over the fragmentation operations to semi-structured data. In [10] horizontal fragmentation has been approached for XML documents.

2 Fundamentals of the OODM

The OODM in [17] is based on an underlying type system. For our purposes here it is sufficient to consider a type system defined as follows:

$$t = b \mid x \mid (a_1 : t_1, \dots, a_n : t_n) \mid \{t\} \mid (a_1 : t_1) \cup \dots \cup (a_n : t_n).$$

Here b represents base types, $(\cdot)$ a record type constructor, $\{\cdot\}$ a set type constructor, and $\cup$ a union type constructor. x is used for type variables, here only for the purpose of defining classes. In order to have object identifiers, we require that among the base types there is at least one type ID, the values of which are the oid-s.

2.1 *Schemata and Databases*

Now consider a type with variables $x_1, \dots, x_n$, but with no occurrence of ID. Replacing all these variables x_i by pairs $r_i : C_i$ with some mutually distinct labels r_i (called *references*) and names C_i (called *class names*) results in a *structure expression*.

A *class* consists of a class name C, a structure expression exp_C, and a set $\{C_1, \dots, C_n\}$ of class names (called the *superclasses*). If we replace all $r_i : C_i$ in exp_C by the base type ID, the resulting parameter-less type T_C is called the *representation type* of the class C.

A *schema* is a finite set of classes that is closed in the sense that all class names appearing in a structure expression or as a superclass must be names of classes defined in the schema.

EXAMPLE 1. Choose the following schema as an example:

`Class` REGION `Struct` (name: *STRING*, central : h: CITY,
cities : { c : CITY })

`Class` CITY `Struct` (name: *STRING*, in : r: REGION, population: *NAT*)

A *database db* for a schema $\mathcal{S}$ assigns to each class $C \in \mathcal{S}$ a finite set $db(C)$ of pairs (i, v), where i is an oid, i.e. a value of type ID, and v is a value of type t_C, such that certain conditions are satisfied. Informally, these conditions are the uniqueness of identifiers, the inclusion of the set of oid-s in a class in the set of oid-s for each of its subclasses (inclusion integrity), the appearance of each oid occurring in a value v as an oid i in the referenced class (referential integrity), and (weak) value-identifiability. For a formal treatment of these conditions we refer to [17].

2.2 A Simple Query Algebra

In order to define a query algebra we could follow the unified and generalised approach in [15], but for our purposes here we can use a simpler algebra. Other operations including nesting and unnesting can be defined explicitly in this section.

The operations of the algebra must allow us to refer to elements of classes and their components. For this we use path expressions. Let C be a class with structure expression exp_C and representation type T_C. *Path expressions* for class C (or exp_C) have one of the following formats:

- *ident* of type ID, or *value* of type T_C,
- If exp_C uses a record type, i.e. $exp_C = (a_1 : exp_1, \dots, a_n : exp_n)$, then we get path expressions
  - $value.a_i$ of type T_i which is the representation type for exp_i and $1 \leq i \leq n$,
  - $value!a_i$ of type T_D if a_i is a reference to class D, i.e. $a_i : exp_i = a_i : D$,
  - $value.a_i.path_i$ where $path_i$ is a path expression for exp_i,
  - $value!a_i.path_i$ where a_i is a reference to class D, i.e. $a_i : exp_i = a_i : D$ and $path_i$ is a path expression for exp_D.
- If exp_C use a union type, i.e. $exp_C = (a_1 : exp_1) \cup \dots \cup (a_n : exp_n)$, then we get path expressions
  - $value.a_i$ of type T_i which is the representation type for exp_i,
  - $value.a_i.path_i$ where $path_i$ is a path expression for exp_i.
- If exp_C uses a set type, i.e. $exp_C = \{exp\}$, then we obtain path expressions
  - *value* of type $T_C = \{exp\}$,
  - $value.path$ where *path* is a path expression corresponding to *exp*.
- If exp_C uses only a reference , i.e. $exp_C = r : D$, then we obtain a path expression $value!r.path$ where *path* is path expression for the class D.

In the OODM each query should result in a set of pairs (i, v), where i is an identifier and v is a value of some proper type. The value v may contain identifiers, which must appear in the database or in the query result.

A *query* Q on $\mathcal{S}$ consists of a structure expression exp_Q called answer schema (with all references pointing to classes in $\mathcal{S}$) and an algebra expression q, i.e. a query Q is defined in the form $Q = (exp_Q, q)$. The operators in the query algebra take one or two class instances as arguments and return another class instance as a result.

Let $\mathcal{S}$ be a database schema, *db* be a database instance over $\mathcal{S}$, *id*(*db*) be the set of all identifiers appearing in *db*, *db*(Q) be the resulting class instance of evaluating query algebra $Q = (exp_Q, q)$. The query algebra operations are defined as follows:

- $q = C$ with C is a class name appearing in $\mathcal{S}$, resulting in $db(Q) = \{(id_w, v) \mid id_w : ID, id_w \notin id(db). \exists id : ID.(id, v) \in db(C)\}$;
- $q = (v : T)$ with a type T and a value v of type T and $Q = (T, q)$, resulting in $db(Q) = \{(id_w, v) \mid id_w : ID \wedge id_w \notin id(db)\}$;

- $q' = \sigma_\varphi(Q)$ (selection) with a selection formulae φ that is defined either as $path_1 = path_2$ or as $path = v$, resulting in $db(Q') = \{(id_w, v) \mid id_w : ID \wedge id_w \notin id(db). \exists id : ID.(id, v) \in db(Q) \wedge \varphi(v) = true\}$;
- $q' = \rho_{a_1 \mapsto b_1, \dots, a_n \mapsto b_n}(Q)$ (renaming) resulting in $db(Q') = \{(id_w, v) \mid id_w : ID \wedge id_w \notin id(db). \exists id : ID.(id, v) \in db(Q)\}$ with b_i as a new name for attribute a_i;
- $q' = \pi_{Q'}(Q)$ (generalised projection) with a new structure expression $exp_{Q'}$ that is a super structure expression of exp_Q, and resulting in $db(Q') = \{(id_w, v') \mid id_w : ID \wedge id_w \notin id(db). \exists id : ID.(id, v) \in db(Q).v' = \pi^Q_{Q'}(v)\}$;
- $q' = q_1 \bowtie_{exp} q_2$ (generalised join) with a common super structure expression, i.e. $exp_{Q_i} \le exp$ for $i = 1, 2$, resulting in

$$db(Q_1 \bowtie_{exp} Q_2) = \{(id_w, v) \mid id_w : ID \wedge id_w \notin id(db). \exists id : ID. \\ (id_1, v_1) \in db(Q_1).(id_2, v_2) \in db(Q_2).\pi^{Q_1}_{exp}(v_1) = \pi^{Q_2}_{exp}(v_2) \wedge \\ \pi^{Q_1 \bowtie Q_2}_{Q_1}(v) = v_1 \wedge \pi^{Q_1 \bowtie Q_2}_{Q_2}(v) = v_2\};$$

- $q' = \nu_a(Q)$ (nest operation) with a as a tuple of components of Q resulting in

$$db(Q') = \{(id_w, w) \mid id_w : ID \wedge id_w \notin id(db). \exists id : ID.(id_1, v_1) \in db(Q). \\ \pi^Q_{Q-exp}(w) = \pi^Q_{Q-exp}(v_1) \wedge \pi^Q_{exp}(w) = \{\pi^Q_{exp}(v_2) \mid (id_2, v_2) \\ \in db(Q) \wedge \pi^Q_{Q-exp}(v_1) = \pi^Q_{Q-exp}(v_2)\}\}$$

  with $exp_{Q'} = (exp_Q - exp) \cup \{exp\}$.
- unnest $q' = \mu_{exp}(Q)$ (unnest operation) with exp as a set attribute of Q resulting in

$$db(Q') = \{(id_w, w) \mid id_w : ID \wedge id_w \notin id(db). \exists id : ID.(id, v) \in db(Q). \\ \pi^Q_{Q-\{exp\}}(w) = \pi^Q_{Q-\{exp\}}(v) \wedge \pi^Q_{exp}(w) \in \pi^Q_{exp}(v)\}$$

  with $exp_{Q'} = (exp_Q - \{exp\}) \cup exp$.
- the usual set operations $\cup$ (union), $-$ (difference), and $\cap$ (intersection).

Queries in this algebra can be rewritten in the usual way as labelled *query trees.*

2.3 *Horizontal Fragmentation*

In [16] three operations for fragmentation of object oriented databases have been discussed: splitting, horizontal fragmentation and vertical fragmentation. In this article we only refer to horizontal fragmentation, which we briefly review here.

According to the definition of databases for an OODM schema, each class will be associated with a set of pairs. Hence, we have the trivial generalisation of relational horizontal fragmentation.

For this let C be some class. Take boolean valued function φ_i such that for each database db we obtain $db(C) = \bigcup_{i=1}^{n} \sigma_{\varphi_i}(db(C))$ with disjoint sets $\sigma_{\varphi_i}(db(C)$.

We then replace C in the schema by n new classes C_i, all with $exp_{C_i} = exp_C$. However, as there may be classes D referencing C, i.e. $r : C$ occurs within exp_D, we have to replace this reference as well. This can be done by replacing $r : C$ in exp_D by $(a_1 : r_1 : C_1, \dots, a_n : r_n : C_n)$ with new pairwise distinct reference names $r_1, \dots, r_n$.

EXAMPLE 2. Take the schema from Example 1 and fragment the class CITY using $\varphi_1 \equiv$ name $\leq$ 'Paraparaumu' and $\varphi_2 \equiv$ name $>$ 'Paraparaumu'.

Then the new schema will be

`Class` REGION `Struct` (name: *STRING*, central : (ltp : h_1: CITY_1) $\cup$ (gtp : h_2: CITY_2), cities : { (ltp : c_1: CITY_1) $\cup$ (gtp : c_2: CITY_2)})

`Class` CITY_1 `Struct` (name: *STRING*, in : r: REGION, population: *NAT*)

`Class` CITY_2 `Struct` (name: *STRING*, in : r: REGION, population: *NAT*)

3 A Heuristic Method for Horizontal Fragmentation

We now present a heuristic approach to horizontal fragmentation in the context of the OODM. The major objective of the approach is to base the fragmentation decision on the efficiency of the most frequent queries. Thus, we start analysing the selection predicates used in these queries. In the general literature [4, 13] the recommended rule of thumb is to consider only the 20% most frequent queries, as these are supposed to account for about 80% of the data access.

3.1 *Normal Predicates*

Let $Q_1, \ldots, Q_k$ be the queries we are interested in. According to our introduction of the query algebra we may assume to be given the query tree for each of them. Furthermore, we may tacitly assume that these query trees are 'optimised' by applying appropriate query optimisation techniques. Then the leaves of these trees correspond to basic queries in the form C for a class name C, and most predecessors of leaves will correspond to a selection query $\sigma_\varphi(C)$. The selection predicates φ in these queries give rise to simple predicates.

In general, a *simple predicate* for a class C has the form *path* θ v with a path expression *path* on C, a value v of the corresponding type, and a comparison operator Θ, which can be one of $=$, $\neq$, $\leq$, $<$, $>$, $\geq$, $\subseteq$, $\supseteq$, $\not\subseteq$, $\not\supseteq$, $\in$, $\notin$, $\ni$, and $\not\ni$.

EXAMPLE 3. Choose the schema of class REGION from example 1:

`Class` REGION `Struct` (name: *STRING*, central : h: CITY,
cities : { c : CITY }).

There is a simple predicate defined on it for searching which region the city 'Paraparaumu' is in:

$$\varphi \equiv \mathit{value!cities.name} \ni \text{'Paraparaumu'}$$

Now let $\Phi = \{\varphi_1, \ldots, \varphi_m\}$ denote a set of simple predicates on a class C. Then the set of *normal predicates* $\mathfrak{N} = \{\mathcal{N}_1, \ldots, \mathcal{N}_n\}$ on class C is the set of all satisfiable predicates of the form $\mathcal{N}_j \equiv \varphi_1^* \wedge \cdots \wedge \varphi_m^*$, where φ_i^* is either φ_i or $\neg\varphi_i$.

EXAMPLE 4. Take the class schema CITY from example 1:

`Class` CITY `Struct` (name: *STRING*, in : r: REGION, population: *NAT*).

A normal predicate can be defined as:

$$\mathcal{N}_j \equiv \underbrace{\mathit{value.name} \leq \text{'Marton'}}_{\varphi_1} \wedge \underbrace{\mathit{value.population} > 50000}_{\varphi_2}$$

Same as for the RDM [13] the normal predicates do not exhaust all possible selection formulae on a class C. However, as the OODM allows cyclic references to be used on a schema, there are infinitely many such selection formulae. Even more, there are path expressions of arbitrary length, so we have no chance to capture the complete variety of selection. On the other hand, we want to define only finitely many fragments. So we do not lose much by restricting ourself to normal predicates.

Of course, the normal predicates derived from the simple predicates on class C can be used to define the horizontal fragments of class C. The idea of our heuristic procedure for horizontal fragmentation is to reduce the number of simple predicates taken into account and thus to reduce the number of horizontal fragments, if this reduces the query processing costs.

3.2 *Size Calculation*

Crucial to the query costs are the sizes of class instances. So let us first look at them. The calculation of the size

for classes is more complicated in the OODM than the calculation of size for relations in the relational model. In the relational model, we only use the record type constructor. Then the size of a relation can be calculated based on the size of a tuple. In the object oriented model, however, other type constructors such as set and maybe some other bulk type constructors are involved. The size calculation for classes based on the record type constructor only will occur as a special case in the OODM.

Let us first consider the simplest situation: a class C with a representation type $T_C = b$, with b indicating a base type. Let n denote the number of objects in a class instance $db(C)$ over T_C, l_v the average space (in bits) for a value of type b, and l_{id} the size of an identifier, i.e. a value of type ID. Then the average size of $db(C)$ is $s_C = n \cdot (l_v + l_{id})$.

Note that this basic case already includes references, which leads to the type $b = ID$ and $l'_v = l_{id}$.

Now let C be a class defined by a record type constructor, i.e. we have $T_C = (a_1 : T_1, \dots, a_n : T_n)$. The calculation of the size of a class is very similar to that in the relational model. Let n denote the number of objects in a class instance $db(C)$ over T_C, $l_0 = l_{id}$ and l_{id} the size of an identifier, and l_j the average space (in bits) for field a_j of T_C. Then the average size of $db(C)$ over exp_C is $s_C = n \cdot \sum_{j=0}^{k} l_j$.

If there is a class C with a structure expression $T_C = \{exp\}$, the calculation of the size of each class instance is different from the first two cases. Let n denote the number of objects in a class instance $db(C)$ over T_C, l_{id} denote the size of an identifier, m the number of values of an object o in a class instance $db(C)$ over T_C, and l_v the average space (in bits) for a value over exp of an object o in a class instance. Then the average size of a object o over T_C is $l_{id} + m \cdot l_v$ and the average size of class over T_C is $s_C = n \cdot (l_{id} + m \cdot l_v)$.

However, the cost model is still not realistic. Even if we accept the inaccuracy resulting from using only mean values, it is hardly the case that a complex value will be stored as a simple unit, especially if sets are involved. It is more likely to use data structures known from the network data model for this.

If T_C involves a set type constructor, i.e. $T_C = (a_1 : T_1, \dots, a_i : \{T_i\}, \dots, a_n : T_n)$, we would store values of type T_i separately as a linked list and just include a pointer to the first element of this list as well as a pointer back from the last element of the list.

This means that instead of l_v we have to consider $l_{id} + l_v$ assuming the l_{id} is the size of a pointer. Then the whole set value needs the space $l_{id} + m \cdot (l_{id} + l_v)$, where m is again the average number of elements in the set. Note that the size of a set is $m \cdot l_v + (m+1) \cdot l_{id}$. It means that we need $(m+1) \cdot l_{id}$ more space to store the values of a set if we treat it as being stored in a linked list with a pointer to an object o of a class C where m is the number of objects in the set.

3.3 *A Query Processing Cost Model*

The calculation of sizes of class instances applies also to the intermediate results of all queries. These intermediate results correspond to the non-leaf nodes of the query trees. In addition, these nodes correspond to a particular operation of the query algebra.

- The size of a selection node σ_φ is $p \cdot s$, where s is the size of the successor node and $100p$ is the average percentage of objects in the successor satisfying φ.
- The size of a projection node π_{exp} is $(1 - c_i) \cdot s \cdot \frac{l_f}{l_o}$ where $l_f(l_o)$ is the average size of an object over $exp(exp_C)$. s is the size assigned to the successor and c_i is the probability that two classes coincide on exp.
- For a join node the assigned size is $\frac{s_1}{l_1} \cdot p \cdot \frac{s_2}{l_2}(l_1 + l_2 - l)$, where s_i are the sizes of the successors, l_i are the corresponding object sizes, l is the size of a tuple over the common attributes and p is the matching probability.
- For a union node the size is $s_1 + s_2 - p \cdot s_1$ with the probability p for an object of C_1 to coincide with an object of C_2.
- For a difference node the assigned size is $s_1 \cdot (1-p)$ with the probability p for an object of C_1 to coincide with an object of C_2.
- For a renaming node the assigned size is exactly the size s assigned to the successor.
- For a nest node the size is $s + n \cdot (m+1) \cdot l_{id} - n \cdot (m-1) l_{un}$ with l_{un} the length of the unnested attributes, m the average number of elements in the nested set, n the number of objects in a class instance $db(C)$ over T_C.
- For a unnest node the size is $s - n \cdot (m+1) \cdot l_{id} + n \cdot (m-1) l_{un}$ with again l_{un} the length of the unnested attributes, m the average number of elements in the nested set, n the number of objects in a class instance $db(C)$ over T_C.

Fragmentation of a class C results in a set of fragments $\{f_1, \dots, f_n\}$ of average sizes $s_1, \dots, s_n$, if the network has nodes $N_1, \dots, N_k$ we have to allocate these fragments to the nodes, which gives rise to a mapping $\lambda : \{1, \dots, n\} \to \{1, \dots, k\}$, which we call a *location assignment*. However, the fragments only appear on the leaves of query trees. More generally, we must associate a node $\lambda(v)$ with each node v in each relevant query tree. $\lambda(v)$ indicates the node in the network, at which the intermediate query result corresponding to v will be stored.

Given a location assignment λ we can compute the total costs of query processing. Let the set of queries be $Q^m = \{Q_1, \dots, Q_m\}$. Query costs are composed of two parts: *storage costs* and *transportation costs*: $costs_\lambda(Q_j) = stor_\lambda(Q_j) + trans_\lambda(Q_j)$. The storage costs give a measure for retrieving the data back from secondary storage, which is mainly determined by the size of the data. The transportation costs provide a measure for transporting between two nodes of the network.

The storage costs of a query Q_j depend on the size of the involved classes or fragments, respectively, and on the assigned locations which decide the storage cost factors.

It can be expressed as $stor_\lambda(Q_j) = \sum_h s(h) \cdot d_{\lambda(h)}$. where h ranges over the nodes of the query tree for Q_j, $s(h)$ are the sizes of the involved class instances, and d_i indicates the storage cost factor for node N_i $(i = 1, \dots, k)$.

The transportation costs of query Q_j depend on the sizes of the involved classes or fragments of classes and on the assigned locations which decide the transport cost factor between every pair of sites. It can be expressed by $trans_\lambda(Q_j) = \sum_h \sum_{h'} c_{\lambda(h')\lambda(h)} \cdot s(h')$. Again the sum ranges over the nodes h of the query tree for Q_j and, h' runs over the predecessors of h in the query tree, and c_{ij} is a transportation cost factor for data transport from node N_i to node N_j $(i, j \in \{1, \dots, k\})$.

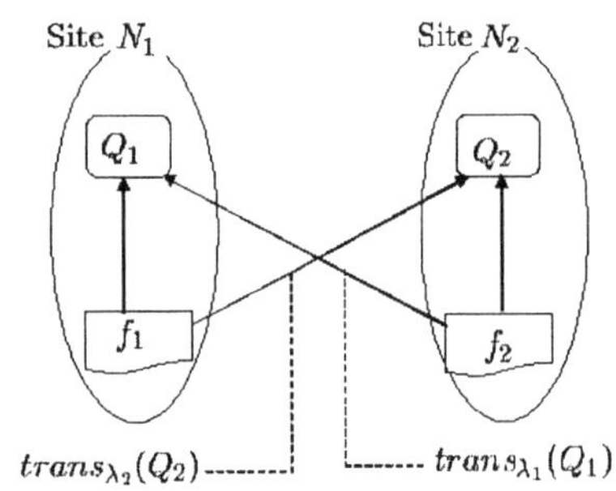

Fig. 1. Allocation of Fragments to Two Sites

For each query Q_j we get a value for its frequency $freq_j$. The total costs of all the queries in Q^m are the sum of the costs of each query multiplied by its frequency. It can be expressed by $cost_\lambda = \sum_{j=1}^{m} cost_\lambda(Q_j) \cdot freq_j$.

3.4 *The Fragmentation Procedure*

Fragmentation and allocation are generally considered as two isolated problems in [13] and [4]. After fragmentation, the fragments are allocated to reside at one node in a distributed management system (not considering replication at this stage in this paper). There is no cost model involved for the fragmentation design in [13]. But we argue that the values of the costs of queries after fragmentation will affect the decision on whether we need to perform fragmentation or not. A cost model should be used to evaluate different fragmentatin solutions. Let us look at the following example to see whether fragmentation of a class and allocating result fragments to different sites will achieve better performance.

EXAMPLE 5. Consider a class being fragmented into two fragments f_1, f_2, and two queries Q_1, Q_2 executing at two different sites N_1, N_2 to access these two fragments remotely. The frequencies of Q_1 and Q_2 are $freq_1$ and $freq_2$, respectively. This design is shown in Figure 1.

If the sizes of f_1 and f_2 are s_1 and s_2, respectively, ignoring storing costs for all the fragments, we have total query costs of:

$$\begin{aligned} cost_{\lambda_1} &= trans_{\lambda_1}(Q_1) + trans_{\lambda_1}(Q_2) \\ &= s_2 \cdot freq_1 \cdot c_{21} + s_1 \cdot freq_2 \cdot c_{12} \end{aligned}$$

with c_{12} and c_{21} as transportation cost factors. Generally, c_{12} should be equal to c_{21}.

If query Q_1 is executed more frequently, say $freq_1 > freq_2$, we do not fragment the class and put the whole class at site N_1, the site that Q_1 is executed. This design is shown in Figure 2.

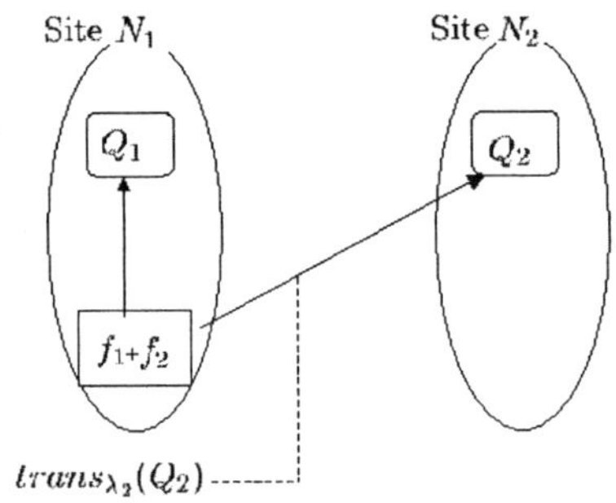

Fig. 2. Allocation of Class to One Site

The total query costs for the second design are:

$$cost_{\lambda_2} = trans_{\lambda_2}(Q_2) = (s_1 + s_2) \cdot freq_2 \cdot c_{12}$$

Comparing costs $cost_{\lambda_1}$ and $cost_{\lambda_2}$ we get $cost_{\lambda_1} > cost_{\lambda_2}$.

It can be concluded that fragmentation of classes does not always minimize the query cost. In general, the distribution could be called optimal, if we find a location assignment λ and a fragmentation solution such that the resulting total query costs are minimal. As this problem is computably intractable, we suggest to use a heuristic approach instead.

For a given database schema $\mathcal{S} = \{C_1, \dots, C_i, \dots C_n\}$, there is a set of queries $\{Q_1, \dots, Q_m\}$ that access the database most frequently or that are used by the most critical transactions. Then the heuristic procedure of horizontal fragmentation includes the following steps:

1. Identify the set of most frequent queries accessing class C_i and rewrite them using the query algebra.
2. Identify the sites at which the queries will be issued. Treat queries that are started at several sites as several different queries. Estimate for each query the frequency of its execution.
3. Sort queries by their frequencies. Then we get a list of queries $[Q_1, \dots, Q_j, \dots, Q_m]$ with the corresponding list of values of frequencies $F^m = [freq_1, \dots, freq_m]$ for Q^m, i.e. we always get $freq_{k-1} \geq freq_k$.
4. Determine optimized query trees for all the queries. Extract simple predicates from these trees. We get a set Φ of simple predicates φ.
5. Construct a usage matrix based on the simple predicates obtained in the previous steps, i.e. determine which simple predicates are used by which queries.
6. From the matrix obtained in the previous step we can get a list Φ_i^b of sorted simple predicates $\Phi_i^b = [\varphi_{i1}, \dots \varphi_{ib}]$ where we always get $freq_{j-1} \geq freq_j$. The number of simple predicates is b. We get a list $X = [0, 1, \dots, x_1, \dots, x_2, \dots, b]$ of indices for the simple predicates.

7. Perform the following steps iteratively to find a reasonable number of simple predicates for horizontal fragmentation and fragment the given class C_i:
   Choose four numbers a, b, x_1, x_2 from X with $a < x_1 < x_2 < b$, including initially the two bounds, i.e. start with $0, x_1, x_2, b$ with $0 < x_1 < x_2 < b$. For each number $x \in \{a, b, x_1, x_2\}$ we calculate the corresponding query costs by the following procedures:

   (a) Choose first x simple predicates in the list Φ_i^b, i.e. $\Phi_i^x = \{\varphi_{i1}, \dots \varphi_{ix}\}$ and build the corresponding set of normal predicates $\mathcal{N}^x$.
   (b) Fragment the class C_i according to the set $\mathcal{N}^x$ of normal predicates obtained in the previous step. Let the fragments be $F_1^x, \dots, F_r^x$.
   (c) Calculate the frequencies $freq_k^j$ of access to the fragments F_k from site j, and use this to determine fragment allocation to sites. Put the fragments to the nodes that access them most frequently.
   (d) Calculate the total query costs $cost_x$ with the cost model introduced above.

   We get four values for the query costs $cost_a, cost_{x_1}, cost_{x_2}, cost_b$. Let min denote the minimal value among these four values. Comparing these query costs we might have the following four situations:

   – If $min = cost_a$ then set $b = x_1$ and choose two new values for x_1, x_2 satisfying $a < x_1 < x_2 < b$. If we can find such two numbers then continue the procedure. If we can only get one number x from the list X, calculate the query cost for it. If $cost_x < cost_a$, then x should be the number of simple predicates that we should use to fragment the class C_i. Thus, let $y := x$.
     If $cost_x > cost_a$ or there is no number left between the new a and b, then a is the number of simple predicates that we will use for fragmentation. Thus, let $y := a$.
   – If $min = cost_{x_1}$, then set $b = x_2$ and choose a new number between a and b so that we get new numbers for x_1, x_2. Continue with the procedure.
     If we can not find a new number between new a and b, then x_1 is the number of simple predicates that we should choose for fragmentation, thus $y := x_1$.
   – If $min = cost_{x_2}$, then set $a = x_1$. Choose a new number between a and b so that we get new values for x_1, x_2 and continue the procedure with the new $a < x_1 < x_2 < b$. If we do not find such a new number, x_2 will be the number of simple predicates that we need for fragmenting the class C_i, i.e. take $y := x_2$.
   – If $min = cost_b$, then set $a = x_2$ and find two new numbers between a and b satisfying the condition $a < x_1 < x_2 < b$. If we can find such numbers, then we continue the iteration with the new values.
     If there is only one number x left between a and b, we calculate the corresponding query cost and compare it with $cost_b$. The number that leads to the minimal cost will be the number of simple predicates that we need for horizontal fragmentation, i.e. for $cost_x < cost_b$ we choose $y := x$, otherwise $y := b$.
     If there is no number between a and b left in X, then b is the number of simple predicates needed for fragmenting C_i. Thus, $y := b$.

   The result is the set $\Phi_i^y = \{\varphi_{i1}, \dots \varphi_{iy}\}$ of simple predicates. Take the corresponding set $\mathcal{N}^y$ of normal predicates for fragmenting C_i.
   Note that a, b, x_1, x_2 are updated in each loop. Searching requires only one or two new points in each loop. Only for these new points, query costs have to be calculated. The procedure stops when the remaining interval (a, b) only contains one number or is empty.

The above procedure is a heuristic one based on the assumption that a reasonable fragmentation schema can be obtained by looking at most frequently used simple predicates. Our approach with the above steps can rapidly search for a reasonable number of simple predicates which results in presumably low total query costs.

3.5 *Analysis*

In this section horizontal fragmentation for object oriented databases has been discussed in more detail. Definitions for simple predicates and normal predicates have been outlined. A heuristic horizontal fragmentation procedure has been proposed based on a cost model. The horizontal fragmentation approach presented in this section paper is superior to that presented in [13] and the one by [6]. Characteristics and potential benefits of the approach presented in this section, and differences to other approaches in the literature can be summarized as follows:

Firstly, the approach can deal with both simple and complex attributes using the same procedure. In [6], simple and complex attributes are treated with different algorithms. Simple attributes refer to the attributes of primitive attribute types only, i.e. those that do not contain other classes as part of them. Complex attributes have the domain of attribute as another class [6]. In our approach, attributes are defined using the underlying object oriented type system. Attributes defined on the abstract identifier type ID are treated as being defined on one of the base types. Our approach is more universal than others in terms of dealing with attributes of different types, and is more easily put into practical use.

Secondly, the expression of simple predicates is extended such that simple predicates can be defined on identifiers as well as on values. The expression of simple predicates on values is also extended to suit various type constructors in the underlying type system of the object oriented model. In [6], the definition of simple predicates is adopted from [13] without any adaption. However, the format of simple predicates in [13] cannot deal with the situation that a simple predicate is defined on a complex type constructor, i.e. a finite set type constructor. In the approach presented, the format of simple predicates is extended in a way that simple predicates can be defined not only on record type constructors or base type constructors but also on finite set type constructors. Note that the domain of the values for the comparison operator θ has been extended to include some set comparing operations. Path expressions have been introduced to refer to any elements of a class.

Thirdly, instead of using minterm predicates as found in [13] and [6], we introduced normal predicates on classes as the satisfiable minterm predicates. Horizontal fragmentation operations based on a set of normal predicates could therefore be guaranteed to satisfy the characteristics of fragmentation discussed in [13]. The approach did not rely on dependencies between simple queries as in [13], because these can hardly be determined. It is very hard to use these dependencies to determine the satisfiability of a conjunction of simple predicates, or to simplify them if they are satisfiable.

Fourthly, the approach applies a cost model for horizontal fragmentation design. In [13], horizontal fragmentation is performed without evaluating the overall system performance. We argue that the larger degree of fragmentation does not necessary lead to the better system overall performance. There exists a cut off point for the degree of horizontal fragmentation that the system has the best performance. However it is computationally intractable to find the optimized fragmentation solution by comparing total costs for all possible fragmentation schemata [13]. The heuristic procedure proposed in this section paper is based on a cost model with which the system performance can be evaluated once a database is being fragmented. One of the characteristics of this procedure is that it can rapidly achieve a horizontal fragmentation schema that is designed to result in low total query cost, or, in other words, the system's overall performance being improved.

4 Conclusion

In this paper we presented a heuristic approach to horizontal fragmentation for object oriented databases. The major objective is to provide a tractable approach to minimising the query processing costs for the most frequent queries. In general, this would require to consider all possible fragmentations and all possible allocations of intermediate query results to the nodes of a network, which is intractable. Instead of this we suggest to consider reducing the number of fragments by determining the right cut point in a list of simple predicates.

Our approach is based on a rather sophisticated data model, which in general needs more than horizontal and vertical fragmentation for distribution design. For this we introduced a third fragmentation operation called splitting. We are currently working on vertical and splitting fragmentation. Experimental evaluation for the proposed procedure is in its early phases. The work presented in this paper is our first attempt to overcome deficiencies of distribution design for object oriented databases. We consider the use of a simplistic data model and the neglection of intermediate query results as the most striking examples of such deficiencies.

References

1. Abiteboul, S., and Kanellakis, P. Object identity as a query language primitive. In *Proc. SIGMOD 1989* (1989), ACM.
2. Apers, P. M. G. Data allocation in distributed database systems. *ACM Trans. Database Syst. 13* (1988), 263–304.
3. Bellatreche, L., Karlapalem, K., and Simonet, A. Algorithms and support for horizontal class partitioning in object-oriented databases. *Distributed and Parallel Databases 8*, 2 (2000), 155–179.
4. Ceri, S., and Pelagatti, G. *Distributed Databases Principles and System.* McGraw-Hill, New York, 1984.
5. Chinchwadkar, G. S., and Goh, A. An overview of vertical partitioning in object oriented databases. *The Computer Journal 42*, 1 (1999).
6. Ezeife, C. I., and Barker, K. A comprehensive approach to horizontal class fragmentation in a distributed object based system. *Distributed and Parallel Databases 3*, 3 (1995), 247–272.
7. Ezeife, C. I., and Barker, K. Distributed object based design: Vertical fragmentation of classes. *Distributed and Parallel Databases 6*, 4 (1998), 317–350.
8. Khalil, N., Eid, D., and Khair, M. Availability and reliability issues in distributed databases using optimal horizontal fragmentation. In *Database and Expert Systems Applications* (1999), T. J. M. Bench-Capon, G. Soda, and A. M. Tjoa, Eds., vol. 1677 of *Lecture Notes in Computer Science*, Springer, pp. 771–780.
9. Ma, H. Distribution design in object oriented databases. Master's thesis, Massey University, 2003.

10. MA, H., AND SCHEWE, K.-D. Fragmentation of XML documents. In *Proceedings XVIII Simpósio Brasileiro de Bancos de Dados (SBBD 2003)* (Manaus, Brazil, 2003), pp. 200–214.
11. MALINOWSKI, E., AND CHAKRAVARTHY, S. Fragmentation techniques for distributing object-oriented databases. In *Conceptual Modeling - ER '97* (1997), D. W. Embley and R. C. Goldstein, Eds., vol. 1331 of *Lecture Notes in Computer Science*, Springer, pp. 347–360.
12. NAVATHE, S. B., AND RA, M. Vertical partitioning for database design: A graphical algorithm. *ACM SIGMOD 14*, 4 (1989), 440–450.
13. ÖZSU, M. T., AND VALDURIEZ, P. *Principles of Distributed Database Systems.* Alan Apt, New Jersey, 1999.
14. RA, M. Horizontal partitioning for distributed database design. In *Advances in Database Research* (1993), M. Orlowska and M. Papazoglou, Eds., World Scientific Publishing, pp. 101–120.
15. SCHEWE, K.-D. On the unification of query algebras and their extension to rational tree structures. In *Proc. Australasian Database Conference* (2001), M. Orlowska and J. Roddick, Eds.
16. SCHEWE, K.-D. Fragmentation of object oriented and semi-structured data. In *Databases and Information Systems II* (2002), H.-M. Haav and A. Kalja, Eds., Kluwer Academic Publishers, pp. 1–14.
17. SCHEWE, K.-D., AND THALHEIM, B. Fundamental concepts of object oriented databases. *Acta Cybernetica 11*, 4 (1993), 49–84.
18. TAMHANKAR, A. M., AND RAM, S. Database fragmentation and allocation: An integrated methodology and case study. *IEEE Transactions on Systems Management 28*, 3 (1998), 194–207.
19. ZHANG, Y. On horizontal fragmentation of distributed database design. In *Advances in Database Research* (1993), M. Orlowska and M. Papazoglou, Eds., World Scientific Publishing, pp. 121–130.

Databases and Information Systems
J. Barzdins and A. Caplinskas (Eds.)
IOS Press, 2005

Successful Database Integration through View Cooperation

Vojtech VESTENICKÝ
Charles University, Malostranské nám. 25, 118 00 Prague 1, Czech Republic
vestenicky@ksi.ms.mff.cuni.cz

Abstract. We present an approach for cooperation between databases modelled as HERM schemas. Cooperation between semantically related parts is based on views. Such views transparently support an information exchange between schemas. View definitions are generated in form of HERM algebra expressions by rules. The rules aim at solving integration conflicts, including constraints, on schema and instance level. The advantage is their simplicity and low complexity (i.e. polynomial). They can be combined into new rules to cover more complicated cases. Resulting views in HERM algebra can be translated into SQL and deployed on any common DBMS.

1 Introduction

Full view integration aims at developing a more general schema and a set of selectors on this schema such that each view is generated by a selector. The drawback is the undecidability of view integration in general and the infeasibility of view integration in most cases [3]. The advantage is that it provides a general solution if there is a solution.

View consistency finds the mechanisms which can be used for the cooperation or co-existence of views. View cooperation enables exchange of information between views. The disadvantage is the need for computational support. The advantage is the existence of view cooperation in situations where view integration is unlikely [11].

Our approach proposes a cooperation strategy. We argue that it has several advantages over the approaches aiming at full view integration:

In global as view approaches (GAV), like COIN [4], MOMIS [1] and IBIS [2], the complexity is high, constraints either simple or omitted and there is no support for complex types.

In local as view approaches (LAV) [7, 5, 6] we face the same problems as in GAV.

The combination of the GAV and LAV introduced in [8] as Both As View (BAV) approach uses reversible schema transformations. The approach concentrates only on schema related conflicts and neglects the constraints.

A well known approach to solving structural conflicts between schemas using view integration has been proposed by Spaccapietra [9]. Transformation rules produce new views that can be used to obtain an integrated schema. Our approach resembles the Spaccapietra's work in some points: we use rules to generate views, we constructs the functions to map values and consider constraints. However, the complexity of generating the views in our approach is lower (especially in rules including paths). We do not modify the views to construct an integrated schema, because it is not always possible and often not desired. We distinguish two types of views: query views and updatable views. This separation allows for maximal information extraction and use of different update strategies. The cooperation relates to interoperable database systems. It is defined by functions which map one database system into another. Two generated views may support different operations, e.g. for an update there can exist mapping only in one direction. This offers much more flexibility. The success of the integration process is closely related to a data model and modeling techniques [12]. Poor modeling techniques result in poor models which are very hard to integrate. We have chosen a Higher-order Entity-Relationship Model (HERM) [10] to represent all participating schemas. HERM has advantages over the traditional ER Model and carries more semantic. Thus, we are able to handle complex types and more constraints than other approaches.

We now briefly introduce some of HERM extensions:

For pairwise different nested attributes $X_1, \ldots, X_n$ is $X(X_1, \ldots, X_n)$ a *nested tuple attribute*.

If Y is a nested attribute then $X\{Y\}$ is a *nested set attribute*.

We denote a set of attributes of some type R (R can be also the whole schema) with $attr(R)$, a set of entity types $ent(R)$ and a set of relationship types $rel_i(R)$, where i is the order of the relationship types (e.g. $ent(R) = rel_0(R)$). A set of components of a relationship type R is $comp(R)$.

A *cluster type* $C = R_1 + R_2 + \ldots + R_k$, $C^C = \bigcup_{i=1}^{k} R_i^C$ (C^C and R^C are instances of a cluster and relationship types respectively), where $\forall R_i^C, R_j^C$, $i \neq j, i, j = 1, \ldots, k$, $R_i^C \cap R_j^C = \emptyset$. If $R_1, \ldots, R_k$ are entity types then C is a cluster of entity types. The cluster type models the super/subtype relationship with more than one subtype. We require that identification types of the components of the cluster are domain-compatible.

The semantic correspondence between R_1 and R_2, entity/relationship types from different schemas, is given by a relation $\mathcal{E}(R_1, R_2)$.

For the rest of the paper we assume two different schemas S_1 and S_2 as an input to the integration process. The relation $\mathcal{E}$ is defined only for the elements of these schemas. We further require that both schemas S_1 and S_2 are connected (see Def. 1).

Definition 1. *Schema $S = \{E_1, \ldots, E_m, R_1, \ldots, R_n\}$ is called connected, if all relationship types have only entity types $E_1, \ldots, E_m$ and relationship types $R_1, \ldots, R_n$ as their components.*

For the definition of views we use *HERM algebra operations*: σ_ϕ (selection) with a selection formula ϕ, $\pi_{A_1,\ldots,A_m}$ (projection) with a generalized subset $\{A_1, \ldots, A_m\}$, ρ_f (renaming) with a renaming function f, $\bowtie_G$(join) with a common generalized subset G, $\cup$ (union), – (difference), $\nu_{X:\{A_1,\ldots,A_n\}}$(nest) and μ_A (unnest) with a set attribute A (see [10] for details).

HERM diagram is a graphical representation of the HERM model. The clusters are represented with a symbol $\oplus$. The edges $E(ntity) \rightarrow A(ttribute)$ can be labelled by $dom(A)$. Other edges could be labelled by integrity constraints, e.g. cardinality constraints. All attributes A of a $R(elationship)$ constituting a key, i.e. $A \in id(R)$, are underlined.

This approach builds on the theory introduced in [10], where the integration process is described as follows: Given the two database schemas S_1, S_2 and the corresponding database states S_1^C, S_2^C.

1. We must find the functions f_{12} and f_{21} which realize the partial embedding of types from schema S_1 to S_2 and S_2 to S_1, i.e. $f_{12} : S_1 \overset{\theta}{\rightarrow} S_2$, $f_{21} : S_2 \overset{\theta}{\rightarrow} S_1$.

2. Then the functions f_{12}^C and f_{21}^C need to be constructed which define a partial embedding of corresponding instances from S_1^C to S_2^C, i.e. $f_{12}^C : S_1^C \twoheadrightarrow S_2^C$, $f_{21}^C : S_2^C \twoheadrightarrow S_1^C$.

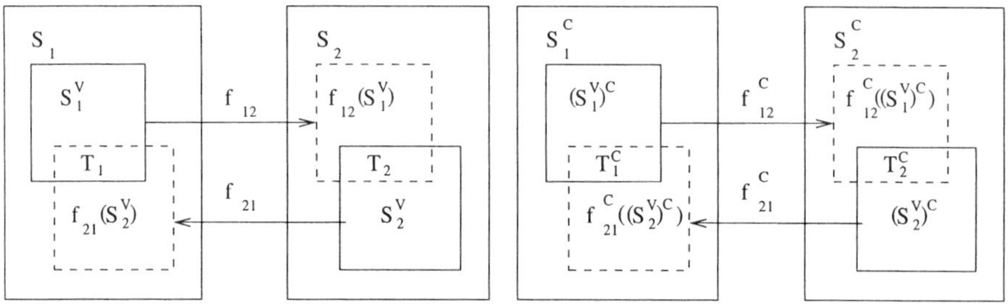

Figure 1: View Cooperation

The two schema morphisms f_{12}, f_{12}^C and f_{21}, f_{21}^C cooperate and define a view cooperation (see Fig. 1) if for each $T_1 \in S_1^V \cap f_{21}(S_2^V)$ and each $T_2 \in S_2^V \cap f_{12}(S_1^V)$, for each pair $T_1^C \in S_1^C$ and $T_2^C \in S_2^C$

- the functions $f_{12}^C(T_1^C)$,$f_{21}^C(f_{12}^C(T_1^C))$,$f_{21}^C(T_2^C)$,$f_{12}^C(f_{21}^C(T_2^C))$ are defined and
- the equalities $f_{21}^C(f_{12}^C(T_1^C)) = T_1^C$, $f_{12}^C(f_{21}^C(T_2^C)) = T_2^C$ are valid.

The functions f_{12} and f_{21} are in our approach represented by the view definitions in HERM algebra.

2 Integration Process

It is comprised of the following main phases:

1. **Obtain views for queries**: application of rules and generation of view definitions using the HERM algebra for each pair of semantically corresponding components in $\mathcal{E}$, no constraints are considered. The result is a maximal set of views containing semantically related components and their attributes.

2. **Obtain views for updates**: rules generate updatable views by adjusting the query views to satisfy constraints.

3. **Obtain mapping functions for instances**: construction of mapping functions for the view instances.

4. **Obtain cooperating views**: searching for overlapping parts in the pairs of the views. These parts represent then the final result of the integration process, cooperating views.

5. **Translation into SQL**: the resulting view definitions are translated into SQL statements that create the actual views in the underlying database.

There are some typical cases which cover common conflicts. They are related not only to the construction of the rules, but also to the construction of the views. View definitions in our approach can be limited to two *view patterns*:

- The views representing the semantic correspondence of two entity/relationship types and all their semantically related attributes are contained within these types, look like $\pi_{attr}(\rho_f(R))$, where $attr$ is a list of semantically related attributes, function f solves the naming inequalities by defining the mapping from one name space to another and R is the relation from which the mapping starts.

- Structural incompatibilities are the main reason for the existence of these view types. If the semantically related components are in different relationship types then it is necessary to access those types and extract the desired information. The view definition must therefore contain a join: $\pi_{attr}(\rho_f(R_1 \bowtie (\ldots \bowtie (R_{n-1} \bowtie R_n))))$, where $\{R_1, \ldots, R_n\}$ is a set of relationship types that are connected and have the semantically related attributes as their components.

2.1 *Obtain Views for Queries*

These views contain the maximal possible set of semantically related components. They aim at information extraction and may not be updatable. They serve as building components for further view definitions. The views satisfying the restrictions posed by different schemas are a subset of these query views, i.e. theirs specialization.

Views are generated by rules. The rules expect as an input two types from different schemas S_1 and S_2. They can be entity/relationship types and must satisfy the condition stated in the beginning of the rule. These conditions allow to handle different types of integration problems. We mention only a few rules for illustration (for others see [13]).

1. **For entity types one of them with nested set attribute** $E_1 \in S_1$ and $E_2, E_3, R \in S_2$ where:

   - $E_1 = (A_1, \ldots, A_n, A\{X\})$, $A_1, \ldots, A_n$ are simple attributes, $A\{X\}$ is a nested set attribute containing the type X,
   - $id(E_1) \in attr(E_1) - \{A\}$, i.e. nested attribute is not a part of the key,
   - $E_2 = (B_1, \ldots, B_k)$, $(E_1, E_2) \in \mathcal{E}$, i.e. E_1 and E_2 are semantically related,
   - $E_3 = (C_1, \ldots, C_l)$, $\exists c \in attr(E_3).(c, A) \in \mathcal{E}$, i.e. for the complex attribute A of E_1 there exists some semantically corresponding attribute c in E_3,
   - $comp(R) = \{E_2, E_3\}$, i.e. E_2 and E_3 are components of the relationship type R.

   This rule handles the case where an entity type in the first schema contains a nested set attribute and in the second schema there exists a decomposition into two entity types one of them holding the equivalent of the nested attribute.

   We explain the situation using the Fig. 2. The entity type $Book$ in the schema S_1 has a nested set attribute $Author$ containing the set of elements of type $Name$. This simple example reflects the fact, that a book may have been written by more authors and we want to keep the track of their names. In schema S_2 can the same situation be modelled using the entity type $Book$ and a separate entity type $Writer$ with the attribute $Name$. Both entity types of schema S_2 are connected through the relationship type $written\_by$. We define the views for this example first. Let us assume $\mathcal{E}$ contains the semantic correspondences between components: $\mathcal{E} = \{(Book, Book), (Author, Name), (Title, Title)\}$. The views are then:

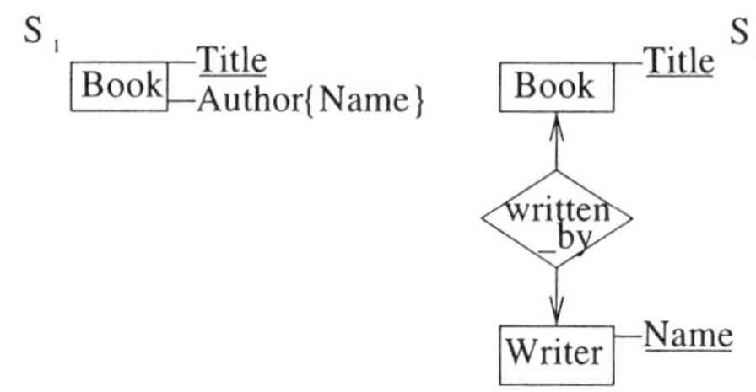

Figure 2: Entity $Book$ with the nested set attribute $Author$ and the decomposed equivalent in schema S_2

- $V_{S_1.Book,S_2.Book} = \pi_{Title,Name}(\rho_{Author \rightarrow Name}(S_1.Book))$,
- $V_{S_2.Book,S_1.Book} = \pi_{Title,Author}(\rho_{Name \rightarrow Author}(S_2.Book \bowtie (written\_by \bowtie Writer)))$.

Note: the attribute $Name$ in the projection part of the view $V_{S_1.Book,S_2.Book}$ is a complex attribute and holds the set of author names. We must therefore transform this complex attribute before we can work with instances, see Sec. 2.3.

Having defined the views in the above example, we turn to the general case and construct:

- *the mapping functions for the attributes*:
  $map_{E_1,E_2} = \{(a,b) | a \in attr(E_1), b \in attr(E_2) \cup attr(E_3), (a,b) \in \mathcal{E}\}$,
  $map_{E_2,E_1} = \{(a,b) | a \in attr(E_2) \cup attr(E_3), b \in attr(E_1), (a,b) \in \mathcal{E}\}$.
- *view definitions* using the HERM algebra:
  - $aset_{E_1,E_2} = \{b | (a,b) \in map_{E_1,E_2}\}$,
    $V_{E_1,E_2} = \pi_{aset_{E_1,E_2}}(\rho_{map_{E_1,E_2}}(E_1))$
  - $aset_{E_2,E_1} = \{b | (a,b) \in map_{E_2,E_1}\}$,
    $V_{E_2,E_1} = \pi_{aset_{R_2,R_1}}(\rho_{E_2,E_1}(E_2 \bowtie (R \bowtie E_3)))$

Other structural conflicts involving complex types can be solved by similar rules.

2. **For two relationship types $R_1 \in S_1$, $R_2 \in S_2$ and semantically related attributes are not contained within the matching relationship type** where:

   - $R_1 = (comp(R_1), A_1, \ldots, A_n)$,$R_2 = (comp(R_2), B_1, \ldots, B_m)$ and $(R_1, R_2) \in \mathcal{E}$.

   Here the attributes can have their semantic equivalents in different relationship/entity types. See Fig. 3, e.g. the semantically corresponding attributes A' are outside the matching type R_1. The dotted lines represent the semantic correspondence between the attributes and relationship types. We determine the set of corresponding attributes for R_1 and R_2:

   $aset^{e\_attr}_{R_1} = \{B_j | A_i \in attr(R_1), i = 1, \ldots, n, (A_i, B_j) \in \mathcal{E}\}$
   $aset^{e\_attr}_{R_2} = \{A_i | B_j \in attr(R_2), j = 1, \ldots, m, (A_i, B_j) \in \mathcal{E}\}$

   Then we find relationship types where these attributes reside, e.g. in Fig. 3 $R_{1,1}$, $R_{1,2}$.

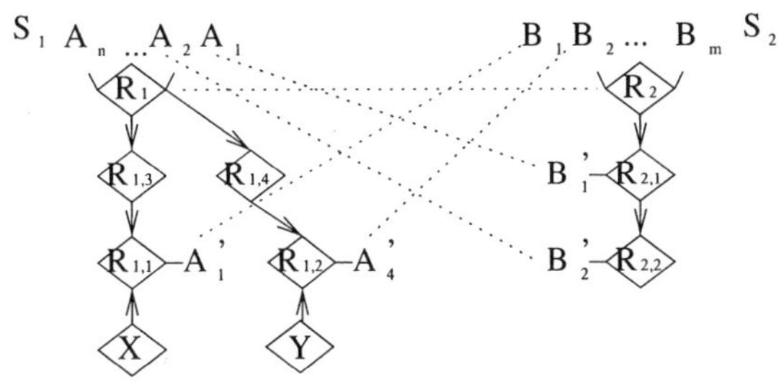

Figure 3: Related attributes are outside the matching types R_1 and R_2

Definition 2. *Function* $f_{rel} : A \rightarrow P$, $A \in attr(S)$, $P \in rel(S)$ *is defined as* $f_{rel}(A) = \{p | a \in A, a \in attr(p)\}$.

Assume we get the sets of relationship types for both attribute sets: $rset_{R_1}^{rel} = f_{rel}(aset_{R_2}^{e\_attr})$ and $rset_{R_2}^{rel} = f_{rel}(aset_{R_1}^{e\_attr})$. For each of the above sets other types must be found, such that all types are connected through some path. Those nodes must exist in each schema because the schemas are connected. If we look at the Fig. 3 we can see the types, e.g. $R_{1,3}^{'}$ and $R_{1,4}^{'}$. We define the function that returns those extended sets using partially an algorithmic description:

Definition 3. *The function* $find : R \times R' \rightarrow R''$, *where* R *is an input set of relationship types,* R' *is a set of all relationship types of schema* S, *and* R'' *is a result set of connected relationship types containing* R.

function *find*$(rset, rschema)$
$Con = \{r | r \in rset, s \in rset - \{r\}, r \in comp(s) \vee s \in comp(r)\}$ `/** already connected`
$Con\_r = \{r | r \in Con, \nexists s \in Con - \{r\}.r \in comp(s)\}$ `/** root types`
$Con\_s = Con - Con\_r$
$rset\_r = rset - Con\_r$
if $(Con\_r = rset\_r)$ **then return** $rset$ `/** check if all connected`
else
begin
$new\_r = \{s | r \in rset\_r, s \in rschema, r \in comp(s)\}$ `/** extend the set`
$new\_input = new\_r \cup rset\_r$
find=$Con\_s \cup$ *find*$(new\_input, rschema)$ `/** check new extended set`
end

Theorem 1. *Function find terminates and its complexity is polynomial, i.e.* $O(3n^3)$, *where* $n = |rschema|$ *(number of schema relationship types).*

The proof is straightforward. The function $find$ may return some types which are not needed to keep the required components connected (see Fig. 3 X and Y). The following function removes unnecessary types from the resulting set.

Definition 4. *Function* $remove : I \times R \rightarrow R'$, *where* I *equals the parameter rset of function find,* R *is the result from find, and* R' *is a resulting set of relationship types without unnecessary types.*

function $remove(Input, Result)$
$Test = Result - Input$
while $|Test| > 0$ **do**
 $t \in Test$
 $IsCon = Result - \{t\}$ `/** try to remove one`
 if $(\forall x, y \in IsCon. x \neq y \Rightarrow (x \in comp(y) \vee y \in comp(x)))$ **then** `/** still connected?`
  $Result = Result - \{t\}$ `/** OK, remove it`
 $Test = Test - \{t\}$
do while
return $Result$

Theorem 2. *Function remove terminates and its complexity is* $O(n^3)$, *where* $n = |Test|$.

Assume we executed the $find$ and $remove$ functions:
$Result_{R_1} = find(rset_{R_1}^{rel}, rel(S_1)), rset_{R_1}^{Con} = remove(rset_{R_1}^{rel}, Result_{R_1}),$
$Result_{R_2} = find(rset_{R_2}^{rel}, rel(S_2)), rset_{R_2}^{Con} = remove(rset_{R_2}^{rel}, Result_{R_2})$
and received the results of the form: $rset_{R_1}^{Con} = \{R_1, R_{11}, \ldots, R_{1c}\}$, $rset_{R_2}^{Con} = \{R_2, R_{21}, \ldots, R_{2d}\}$. Notice that each of the previous sets have the matching relationship type R_1 or R_2 as their element. We further need *root types* of these sets.

Definition 5. *Function* $root\_types : R \rightarrow R'$, *where* R *is a set of relationship types and* R' *is a set of root types,* $root\_types = \{r | r \in R, \nexists s \in R - \{r\}. r \in comp(s)\}$.

Theorem 3. *Function* $root\_types$ *runs in* $O(n^2)$, *where* $n = |rset|$.

Suppose we executed the function: $rset_{R_1}^{Root} = root\_types(rset_{R_1}^{Con})$, $rset_{R_2}^{Root} = root\_types(rset_{R_2}^{Con})$ and obtained the sets: $rset_{R_1}^{Root} = \{R_1, R_{11}, \ldots, R_{1e}\}$, $rset_{R_2}^{Root} = \{R_{21}, \ldots, R_{2f}\}$.

We now have the root types in $rset_{R_i}^{Root}$ and the connected types in $rset_{R_i}^{Con}$, for $i = 1, 2$. It is further valid that $rset_{R_i}^{Root} \subseteq rset_{R_i}^{Con}$. We separate the non-root types out of the connected set of types: $rset_{R_i}^{\overline{Root}} = rset_{R_i}^{Con} - rset_{R_i}^{Root}$. Let those respected sets be: $rset_{R_1}^{\overline{Root}} = \{R_{1,p}, \ldots R_{1,r}\}$, $rset_{R_2}^{\overline{Root}} = \{R_{2,s}, \ldots R_{2,t}\}$. The root types will be used in the join part of the view first, then come the non-root types. This is based on the assumption that the root types possess also the keys of the non-root types.

Theorem 4. *For the root types in a set of* $rset_{R_i}^{Root}$ *there exists a sequence* $Seq = \langle r_1, \ldots, r_n \rangle$, *where* $n = |rset_{R_i}^{Root}|$ *such that:*
$\forall r_j \in rset_{R_i}^{Root}, j = 1, \ldots, n, i = 1, 2.((id(r_1) \subseteq id(r_2)) \vee (id(r_2) \subseteq id(r_1))) \wedge ((id(r_2) \subseteq id(r_3)) \vee (id(r_3) \subseteq id(r_2))) \wedge \ldots \wedge ((id(r_{j-1}) \subseteq id(r_j)) \vee (id(r_j) \subseteq id(r_{j-1}))) \Rightarrow \langle r_1, \ldots, r_n \rangle$.

Proof: the existence of a sequence Seq is guaranteed by the connectedness of the input schemas.

We need this sequence for the join part of the view where the relationship types being joined should have some common set of joining attributes.

Definition 6. *Function* $root\_seq : R \rightarrow (R', I)$ *gets as an input parameter the set of root types* R *and returns the set of pairs* (R', I) *where* R *is a root type and* I *is the order in the sequence.*

```
function root_seq(Root)
r ∈ Root
Key = id(r) /** current set of keys
i = 1
Result = Result ∪ {(r, i)} /** first in sequence
Root = Root − {r}
while (|Root| > 0) do
 for s ∈ Root do /** find next
 if ((Key ∩ id(s)) ≠ ∅) then /** check if common keys exist
 begin
  i = i + 1
  Result = Result ∪ {(s, i)} /** next in the sequence found
  Key = Key ∪ id(s) /** extend the key set
  Root = Root − {s}
 end
 do for
do while
return Result
```

Theorem 5. *Complexity of the function* $root\_seq$ *is* $O(n^2)$, *where* $n = |Root|$.

We now have all what we need to define the views:

- *the mapping functions for the attributes*:
  $map_{R_1,R_2} = \{(a, b)|a \in aset_{R_2}^{e\_attr}, b \in R_2, (a, b) \in \mathcal{E}\}$,
  $map_{R_2,R_1} = \{(a, b)|a \in aset_{R_1}^{e\_attr}, b \in R_1, (a, b) \in \mathcal{E}\}$,
  the attribute sets for the projection part:
  $aset_{R_1} = \{b|(a, b) \in map_{R_1,R_2}\}$,$aset_{R_2} = \{b|(a, b) \in map_{R_2,R_1}\}$,
- *view definitions*:
  Suppose we executed the function *root_seq* and got the following sets:
  $Seq_{R_1}^{Root}$ =root_seq($rset_{R_1}^{Con}$),$Seq_{R_1}^{Root} = \{(R_{1,1}, 1), \ldots, (R_{1,g}, g)\}$,
  $Seq_{R_2}^{Root}$ =root_seq($rset_{R_2}^{Con}$),$Seq_{R_2}^{Root} = \{(R_{2,1}, 1), \ldots, (R_{2,h}, h)\}$.
  The types with the lower index are used in the join definition first. Then we have non-root types: $rset_{R_1}^{\overline{Root}} = \{R_{1,p}, \ldots R_{1,r}\}$, $rset_{R_2}^{\overline{Root}} = \{R_{2,s}, \ldots R_{2,t}\}$.
  - $V_{R_1,R_2} = \pi_{aset_{R_1}}(\rho_{map_{R_1,R_2}}(R_{1,p} \bowtie (\ldots \bowtie (R_{1,r} \bowtie (R_{1,g} \bowtie (\ldots \bowtie (R_{1,2} \bowtie R_{1,1})))))))$
  - $V_{R_2,R_1} = \pi_{aset_{R_2}}(\rho_{map_{R_2,R_1}}(R_{2,s} \bowtie (\ldots \bowtie (R_{2,t} \bowtie (R_{2,g} \bowtie (\ldots \bowtie (R_{2,2} \bowtie R_{2,1})))))))$

Rules solving other conflicts, e.g. in IsA hierarchies, can be easily derived from this rule.

3. **Entity/relationship type with entity/relationship type containing cluster** $R_1 \in S_1$ and $R_2 \in S_2$ where:

   - $R_1 = (comp(R_1), A_1, \ldots, A_n)$, if $comp(R_1) = \emptyset$ then R_1 is an entity type,
   - $R_2 = (comp(R_2), R_{2,1} + \ldots + R_{2,s}, B_1, \ldots, B_m)$, $R_2 \in rel_i(S_2)$, R_2 contains cluster $R_{2,1} + \ldots + R_{2,s}$, where $\{R_{2,1} + \ldots + R_{2,s}\} \in rel_j(S_2), j < i$,
   - $(R_1, R_2) \in \mathcal{E}$, i.e. R_1 is semantically related to R_2

   These conflicts can arise, if one designer uses just one type or different degree of abstraction, and another prefers super/subtype relationships using clusters. The result of such process may look like in Fig. 4.

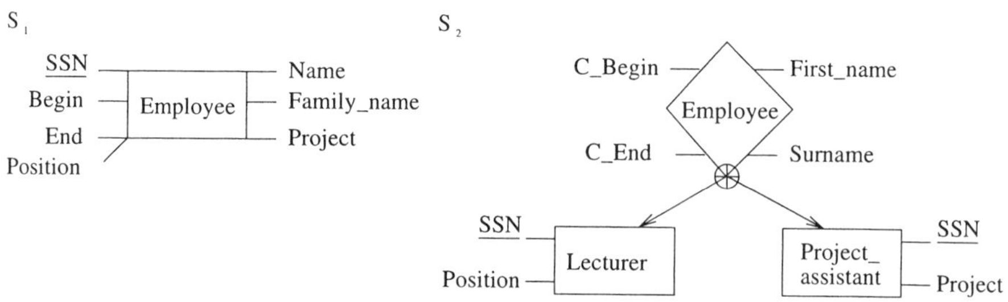

Figure 4: The entity type $Employee$ in S_1 is modelled as cluster in S_2

Here the information about employees has been differently modelled. In schema S_1 only one entity type has been chosen: $Employee(SSN, Name, Family\_name, Begin, End, Position, Project, id\{SSN\})$. Schema S_2 contains the relationship type $Employee$ with the cluster: $Employee(Lecturer + Project\_assistant, First\_name, Surname, C\_Begin, C\_End, id\{Lecturer + Project\_assistant\})$. The key SSN from the cluster has been inherited.

Let us assume that the semantic correspondences are defined by the relation $\mathcal{E} = \{(SSN, SSN), (Position, Type), (Project, Project), (Name, First\_name), (End, C\_End), (Family\_name, Surname), (Begin, C\_Begin)\}$.

The cluster is disjoint union of types. We will therefore define the views for every pair $Employee$ and its cluster component of schema S_2, i.e. $(Employee, Lecturer)$ and $(Employee, Project\_assistant)$. The view definitions are then:

- $V_{S_1.Employee, S_2.Employee-Lecturer} = \pi_{First\_name, Surname, C\_Begin, C\_End, Type}$ $(\rho_{Name \to First\_name, Family\_name \to Surname, Begin \to C\_Begin, End \to C\_End, Pos. \to Type}$ $(S_1.Employee))$,
- $V_{S_2.Employee-Lecturer, S_1.Employee} = \pi_{SSN, Name, Family\_name, Begin, End, Position}$ $(\rho_{First\_name \to Name, Surname \to Family\_name, C\_Begin \to Begin, C\_End \to End, Type \to Pos.}$ $(Lecturer \bowtie S_2.Employee))$,
- $V_{S_1.Employee, S_2.Employee-Project\_assistant} = \pi_{First\_name, Surname, C\_Begin, C\_End, Project}$ $(\rho_{Name \to First\_name, Family\_name \to Surname, Begin \to C\_Begin, End \to C\_End}(S_1.Employee))$,

- $V_{S_2.Employee-Project\_assistant,S_1.Employee} = \pi_{SSN,Name,Family\_name,Begin,End,Project}$ $(\rho_{First\_name \to Name,Surname \to Family\_name,C\_Begin \to Begin,C\_End \to End}$ $(Project\_assistant \bowtie S_2.Employee))$.

The view definitions for a general case must also reflect the different choices:

- *the mapping functions for the attributes*, where $t = 1, \ldots, s$:
  $map_{R_1,R_2-R_{2,t}} = \{(a,b)|a \in attr(R_1), b \in attr(R_2) \cup attr(R_{2,t}), (a,b) \in \mathcal{E}\}$,
  $map_{R_2-R_{2,t},R_1} = map_{12}^{-1}$,
- *view definitions*,where $t = 1, \ldots, s$:
  - $aset_{R_1,R_2-R_{2,t}} = \{b|(a,b) \in map_{R_1,R_2-R_{2,t}}\}$,
    $V_{R_1,R_2-R_{2,t}} = \pi_{aset_{R_1,R_2-R_{2,t}}}(\rho_{map_{R_1,R_2-R_{2,t}}}(R_1))$,
  - $aset_{R_2-R_{2,t},R_1} = \{b|(a,b) \in map_{R_2-R_{2,t},R_1}\}$,
    $V_{R_2-R_{2,t},R_1} = \pi_{aset_{R_2-R_{2,t},R_1}}(\rho_{map_{R_2-R_{2,t},R_1}}(R_{2,t} \bowtie R_2))$.

2.2 *Obtain Views for Updates*

In this phase of the integration process another rules check the views against the most important constraints, key constraints, attribute constraints and domain incompatibilities (other can be easily added) to make the views updatable. We select the rule from Sec. 2.1 and consider the updatability.

For two relationship types where semantically related attributes are not contained within the matching relationship type. The query view definitions are characterized by two types of sets of relationship types used in the join parts of the views: root types and non-root types.

By the root types plays the ordering an important role: $Seq_{R_1}^{Root} = \langle(R_{1,1}, 1), \ldots, (R_{1,g}, g)\rangle$, $Seq_{R_2}^{Root} = \langle(R_{2,1}, 1), \ldots, (R_{2,h}, h)\rangle$, $x \in Seq_{R_i}^{Root}$ means that an element x is a part of the sequence $Seq_{R_i}^{Root}$.

Let the non-root types be: $rset_{R_1}^{\overline{Root}} = \{R_{1,p}, \ldots R_{1,r}\}$, $rset_{R_2}^{\overline{Root}} = \{R_{2,s}, \ldots R_{2,t}\}$. The view definitions are:

- $V_{R_1,R_2} = \pi_{aset_{R_1}}(\rho_{map_{R_1,R_2}}(R_{1,p} \bowtie (\ldots \bowtie (R_{1,r} \bowtie (R_{1,g} \bowtie (\ldots \bowtie (R_{1,2} \bowtie R_{1,1})))))))$,
- $V_{R_2,R_1} = \pi_{aset_{R_2}}(\rho_{map_{R_2,R_1}}(R_{2,s} \bowtie (\ldots \bowtie (R_{2,t} \bowtie (R_{2,g} \bowtie (\ldots \bowtie (R_{2,2} \bowtie R_{2,1})))))))$.

We follow the identifiability property and turn to the key attributes first. In order to update the views, the domains of these keys must be compatible.

Definition 7. *The function* $f^{comp\_keys} R_1 \times R_2 \to A$, $f^{comp\_keys}(R_1, R_2) = \{b|a \in id(R_1), b \in id(R_2), (a,b) \in \mathcal{E}, dom(a) = dom(b)\}$ *returns for the relationship type* R_1 *the set of semantically equal and domain compatible key attributes of relationship type* R_2.

There are two update strategies for these views:

1. If not all of the keys in R_1 have their semantically related and domain compatible pairs in R_2, i.e. $id(R_1) \supset f_{R_2,R_1}^{comp\_keys}$ (see Def. 7), then we say that the view V_{R_2,R_1} is not updatable. The above condition is also valid for the proper parameters of view V_{R_1,R_2}. It means, it may happen that only one of the views is updatable while the other is not. Both views are updatable if the number of key attributes in both types R_1 and R_2 is equal and that is not the general case.

2. If all keys of R_1 have their semantically related and domain compatible pairs in R_2, i.e. $id(R_1) = f^{comp\_keys}_{R_2,R_1}$, then we say that the view V_{R_2,R_1} (respectively V_{R_1,R_2}) is updatable. But only for the attributes that reside in R_1 or their semantic counter parts in R_2. Attributes outside these types must be accessed through some relationship type. If we look at the Fig. 3, the components of the R_1 and R_2 (e.g. $R_{1,3}$,$R_{1,4}$) are identifiable, as their key sets are subsets of keys in R_1 and R_2. The problem is posed by other types, e.g. X and Y. If we want to update the attributes residing in these types, we must have the corresponding keys. We take the set of attributes, e.g. $aset_{R_1}$, from the original projection part of the view definition V_{R_1,R_2} (see above). This set contains the equivalent attribute pairs in schema S_2. We use the existing mapping map_{R_1,R_2} and extract the attributes $aset_{S_1}$ in schema S_1: $aset_{S_1} = \{a|(a,b) \in map_{R_1,R_2}\}$. So if we wanted to update the view V_{R_1,R_2} this is a set of attributes in schema S_1 which must be sufficient for the identification of all in the join definition participating relationship types. We do the same for the set $aset_{R_2}$ and get $aset_{S_2} = \{a|(a,b) \in map_{R_2,R_1}\}$. It remains to check for which relationship types these sets provide the identifiability. We consider root types first (keys of other types are included in root types).

Definition 8. *The function $f^{ident}_{rel} : R \times A \rightarrow R'$ for the set of relationship types R and attributes A returns the set of relationship types R' whose all key attributes are included in A, i.e. $f^{ident}_{rel}(R, A) = \{r|r \in R, A_{id} \subseteq A, id(r) = A_{id}\}$.*

We use the function f^{ident}_{rel}: $rset^{Root}_{R_1} = \{a|(a,b) \in Seq^{Root}_{R_1}\}$, $rset^{Root}_{R_2} = \{a|(a,b) \in Seq^{Root}_{R_2}\}$, $rset^{Root\_ident}_{R_1} = f^{ident}_{rel}(rset^{Root}_{R_1}, aset_{S_1})$, $rset^{Root\_ident}_{R_2} = f^{ident}_{rel}(rset^{Root}_{R_2}, aset_{S_2})$. While $rset^{Root\_ident}_{R_1} \subseteq rset^{Root}_{R_1}$ and $rset^{Root\_ident}_{R_2} \subseteq rset^{Root}_{R_2}$ holds, we must remove not only the components of unidentifiable root types but also those relationship types which are no longer connected with parts where R_1 respectively R_2 resides.

Definition 9. *The function $f^{components}_{rel} : R \times R' \rightarrow R''$ where R is a relationship type from which we want to determine the components (recursively to all sub-components), R' is a set of relationship types where the components of R may come from, and R'' is a resulting set containing the components of R.*

function $f^{components}_{rel}(R, R')$
for all $x \in R'$ **do**
 if $x \in comp(R)$ **then return** $\{x\} \cup f^{components}_{rel}(x, R' - \{x\})$
do for

Definition 10. *The function $f^{new\_con}_{rel} : R \times R_1 \times R_2 \rightarrow R_3$, where R is a relationship type, R_1 is a set of root types, R_2 is a set of non-root types and R_3 is a new set of connected relationship types.*

function $f^{new\_con}_{rel}(R, root\_ident, non\_root)$
for all $r \in root\_ident$ **do**
 $P = P \cup f^{components}_{rel}(r, non\_root)$
 if $(f^{components}_{rel}(r, non\_root) \cap \{R\}) \neq \emptyset$ **then** $P_R = P_R \cup f^{components}_{rel}(r, non\_root)$
do for
$P_Z = P - P_R$
$Result = \emptyset$
while $|P_Z| > 0$ **do**

$q \in P_Z$
if $((\{q\} \cup Result) \cap P_R) \neq \emptyset$ **then** $Result = Result \cup \{q\}$
$P_Z = P_Z - \{q\}$
do while
return *Result*

We illustrate the usage of the above function. We have the sets of root types $rset_{R_1}^{Root\_ident}$ and $rset_{R_2}^{Root\_ident}$ which can be identified (we have all the keys) with the sets of attributes $aset_{S_1}$ and $aset_{S_2}$ respectively. We have sets of non-root relationship types $rset_{R_1}^{\overline{Root}}$ and $rset_{R_1}^{\overline{Root}}$. If we use these sets as input parameters to the function $f_{rel}^{new\_con}$ we get the following:

$rset_{R_1}^{new\_con} = f_{rel}^{new\_con}(R_1, rset_{R_1}^{Root\_ident}, rset_{R_1}^{\overline{Root}})$,
$rset_{R_2}^{new\_con} = f_{rel}^{new\_con}(R_2, rset_{R_2}^{Root\_ident}, rset_{R_2}^{\overline{Root}})$.

The results are two new connected sets for R_1 and R_2 under consideration of identifiability of root types using the semantically corresponding attributes. The ordering of root types for the join can be achieved by using the function $root\_seq$ for $rset_{R_1}^{Root\_ident}$ and $rset_{R_2}^{Root\_ident}$. Let those result sets be: $Seq_{R_1}^{Root\_ident} = \langle R_{1,1}, \ldots, R_{1,k} \rangle$, $Seq_{R_2}^{Root\_ident} = \langle R_{2,1}, \ldots, R_{2,l} \rangle$. We need new non-root types: $rset_{R_1}^{\overline{Root}} = rset_{R_1}^{new\_con} - rset_{R_1}^{Root\_ident}$, $rset_{R_2}^{\overline{Root}} = rset_{R_2}^{new\_con} - rset_{R_2}^{Root\_ident}$. Let those sets be: $rset_{R_1}^{\overline{Root}} = \{R_{1,p}, \ldots, R_{1,q}\}$, $rset_{R_2}^{\overline{Root}} = \{R_{2,r}, \ldots, R_{2,s}\}$.

The new sets of attributes for the projection part and mappings:

$aset_{R_1,R_2}^{new} = \{b | a \in attr(r), r \in rset_{R_1}^{new\_con}, (a,b) \in \mathcal{E}\}$,
$aset_{R_2,R_1}^{new} = \{b | a \in attr(r), r \in rset_{R_2}^{new\_con}, (a,b) \in \mathcal{E}\}$,

$map_{R_1,R_2}^{new} = \{(a,b) | a \in attr(r), r \in rset_{R_1}^{new\_con}, (a,b) \in \mathcal{E}\}$,
$map_{R_2,R_1}^{new} = \{(a,b) | a \in attr(r), r \in rset_{R_2}^{new\_con}, (a,b) \in \mathcal{E}\}$,

The new view definitions are of the form:

- $V_{R_1,R_2} = \pi_{aset_{R_1,R_2}^{new}}(\rho_{map_{R_1,R_2}^{new}}(R_{1,p} \bowtie (\ldots \bowtie (R_{1,q} \bowtie (R_{1,k} \bowtie (\ldots \bowtie (R_{1,2} \bowtie R_{1,1})))))))$
- $V_{R_2,R_1} = \pi_{aset_{R_2,R_1}^{new}}(\rho_{map_{R_2,R_1}^{new}}(R_{2,r} \bowtie (\ldots \bowtie (R_{2,s} \bowtie (R_{2,l} \bowtie (\ldots \bowtie (R_{2,2} \bowtie R_{2,1})))))))$

2.3 Obtain Mapping Functions for Instances

In the previous sections we have constructed rules which actually map parts of one schema onto the parts of another schema using the view definitions. In the Fig. 1 it is the left part with the functions f_{12} and f_{21}. By, for example, domain incompatibilities the instances must be mapped between the schemas, we need the functions like f_{12}^C and f_{21}^C from the right-part of Fig. 1. We show some rules for illustration:

- **For two entity/relationship types** $R_1 \in S_1$ and $R_2 \in S_2$ where semantically equal attributes are contained within the matching entity/relationship type we have simple view definitions: $V_{R_1,R_2} = \pi_{aset_{R_1,R_2}}(\rho_{map_{R_1,R_2}}(R_1))$, $V_{R_2,R_1} = \pi_{aset_{R_2,R_1}}(\rho_{map_{R_2,R_1}}(R_2))$.

The update condition was the domain-compatibility of the key attributes of both types. This can be weakened, if there exist some function that maps the values from one domain into the values of another domain. We extend this principle to all attributes from the projection part of the view.

We find the sets of domain incompatible attributes for both views:

$$rset_{R_1,R_2}^{dom\_incomp} = \{(a,b)|(a,b) \in map_{R_1,R_2}, dom(a) \neq dom(b)\},$$
$$rset_{R_2,R_1}^{dom\_incomp} = \{(a,b)|(a,b) \in map_{R_2,R_1}, dom(a) \neq dom(b)\}.$$

For these domain incompatible pairs we find the set of functions:

$$fset_{R_1,R_2} = \{f|(a,b) \in rset_{R_1,R_2}^{dom\_incomp}, \exists f.f(dom(a)) = dom(b)\},$$
$$fset_{R_2,R_1} = \{f|(a,b) \in rset_{R_2,R_1}^{dom\_incomp}, \exists f.f(dom(a)) = dom(b)\}.$$

To guarantee the updatability of the views, we require that for all domain incompatible pairs of attributes from $rset_{R_1,R_2}^{dom\_incomp}$ and $rset_{R_2,R_1}^{dom\_incomp}$ such functions exist:

$$\forall (a,b) \in rset_{R_1,R_2}^{dom\_incomp} \exists f \in fset_{R_1,R_2}.f(dom(a)) = dom(b),$$
$$\forall (a,b) \in rset_{R_2,R_1}^{dom\_incomp} \exists f \in fset_{R_2,R_1}.f(dom(a)) = dom(b).$$

If some of the above conditions are not valid we must remove those attributes from the view definition for which there is no mapping function. The set of domain incompatible attributes of R_1 and R_2: $rset_{R_1}^{di} = \{a|(a,b) \in rset_{R_1,R_2}^{dom\_incomp}\}$, $rset_{R_2}^{di} = \{a|(a,b) \in rset_{R_2,R_1}^{dom\_incomp}\}$.

Let the sets of incompatible attributes for which there exist mapping functions be: $rset_{R_1}^{di} = \{A_j, \ldots, A_{j+k}\}$ and $rset_{R_2}^{di} = \{B_l, \ldots, B_{l+m}\}$. Let the functions for $rset_{R_1}^{di}$, $rset_{R_2}^{di}$ be: $fset_{R_1,R_2}^{dmap} = \{f_{11}, \ldots, f_{1k}\}$ and $fset_{R_2,R_1}^{dmap} = \{f_{21}, \ldots, f_{2m}\}$.

The mapping functions for the instances can then be defined like this:

$$f_{R_1,R_2}^C : R_1^C[A_j] \times \ldots \times R_1^C[A_{j+k}] \rightarrow f_{11}(R_1^C[A_j]) \times \ldots \times f_{1k}(R_1^C[A_{j+k}]),$$
$$f_{R_2,R_1}^C : R_2^C[B_l] \times \ldots \times R_2^C[B_{l+m}] \rightarrow f_{21}(R_2^C[B_l]) \times \ldots \times f_{2m}(R_2^C[B_{l+m}]).$$

- **For entity types one of them with nested set attribute** $E_1 \in S_1$ and $E_2, E_3, R \in S_2$ where the view definitions were (see Sec. 2.1):

  - $aset_{E_1,E_2} = \{b|(a,b) \in map_{E_1,E_2}\}$, $V_{E_1,E_2} = \pi_{aset_{E_1,E_2}}(\rho_{map_{E_1,E_2}}(E_1))$,
  - $aset_{E_2,E_1} = \{b|(a,b) \in map_{E_2,E_1}\}$, $V_{E_2,E_1} = \pi_{aset_{R_2,R_1}}(\rho_{E_2,E_1}(E_2 \bowtie (R \bowtie E_3)))$.

The entity type E_1 has a nested set attribute A. The entity type E_3 in other schema holds the semantically equal simple attribute C such that $(A, C) \in \mathcal{E}$. We try to solve the structuring problems with some suitable mapping between instances. We use the HERM algebra operations nest ν and unnest μ (for details refer to [10]). The schema elements contained in the view V are denoted by $\mathcal{V}$. We can now introduce the mappings:

$$f_{E_1,E_2}^C : \mu_C \pi_C (V_{E_1,E_2}^C) \rightarrow E_3^C[C],$$
$$f_{E_2,E_1}^C : \nu_{A:\{C\}} \pi_C \rho_{A \rightarrow C}(V_{E_2,E_1}^C) \rightarrow E_1^C[A].$$

2.4 Obtain Cooperating Views

We have defined the conditions for the view cooperation in Sec. 1. We apply this criteria on the view definitions we have got by using the previous rules. We do not go into detail and suppose the views satisfy all constraints. We focus on the mappings. From the mappings we can decide, whether the current view definitions are cooperating or not. If not, we define the new views which are subsets of the old ones and satisfy the conditions for cooperation.

- **For two entity/relationship types** $R_1 \in S_1$, $R_2 \in S_2$ we had simple view definitions (see Sec. 2.3) and these mapping functions:
  $map_{R_1,R_2} = \{(a,b) | a \in attr(R_1), b \in attr(R_2), (a,b) \in \mathcal{E}\}$,
  $map_{R_2,R_1} = map^{-1}_{R_1,R_2}$,where $map^{-1} = \{(b,a) | (a,b) \in map\}$.

  The set of attributes in the projection part of view V_{R_1,R_2} is $aset_{R_1,R_2}$ and for V_{R_2,R_1} it is $aset_{R_2,R_1}$.

  Theorem 6. *The views V_{R_1,R_2} and V_{R_2,R_1} cooperate if the following holds:*
  $map_{R_2,R_1}(map_{R_1,R_2}(aset_{R_2,R_1})) = aset_{R_2,R_1} \wedge$
  $map_{R_1,R_2}(map_{R_2,R_1}(aset_{R_1,R_2})) = aset_{R_1,R_2}$.

  If we consider the current case, the mapping $map_{R_2,R_1} = map^{-1}_{R_1,R_2}$ which means, that both mappings are in symmetry and Theorem 6 holds. The same is valid in the rule for *two entity/relationship types with nested tuple attributes.*

- **For two relationship types R_1, R_2 and semantically related attributes are not contained within the matching relationship type** have mappings the form:
  $map_{R_1,R_2} = \{(a,b) | a \in aset^{e\_attr}_{R_2}, b \in R_2, (a,b) \in \mathcal{E}\}$,
  $map_{R_2,R_1} = \{(a,b) | a \in aset^{e\_attr}_{R_1}, b \in R_1, (a,b) \in \mathcal{E}\}$.
  The view cooperation must not necessarily take place because the mappings are not symmetric in general. The same conditions for the cooperation of views apply as in the Theorem 6.

  We determine the set of view attributes upon the both views V_{R_1,R_2} and V_{R_2,R_1} cooperate:
  $aset^{coop}_{R_1} = map_{R_2,R_1}(map_{R_1,R_2}(aset_{R_2,R_1}))$, $aset^{coop}_{R_2} = map_{R_1,R_2}(map_{R_2,R_1}(aset_{R_1,R_2}))$.

  We adjust the set of attributes in each view definition:

  $aset_{R_1} = aset^{coop}_{R_2}$ and $aset_{R_2} = aset^{coop}_{R_1}$

  Finally we modify the original mappings:

  $map_{R_1,R_2} = \{(a,b) | a \in aset_{R_2}, b \in aset_{R_1}, (a,b) \in \mathcal{E}\}$,
  $map_{R_2,R_1} = \{(a,b) | a \in aset_{R_1}, b \in aset_{R_2}, (a,b) \in \mathcal{E}\}$.

  The equation $map_{R_1,R_2} = (map_{R_2,R_1})^{-1}$ now holds.

2.5 Translation into SQL

The view definitions we gained through the integration process can be translated to the desired SQL statements that create the views in the underlying database. The SQL standards SQL1999 and SQL2003 contain the support of complex data types with operations nest and unnest, user defined casting functions between different data types, and other extensions that enable the interoperability. At least these features have to be supported by the underlying DBMS (for example, DB2 V.8, Oracle 10) to create the views with the supporting functionality, e.g. instance mappings.

The translation of query views can be realized on the majority of today's DBMS. Update views, however, can not be directly translated. The SQL standards are continually extending the area of updatable views, but it is still not sufficient.

More complex update views must be broken into a set of simple views (presented functions in Sec. 2.2 are an option) and updated as a sequence in a transaction manner. An extensive computational support might be necessary.

3 Conclusion

Cooperation can succeed where integration is unlikely. We have considered cooperation between databases represented with HERM schemas using views. They work as functions that map the types on one another. The advantage is, if there doesn't exist an update view we can still query the relevant information. The user accesses the view in the same way as he would access the entity/relationship type in his schema. Not all constraints and conflicts can be handled on schema level. The importance of instance mapping functions is vital. The views are generated by rules. The aim of these rules is their simplicity and low complexity. Yet the range of conflicts they solve is broad. The updatability of the views in this paper has been reduced to key constraints, but the others can be easily added [13].

Despite the benefits mentioned above, our approach has to be extended. The rules were designed as building blocks for other rules, e.g. integration of two relationship types containing nested tuple attributes and nested set attributes. Inference rules can be constructed that allow derivation of such new rules. Strategies for the generation of SQL statements and execution plans especially for update views are necessary. Update strategies must be developed to address more complex constraints. These strategies can be defined as compositions of mapping functions.

References

[1] Sonia Bergamaschi, Silvana Castano, Maurizio Vincini, and Domenico Beneventano. Semantic integration of heterogeneous information sources. *Data and Knowledge Engineering*, 36(3):215–249, 2001.

[2] Andrea Cali, Diego Calvanese, Giuseppe De Giacomo, Maurizio Lenzerini, Paolo Naggar, and Fabio Vernacotola. Ibis: Semantic data integration at work. *Lecture Notes in Computer Science*, 2681:79–94, 2003.

[3] B. Convent. Unsolvable problems related to the view integration approach. *Proc. of International Conference on Database Theory*, pages 141–156, September 1986.

[4] Cheng Hian Goh, Stt'ephane Bressan, Stuart E. Madnick, and Michael D. Siegel. Context interchange: New features and formalisms for the intelligent integration of information. *ACM Trans. on Information Systems*, 17(3):270–293, 1999.

[5] Alon Y. Halevy. Answering queries using views. *Very Large Databases*, 10(4):270–294, 2001.

[6] Maurizio Lenzerini. Data integration: A theoretical perspective. In *21st ACM SIGACT SIGMOD SIGART Symp. on Principles of Database Systems*, pages 233-246, 2002.

[7] A. Levy. Logic-based techniques in data integration. In J. Minker, editor, *Logic Based Artificial Intelligence*. Kluwer Academic Publishers, 2000.

[8] P.McBrien and A.Poulovassilis. Data integration by bi-directional schema transformation rules. In *ICDE'03*, pages 227–238, Bangalore, March 2003.

[9] Stefano Spaccapietra and Christine Parent. View integration: A step forward in solving structural conflicts. *IEEE Transactions on Data and Knowledge Engineering*, 6(2):258–274, 1994.

[10] B. Thalheim. *Entity-relationship modelling - Fundamentals of database technology.* Springer, Berlin, 2000.

[11] B. Thalheim. Abstraction layers in database structuring: The star, snowflake and hierarchical structuring. Technical Report 13-01, BTU, Cottbus, 2001.

[12] V. Vestenický, J. Lewerenz, and T. Feyer. Modelling the modification component of an information service. *Proc. of Challenges of the Symposium on Advances in Databases and Information Systems - ADBIS-DASFAA'2000*, (ISBN 80-85863-56-1):195–204, September 2000.

[13] Vojtech Vestenický. Schema integration as view cooperation. Technical Report I-12-2003, BTU, Cottbus, 2003.

Data Models and Query Processing

Databases and Information Systems
J. Barzdins and A. Caplinskas (Eds.)
IOS Press, 2005

A Matrix Model for XML Data

Jaroslav POKORNY, Vladimir REJLEK
Charles University, Faculty of Mathematics and Physics
Malostranske nam. 25, Praha, Czech Republic

Abstract. Many modern applications produce and process XML data, which is queried in its both structural and textual component. To ensure such functionality, it is required to approach this data with both database and information retrieval methods. We start with a convenient vector space model and extend it with information about structural properties of an XML data collection *C*. According to the occurrences of a term *t* in the XML structure of *C* we represent *t* by a vector of weights. To reduce its lengths, a form of a DataGuide for *C* is considered with a path as a structure unit. Then, an XML document *D* is represented by a matrix ***D*** of weights. In the paper, we discuss possibilities of ranking in this model and adjusting ***D*** to reflect associations among paths in the DataGuide for *C*. Experimental results for a standard test collection are given and discussed.

Introduction

Many modern applications produce and process XML [15] data, which is queried in its both structural and textual component. This is especially useful if we consider a casual user who looks for information in web-based database systems or intranets containing XML data, like online shops, airline reservations, digital libraries catalogues or any other, and does not expect an exact answer. It is anticipated even that in the future, many websites will be built from XML documents. A remarkable characteristic of such XML collections is that they are mostly heterogeneous, e.g. they contain data from on domain but, possibly, valid w.r.t. different DTDs or XML schemes. A part of these collections will be managed as XML databases [4] and will also provide a way for users to search their contents. To ensure such functionality, it is required to approach these websites with both database and information retrieval (IR) methods.

It is clear that IR approaches will be useful especially in cases of text-rich XML documents. As it is observed recently [9], current XML query languages like XPath and XQuery are rather data-centric than document-centric. In other words, they are no longer appropriate for searching in such environments as they can not cope with the diversity of data. Hence, a research of integration of database querying and IR in context of XML is undoubtedly interesting and promising trend. Despite of the fact that a variety of systems that support such methods have been proposed, conventional IR techniques [14] cannot be employed directly. The reason for it is that two types of queries should be dealt with: content-only queries, i.e. the traditional ones in IR, and content-and-structure queries.

Particularly, a number of techniques to extend the vector space model have been designed, e.g. [1,6,7,9,10]. In [9] so-called *indexing nodes* are explicitly indicated. They are disjoint and group some XML tree nodes. These structures are the basic units of indexing. The approach [6] uses indexing the pairs (*t*, *c*) through weights, where the term *t* is qualified by the context *c* in which it appears. The *c* depends on the path to *t*. Measuring of a context resemblance introduces another coefficients into query processing. In [10], a static index is built for basic indexing nodes and then, during query processing, vector space statistics are generated. A common property of these approaches is that these statistics are used to compute a similarity value for an XML document and a query and a subsequent ranking of these values.

Authors of [7] have proposed to use an extension of the traditional vector space model given by Fox [8]. Fox developed a method for representing in single, extended vector different classes of information about a document, such as author name, terms etc. In [5] a weighting is applied to document fragment sequences that are manipulated by XQuery queries. The associated weighting algorithms are then fundamental for a rank operator.

A usual critique of the mentioned approaches is that they not sufficiently reflect the structure of XML documents. A more advanced, two-phase evaluation schema is proposed in [1]. First, a modified vector space model is employed to obtain similarity scores for the textual nodes of XML trees. Then, the scores are propagated upward in the XML-trees with a possible modification and possibly new scores of other nodes are generated.

We use a vector space ranking technique in context of the entire collection C of XML data. A document D is represented by a matrix $\boldsymbol{D}$, whose each row vector w^t associated with a term t contains the weights of t for each path occurring in C. For the path representation C we use a modification of well-known DataGuide technique [11]. Obviously, some weights in w^t will be equal to 0 dependent on whether t is contained in path or not. The query Q is also associated with a matrix $\boldsymbol{Q}$. The matrix model proposes to evaluate the degree of similarity of D with regard to the Q as the correlation between the matrices $\boldsymbol{D}$ and $\boldsymbol{Q}$. Experiments have shown that it is not possible to rely only on this score. Instead we adjust the matrix $\boldsymbol{D}$ by an additional data structure, so called a path transform matrix, which reflects relationships among paths in the DataGuide for C. Then, the resulted transformed matrices $\boldsymbol{TD}$ and $\boldsymbol{TQ}$ are used for query processing.

The rest of the paper is organized as follows. Section 1 repeats some basics about the vector space model. Section 2 describes its extension to the matrix model. A tree representation of the entire collection of XML data is defined here and associated weighting and ranking system are proposed. The notion of a path transform matrix is introduced and discussed in this section. In Section 3, we discuss some issues connected with the matrix model. In Section 4, we report our experiences in using the matrix model in the application domain given by the well-known Shakespeare's collection. Finally, we conclude in Section 5.

1. Vector Space Model

Suppose a collection C of documents. In the vector space model, documents are treated as points in a high-dimensional space. As document $D_i \in C$ is represented by a vector of term weights, i.e.

$$\boldsymbol{D}_i = [w_{i,1}, w_{i,2},\ldots, w_{i,m}] \in \langle 0,1\rangle^{\mathrm{m}}$$

where m is the number of terms considered in C and i is the document number. The weight $w_{i,j}$ is usually calculated as $TF_{i,j} * \log(N/n_i)$, where TF_{ij} is the number of times the term t_j occurs in a document D_i, n_i is the number of documents in which t_j occurs, and N is the total number of documents in the C. Similarly, a query Q can be expressed as

$$\boldsymbol{Q} = [q_1, q_1,\ldots, q_m] \in \langle 0,1\rangle^{\mathrm{m}}$$

To compare a query with a document, the most commonly used measure is the so-called *cosine similarity*, i.e., the cosine value of the angle between both vectors. So, we can define the similarity between a document D_i and a query Q as

$$Sim(D_i, Q) = \frac{\sum_{j=1}^{m} w_{i,j} * q_j}{\sqrt{\sum_{j=1}^{m} (w_{i,j})^2 * \sum_{j=1}^{m} (q_j)^2}}$$

2. Matrix Model

We will now consider a collection C of XML documents containing large portions of text. Our goal is to model such data not only as XML trees, as usually, but also with respect to their text characteristics as it is usual in IR. Clearly, a pure vector model is not enough, as it does not reflect a structure of XML data.

We will use a lightweighted tree-like data model for handling XML documents, in which we create nodes for elements and attributes. Edges model the containment relationship between elements. The textual content of a element E with subelements (so called mixed content) is represented as a set of child nodes of the E node. Leafs nodes in the tree contain a piece of text (possibly empty). We neglect a difference between elements and attributes. In other words, we deal with attributes as with elements in the model.

Our matrix model is based on the vector space model. It only changes the meaning of the term weight in a document. In the vector space model a weight is expressed by a real number specifying a term weight for the entire document. In the matrix model, the term weight is expressed by a vector. Such a weight should reflect a distribution of the term in the XML structure of the document. Then a document D_i is represented as a matrix

$$\boldsymbol{D}_i = \begin{bmatrix} w_{i,1,1} & w_{i,1,2} & \cdots & w_{i,1,k} \\ w_{i,2,1} & w_{i,2,2} & \cdots & w_{i,2,k} \\ \cdots & \cdots & \cdots & \cdots \\ w_{i,m,1} & w_{i,m,2} & \cdots & w_{i,m,k} \end{bmatrix} \in <0,1>^{m,k}$$

where k denotes the number of paths in the XML structure of the entire document collection C. The value $w_{i,j,s} \in <0,1>$ is the weight of the term t_j on the path s in the document D_i.

2.1 DataGuide for Collection

To apply the above ideas for a collection C of XML data, we need a data model of C. A tree representation can be used also in this case. Such tree can be modelled by the well-known DataGuide [11]. The DataGuide has been developed for the OEM data model originally [2]. Transferring the notion of DataGuide to the XML data model is in principle straightforward. One must however decide whether the DataGuide describes only the primary structure or the secondary structure as well, i.e., constructs such as IDREF. We easily will not assume mutual references among XML documents in our approach.

The idea of a DataGuide is to provide a summary of the structure, generated from the collection C. A *DataGuide for a collection* C, DG_C, of XML documents is a tree T_C fulfilling the following conditions:

- each path in C exists in T_C,
- each path in T_C exists in C,
- paths in T_C are unique (but not in C).

Notice, that in the case of non-existence of a common root it is possible to add formally such node to the forest of trees and get the required tree. Every leaf node of the tree is *annotated* with the IDs of those document leafs it represents (i.e. those reached by the same label path as the index node). In fact, a leaf of the tree is associated with a bag of element contents (PCDATA). Remind that DG_C is not given uniquely in general and does not support ordering in the XML documents in C.

Example 1: Consider a collection C_1 with three documents in Fig. 1. We index terms "David", "Morrell", "Caine". The associated DataGuide DG_{C1} and the list of its paths are depicted in Fig. 2. ◊

The resulting tree is usually much smaller than the forest of document trees constructed from C (although theoretically its size is linear in that of the document forest). Hence it can be supposed to be held in main memory even for large document collections.

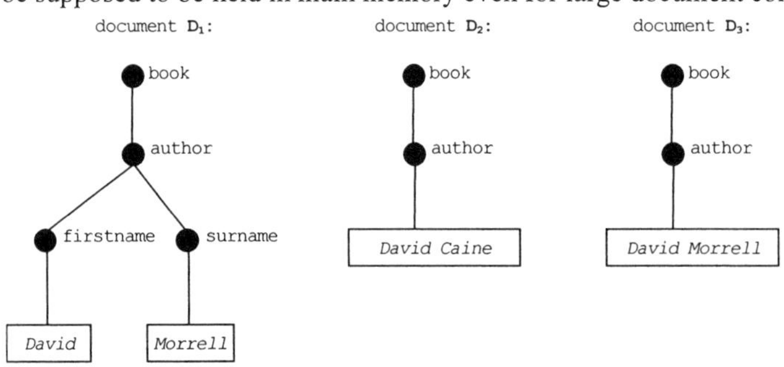

Figure 1. Collection C_1 of XML documents

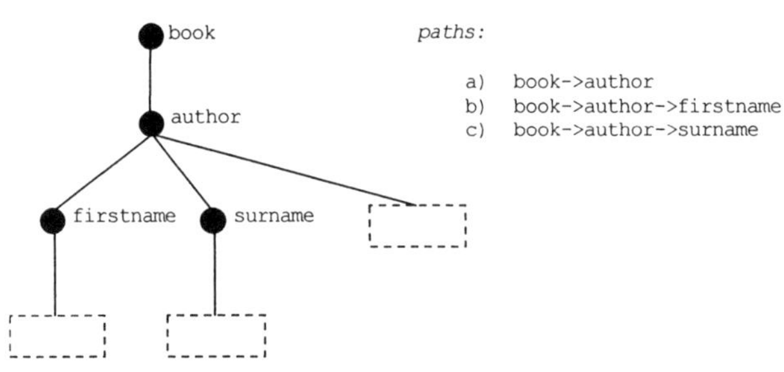

Figure 2. DataGuide for C_1

2.2 Weighting and Ranking

Assuming a DG_C we now try to construct a matrix. Given a term t, path p, and document D_i, the term weight $w_{i,t,p}$ is defined as

$$TF_{t,p} / m_p$$

where $TF_{t,p}$ is the number of t occurrences on p and m_p is the number of all terms associated with p. Our matrix model will be used for tree queries, which cover the core XPath [16] constructs. Consequently, we express such queries in a similar way as documents, i.e. as matrices. For an evaluation of queries we will inspire by the original vector space model and introduce a similarity Sim_1 as

$$Sim_1(D_i, Q) = \sum_{l=1}^{m} \sum_{j=1}^{k} d_{i,l,j} * q_{l,j} \qquad (1)$$

where i, m, and k have the same meaning as above.

Example 1 (continued): The (3-dimensional) matrix $\boldsymbol{D}$ for collection C_1 can be represented in the following form.

	a	b	c	a	b	c	a	b	c
D_1:	[(0	**1**	0)	(0	0	**1**)	(0	0	0)]
D_2:	[(**0,5**	0	0)	(0	0	0)	(**0,5**	0	0)]
D_3:	[(**0,5**	0	0)	(**0,5**	0	0)	(0	0	0)]
		"david"			"morrell"			"caine"	

We evaluate the queries in Fig. 3 in this model.

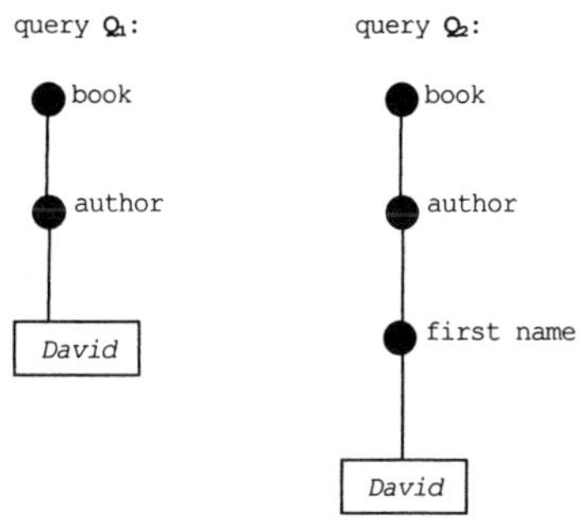

Figure 3. Queries Q_1 and Q_2

Thus, we obtain matrices

	a	b	c	a	b	c	a	b	c
Q_1:	[(1	0	0)	(0	0	0)	(0	0	0)]
Q_2:	[(0	1	0)	(0	0	0)	(0	0	0)]
		"david"			"morrell"			"caine"	

Then, we get the results

$$Sim_1(D_1, Q_1) = \mathbf{0}$$
$$Sim_1(D_2, Q_1) = \mathbf{0.5}$$
$$Sim_1(D_3, Q_1) = \mathbf{0.5}$$
$$Sim_1(D_1, Q_2) = \mathbf{1}$$
$$Sim_1(D_2, Q_2) = \mathbf{0}$$
$$Sim_1(D_3, Q_2) = \mathbf{0}$$

◊

The results based on the Sim_1 are not too successful in Example 1. Particularly, $Sim_1(D_1, Q_1) = Sim_1(D_2, Q_2) = Sim_1(D_3, Q_2) = 0$ is not in accordance with our intuition. The reason is that for different XML structures their similarity is equal to 0. Since our goal is to process documents from heterogeneous sources, e.g. with different DTDs, it is necessary to add another additional structure to our considerations which will enable to describe associations (similarity) among paths in DG_C.

2.3 Path Transform Matrix

By a path transform matrix we *understand a square* matrix **A** of dimension k × k, where k is the number of paths in DG_C, $a_{i,j} \in$ <0,1>, and $a_{i,i} = 1$. For each path the matrix **A** contains a vector that expresses a relationship (similarity) of the path to the other ones. This similarity is expressed by a real number from <0,1>, where 0 means that paths have nothing common and 1 means that the paths are the same.

Example 1 (continued): The path transform matrix could have a form

A_1	a	b	c
a	1	0	0
b	0,5	1	0
c	0,5	0	1

We can observe, that the weights on the path `book→author→firstname` have to be 0.5 times transformed on the path `book→author`. Similarly weights on the path `book→author→surname` are transformed 0.5 times on the path `book→author`. Recalculating all weights (see (2)) we obtain matrices ***TD*** and ***TQ***

	a	b	c	a	b	c	a	b	c
TD_1:	[(0,5	1	0)	(0,5	0	1)	(0	0	0)]
TD_2:	[(0,5	0	0)	(0	0	0)	(0,5	0	0)]
TD_3:	[(0,5	0	0)	(0,5	0	0)	(0	0	0)]
TQ_1:	[(1	0	0)	(0	0	0)	(0	0	0)]
TQ_2:	[(0,5	1	0)	(0	0	0)	(0	0	0)]
		"david"			"morrell"			"caine"	

and associated values of Sim_1

```
Sim1(TD1,TQ1) = 0.5
Sim1(TD2,TQ1) = 0.5
Sim1(TD3,TQ1) = 0.5
Sim1(TD1,TQ2) = 1.25
Sim1(TD2,TQ2) = 0.25
Sim1(TD3,TQ2) = 0.25
```

◊

Since $Sim_1(TD_2, TQ_1) = Sim_1(TD_3, TQ_1) = 0.5$ we get the same result as without ***A***, but due to the paths transformation, the output also contains D_1 and $Sim_1(TD_1,TQ_1)$ is also equal to 0.5, although we work with documents of not the same structure. The highest $Sim_1(TD_1,TQ_2)$=1.25 concerns the document D_1. It is caused by the fact, that query Q_2 is a subtree of D_1. For the first query the structures of D_2 and D_3 are similar, but their element contents match only partially ("david" vs. "david caine" and "david morrell"), i.e. with the result 0.5.

The high similarity of D_1 a Q_2 is caused also by the path transform matrix. This matrix favours documents with more detailed structure (i.e. in path of form `book→author→firstname` and `book→author→surname`). It is due to the transformation `book→author→firstname` 0.5 times on `book→author` and `book→author→surname` 0.5 times on `book→author`. In contrary the inverse transformation from `book→author` to `book→author→firstname` is equal to 0 as well as the transformation `book→author` to `book→author→surname`.

Now we describe in detail, how the transformation is processed. Let ***D*** be a matrix of dimension $m \times k$, where m is the number of all terms of C, k is the number of all paths in DG_C, and $w_{i,j} \in <0,1>$. Let ***A*** be a path transform matrix of dimension k, where $a_{i,j} \in <0,1>$ and $a_{i,i}=1$. Then a one step transformation is a mapping 1st_tr(***D***, ***A***) returning again a matrix of dimension $m \times k$, ***TD***, whose items $w_{i,j}$ are defined as

$$w_{l,i} = \max(w_{l,i}, \max_{j=1,\dots,k}(a_{j,i} * w_{l,j})) \tag{2}$$

A transformation *of the document matrix* ***D*** by the path transform matrix ***A*** is the transitive closure of 1st_tr(***D***, ***A***).

Example 1 (continued): We will try to calculate similarities for a more complicated path transform matrix.

A_2	a	b	c	
a	1	0,2	0,2	book→author
b	0,5	1	0	book→author→firstname
c	0,5	0	1	book→author→surname

We obtain the matrices ***TD*** and ***TQ***

	a	b	c	a	b	c	a	b	c
TD_1:	[(0,5	1	0,1)	(0,5	0,1	1)	(0	0	0)]
TD_2:	[(0,5	0,1	0,1)	(0	0	0)	(0,5	0,1	0,1)]
TD_3:	[(0,5	0,1	0,1)	(0,5	0,1	0,1)	(0	0	0)]
TQ_1:	[(1	0,2	0,2)	(0	0	0)	(0	0	0)]
TQ_2:	[(0,5	1	0,1)	(0	0	0)	(0	0	0)]
		"david"			"morrell"			"caine"	

and associated values of Sim_1.

```
Sim1(TD1,TQ1) = 0.72        Sim1(TD1,TQ2)=1.26
Sim1(TD2,TQ1) = 0.54        Sim1(TD2,TQ2)=0.36
Sim1(TD3,TQ1) = 0.54        Sim1(TD3,TQ2)=0.36
```

◊

Since the current form of function Sim_1 favours a little the document D_1 because of 1s in its matrix, we normalize the vectors used in definition of Sim_1. Most approaches to IR scale usually document vectors so that they all have unit length. The length of a vector is calculated by summing the squares of all the components and taking the square root of the answer. Consequently, we obtain a score function based on the following formula:

$$Sim_2(D_i, Q) = \sum_{l=1}^{m} \frac{\sum_{j=1}^{k} w_{i,l,j} * q_{l,j}}{\sqrt{\sum_{j=1}^{k} (w_{i,l,j})^2 * \sum_{j=1}^{k} (q_{l,j})^2}}$$

Recalculating our example,

```
Sim2(TD1, TQ1) = 0.62     Sim2(TD1,TQ2)=1
Sim2(TD2, TQ1) = 1        Sim2(TD2,TQ2)=0.62
Sim2(TD3, TQ1) = 1        Sim2(TD3,TQ2)=0.62
```

the resulted ranking of documents answered for the query Q_1 can be $\{D_2, D_3, D_1\}$. For the query Q_2 we can get $\{D_1, D_2, D_3\}$.

3. Some Results and Issues

Statement 1: The similarity Sim_2 is (a) irreflexive, (b) non-negative, (c) commutative, and (d) not transitive.
Proof: It is easy to verify the (c) property. The proof of (a) can be shown by example of $Sim_2(TD_3,TD_3) = 2$. In general, the result of Sim_2 can be from $<0,m>$, i.e. (b) holds. For example, it is 0 in the case, if the sets of terms with non-zero weights have empty intersection for documents D_i, D_j, $i \neq j$. Consequently, from the result of $Sim_2(D_1,D_2)$ it is not possible to decide that D_1 a D_2 are the same. The (d) property follows, e.g., from the fact that $Sim_2(D_3, D_1)+Sim_2(D_1, D_2) = 0 < Sim_2(D_3, TD_2) = 1$ for $D_1 = (0, 0, 1)$, $D_2 = (0, 1, 0)$, and $D_3 = (0, 1, 0)$, i.e. the triangular inequality does not hold. Thus, the similarity Sim_2 is not a metric.

Example 2: In Fig. 4 the document D_2 in collection C_2 distinguishes from D_1 and D_3 in its structure, since it uses the element `<title>` instead of `<name>`. Obviously, D_2 comes from other source. Fig. 5 shows an associated DataGuide. Since we want to work with all such documents evenly we set up the similarity between paths `book→name` and `book→title` as 1 in $\boldsymbol{A}$.

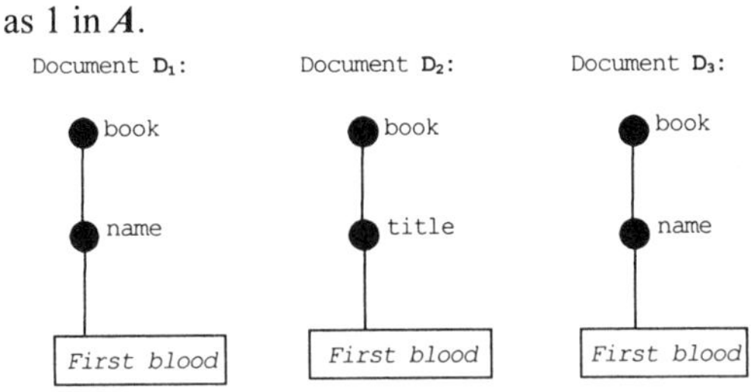

Figure 4. Collection C_2 of XML documents

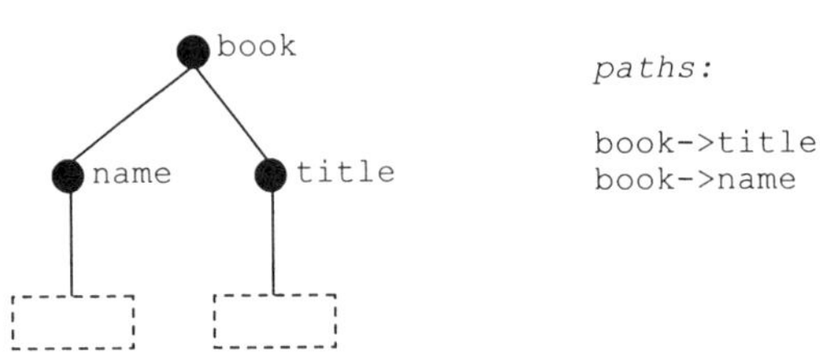

Figure 5. DataGuide for C_2

The matrix $\boldsymbol{A}$

A_3	a	b	
a	1	1	`book→name`
b	1	1	`book→titul`

and the modified document matrix based on DG_{C2}

	a	b	a	b
TD_1: [	(0,5	0,5)	(0,5	0,5)]
TD_2: [	(0,5	0,5)	(0,5	0,5)]
TD_3: [	(0,5	0,5)	(0,5	0,5)]
	"first"		"blood"	

used for the calculation of Sim_2 provide the following results:

$$\mathtt{Sim_2(TD_1,TD_1)=Sim_2(TD_1,TD_2)=Sim_2(TD_1,TD_3)=Sim_2(TD_2,TD_3)=2}$$

Due to the transformations specified in ***A***, the similarity of a document to itself is the same for all documents. It is in accordance with our requirements. We can conclude that the impossibility to determine the sameness from the results of Sim_2 is rather contradictory.

◊

Consider the example of transformations between
`book→author`
and
`book→author→firstname, book→author→surname.`

We can observe that asymmetric path transform matrix favours some structures. Asymmetry of transformation in one direction and vice verse between two paths causes that the path with better transformation contributes more to the resulted similarity, i.e. documents with this structure will be favoured. For usual situations we can assume that the similarity between two paths expressed with one number is enough. Thus, the path transform matrix can be symmetric. On the other hand, a favouring some path can be beneficial as it is documented in Example 1.

We conclude with an issue how to deal with a mixed content in this approach. In the matrix model an element with a mixed content will be indexed as the union of its partial text parts. For example, the element

<x>text$_1$<z>...</z>text$_2$<y>...</y><x>

generates paths `x→z`, `x→y`, and `x`, where `x` refers to text$_1$ ∪ text$_2$.

4. Experiments

To use the matrix model in practice for a XML data collection *C*, it is necessary to explore

- set of terms,
- list of collection paths,

and to set up the associated path transform matrix. The first two data structures are partially generated during lexical analysis of entering the XML documents. With paths it is possible to construct a DG_C. A considerable theoretical foundation behind DataGuides can be found in [12], which proved that creating a DataGuide over a source database is equivalent to a well-studied problem concerning conversion of a nondeterministic finite automaton to a deterministic finite automaton. In our case, when the source database is a tree, this conversion takes linear time.

A computation of the transitive closure of ***A*** is relatively complex. In [13], a greedy algorithm is considered with complexity $O(k^4)$, where *k* is the number of paths in DG_C. In the second step we calculate the transformation of ***D***. Its complexity is $O(m*k^2)$. Observe, these calculations run only once. Assuming a naïve implementation of query evaluation, we obtain the complexity $O(m*k)$.

Our algorithm has been implemented in Java. We did a comparison between vector and matrix model. For our first experiments we used the well-known collection of Shakespeare's plays [3] and synthetic data generated by a widely used database benchmark XBench [17]. The former dataset, C_1, contains 37 plays of the total size 7712KByte. We identified 14201 measurable terms 39 paths there. Documents of the latter collection have been sized 9722 KByte. This collection, C_2, contained 26 documents with 22 paths in DG_C and 2712 measurable terms. Since there is no difference between a document and query in our approach we used some members of collection directly as queries. As the efficiency has been not the main goal our experiments we pay a rather high price to more complex calculations in the matrix model comparing to the vector space model. Memory representations of both type of indices are not too different.

For experiments with different queries we started with "hamlet.xml" query against all plays in both models. (Fig. 6 and 7) and the following paths:

`play→personae→pgroup→persona`

to

`play→personae→persona`

by constant 0.2 and

`play→title`

to

`play→playsubt`

by 0.5, and back by constant 0.5 too.

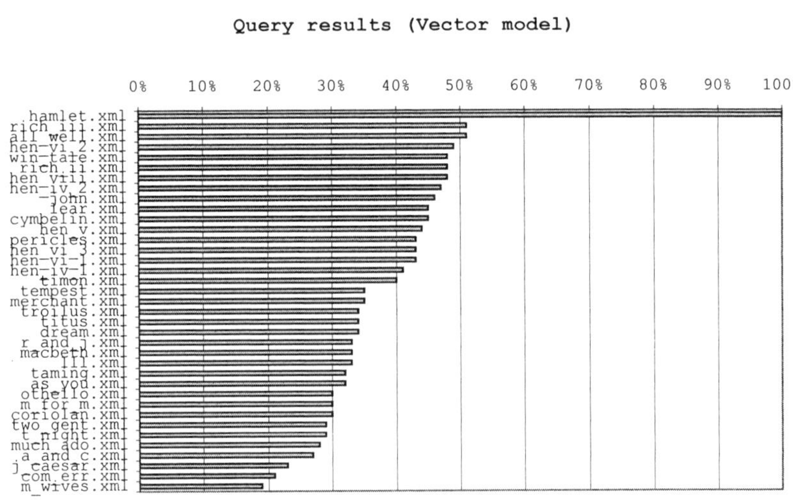

Figure 6. Result of "hamlet.xml" query in the vector space model

We can observe from the graphs in Fig. 6-8 that the matrix model has more balanced results, the values are less dispersed. The reason lies in the document structures that distinguish minimally for particular plays.

The vector space model gives some results of lower quality. For example the tragedies Othello, Mackbeth, and Romeo and Juliet are of less relevance comparing to the comedy "A Midsummer Night's Dream". For finding out a play similar to the tragedy Hamlet it is not too good result. Othello is even under the 30%. Here the matrix model uniquely wins since all mentioned plays have the similarity higher than 40%.

Bigger differences are also observable with Julius Cesar, The Comedy of Errors and The Merry Wives of Windsor However Julius Cesar did not obtain better ranking in the matrix model too. The matrix model also gives more balanced similarities for plays about the king Henry.

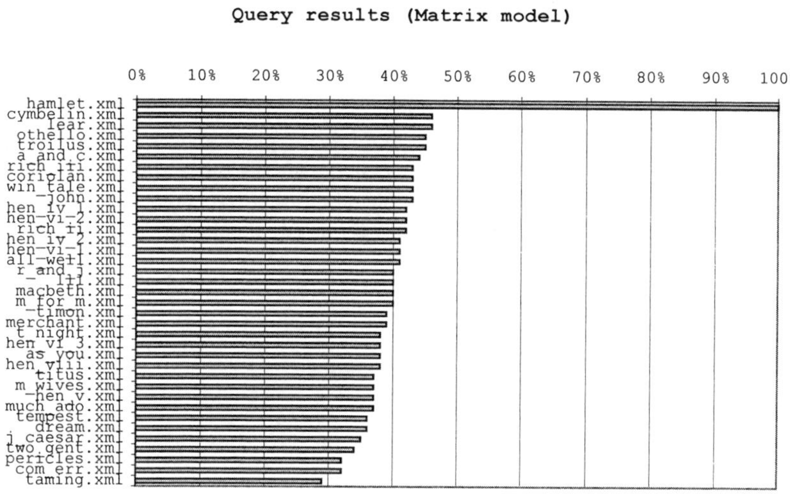

Figure 7. Result of "hamlet.xml" query in the matrix model

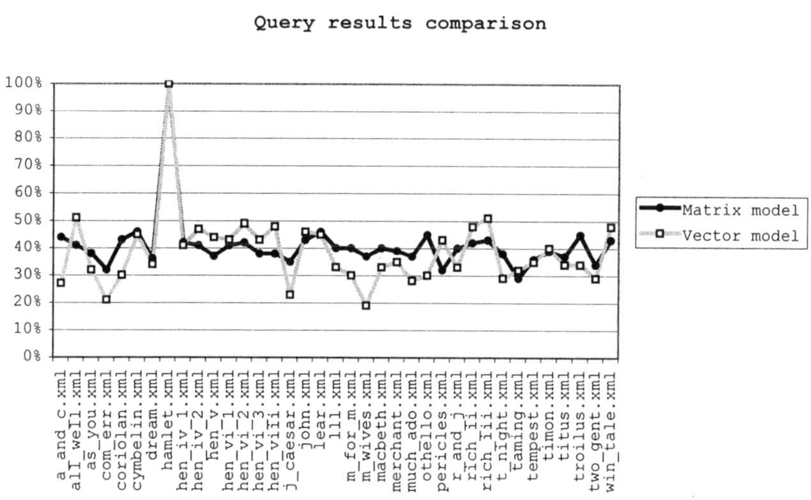

Figure 8. Comparison of the query results in the vector and matrix models

5. Conclusions

The matrix model solves problems connected with adding the structure of documents to convenient vector space model known from IR. Its goal is to enhance possibilities of integrating heterogeneous sources of XML data.

In principle, the matrix model does not distinguish between a query and a document. It is designed for finding out the similarity of two XML documents. The model can be extended for a multi-lingual information sources. Each path in the path transform matrix could be set up for a different language according to the requirements on indexing.

Similarly to other approaches based on transformations, a manual approach to tuning the A matrix is important. Although this could be conceived as a drawback of the method, we believe that any improving of today's information retrieval methods requires such approach, as only poor statistical data is not enough for achieving this goal. Obviously, the use of machine learning techniques and probabilistic estimation based on training sets of appropriate benchmarks may also be beneficial.

Since nowadays XML is increasingly used as a replacement for HTML on the Web, documents usually have XLinks or XPointers to data in other documents. Such inter-documents links could be also added to the searching their contents. Our future work will aim to use the matrix model also in the context of such applications.

Acknowledgement

This research was supported in part by GACR grant 201/03/0912.

References

[1] Anh, V.N., Moffat, A.: Compression and an IR Approach to XML Retrieval. In: *Proc. of the First Workshop of INEX*, Dagstuhl, Germany, December 2002, pp. 99-104.

[2] Abiteboul, S., Quass, D., McHugh, J., Widom, J., and. Wiener, J. L.: The Lorel Query Language for Semistructured Data. *International Journal on Digital Libraries*, 1(1):68–88, 1997.

[3] Bosak, J.: Shakespeare 2.00. *Los Altos, California,* http://www.ibiblio.org/bosak/, 1999

[4] Bourret, R.: XML and Databases, http://www.rpbourret.com/xml/XMLAndDatabases.htm.

[5] Bremer, J.-M., Gertz, M.: XQuery/IR: Integrating XML Document and Data Retrieval In: *Proceedings of the 5th International Workshop on the Web and Databases (WebDB)*, June 2002, pp. 1-6.

[6] Carmel, D., Efraty, N., Landau, G.M., Maarek, Y., Mass, Y.: An Extension of the Vector Space Model for Querying XML Documents via XML Fragments. In: *Proc. of XML and Information Retrieval (Workshop) Tampere*, Finland, 2002, pp. 14-25.

[7] Crouch, C.J., Apte, S., Bapat, H.: Using the Extended Vector Model for XML Retrieval. In: *Proc. of the Second INEX 2002 Workshop*, Dagstuhl, Germany, December 2003, pp. 95-98.

[8] Fox, E.A.: Extending the Boolean and Vector Space Models of Information Retrieval with P-Norm Queries and Multiple Concept Types. PhD thesis, Cornell University Department of Computer Science, 1983.

[9] Fuhr, N., Großjohann, K.: XIRQL: A Query Language for Information Retrieval. In: *Proc. of ACM-SIGIR*, New Orleans, 2001, pp. 172-180.

[10] Grabs, T., Schek, H.: Generating vector spaces on-the-fly for flexible XML retrieval. In: *Proc. of XML and Information Retrieval (Workshop)*, Tampere, Finland, ACM Press, 2002, pp. 4-13.

[11] Goldman, R., Widom, J.: DataGuides: Enabling Query Formulation and Optimization in Semistructured Databases. In: *Proc. of the 25th VLDB Conf.*, 1997, pp. 436-445.

[12] S. Nestorov, J. Ullman, J. Wiener, and S. Chawathe. Representative objects: Concise representations of semistructured, hierarchical data. In: *Proceedings of the 13th Int. Conf. on Data Engineering, Birmingham*, England, April 1997, pp. 79-90.

[13] Rejlek, V.: A similarity of XML documents. Master Thesis, Dep. of Software Engineering, Charles University, 2003, (in Czech).

[14] Salton, G. (editor), *The SMART Retrieval System*, Prentice Hall, New Jersey, 1971.

[15] Extensible Markup Language (XML) 1.0., http://www.w3.org/TR/REC-xml, 1998.

[16] XML Path Language (XPath), Version 1.0, http://www.w3.org/TR/xpath, 1999.

[17] Yao, B.B., Özsu, M.T., Keenleyside, J.: XBench - A Family of Benchmarks for XML DBMSs. *School of Computer Science University of Waterloo*, http://db.uwaterloo.ca/~ddbms/projects/xbench/index.html, 2002.

Databases and Information Systems
J. Barzdins and A. Caplinskas (Eds.)
IOS Press, 2005

Enhancing Hierarchical Queries in Relational Databases with the Nested Set Representation

Lauri PIETARINEN, Boris NOVIKOV
Relational Consulting Oy
Vuorimiehenkatu 15 A 3,00140 Helsinki, Finland
lauri.pietarinen@relational-consulting.com
University of St.-Petersburg
Universitetsky prosp. 28, 198904 St.-Petersburg, Russia
borisnov@acm.org

Abstract. Queries on hierarchical structures have traditionally been the sore point of relational databases because of missing functionality in SQL. Simple tree traversal is not possible without procedural code or vendor specific extensions. Using the nested set representation to represent trees vendor neutral SQL can be used to express ancestor and descendant queries. In our paper we propose enhancements to the original nested set representation to speed up tree traversal queries and enable extra functionality, such as 'by level' queries. We also introduce an algorithm for inserting new nodes into the tree without having to renumber existing ones. We also show how ancestor queries can be optimized using our enhancements. Finally we compare the performance of this new representation with that of the original representation and to the recursive union as implemented in DB2.

1. Introduction

Hierarchical structures are ubiquitous in information systems. We can distinguish between hierarchies in schema, such as a schema involving the entities department, group, team and person, and hierarchies in which the nodes are of the same kind. We can further distinguish between hierarchies with potential cycles, such as bill-of-material type hierarchies in which children can have more than one parent and strict hierarchies in which each node has exactly one parent except for the root node which has no parents [11].

Examples of strict hierarchies include organizational hierarchies, product group hierarchies, newsgroup and email threads, web site maps, file folder structures, project task hierarchies, word hierarchies in thesauri [4] and XML-hierarchies [7],[9].

The traditional and obvious way to represent hierarchies (and arbitrary graphs) in relational databases is the adjacency list approach in which the child node refers to its parent node with its primary key. In spite of this, SQL has not traditionally provided good support for querying hierarchical (or other recursive) structures apart from some vendor specific extensions such as Oracle's Connect By.

The nested set representation [1], [2], [6] to represent hierarchies uses a partly materialized non-recursive structure to improve query performance with a degrading of update performance. While it is not as general as the adjacency list representation, which can be used for arbitrary graphs, it is a good alternative for situations where strict, possibly ordered trees are used. It is based on a numbering scheme in which the nodes are tagged with their postorder and preorder (or visitation order) in the hierarchy.

Because of the non-recursiveness of the nested set representation, queries such as

- get nodes of subtree in preorder

- get ancestors of node
- get closure of tree

are easy to express and efficient to execute using standard SQL available on any SQL-based database.

In this paper we propose modifications to the nested set representation for enhanced performance and functionality and compare this enhanced representation with the recursive union queries available in DB2.

We also introduce new queries to overcome deficiencies in the original nested set approach, especially in ancestor searches over large hierarchies.

The paper proceeds as follows. The next section provides an overview of related work. In Section 3 we describe the original nested set model giving examples in SQL. In section 4 we describe our enhancements for the model. In Section 5 we present some typical use cases with an eye on indexes for optimizing performance. In Section 6 we compare the functionality and performance of our approach with that of DB2 Recursive Union and then conclude with some further suggestions.

2. Related Work

The best known technique to handle hierarchies is based on the edge representation of graphs, where an edge of a graph is stored as a pair of starting and ending nodes. Extensive research in this area has been done, and several efficient query evaluation techniques were proposed in the context of deductive databases [11].

From the practical perspective, the disadvantages of this structure are the following:

- it is difficult to represent the order of nodes (say, child nodes), and
- processing of the hierarchies of indefinite depth requires recursive queries.

Although recursive queries are now available in SQL:1999, this feature is not yet widely supported by DBMS vendors. Additionally, since the result of the recursive union query does not contain order information it is not possible to further join the result without losing the tree structure. Further, the recursive union construct does not enable the output of ordered hierarchies, such as XML documents or newsgroup threads.

The nested set representation is described in [1] in general terms without any references to databases. The application to relational databases can be found in [2] and [6].

Although the nested set representation has turned out to be useful in practice [4] not much attention has been given to it in the scientific community and hence our primary references are to books or vendor user group presentations.

Similar representations are discussed in [7], [8] and [12] but they are either more complex or rely on vendor specific optimization techniques, such as index-anding, or propose modifications to the database optimizer and engine.

In [13] a variation of the nested set representation is described that uses a binary rational numbering.

3. The Nested Set Representation

We now describe the background ideas of the nested set representation and continue with an SQL-based example. We then present some SQL-queries that make use of that example.

3.1 Fundamentals

One way to represent hierarchies is to use nested sets [1]. Each node of the hierarchy is represented as a set. Relationships between nodes are modeled using set inclusions. Sets corresponding to leaf nodes should be disjoint. The elements of these sets are not important for us. For simplicity we assume that sets corresponding to leaf nodes contain single elements, which are all distinct.

The set containing all the other sets is the root of the tree and the sets containing no subsets are the leaf nodes of the tree. The ancestors of a node are the nodes that contain this node and the descendants of a node are the nodes that this node contains. We require that for any two different sets, either one is the proper subset of the other or their intersection is empty. Such a hierarchy is clearly unordered, i.e. the children under a given parent are not in any specific order.

The example hierarchy in Figure 1 could be represented with nested sets as in Figure 2.

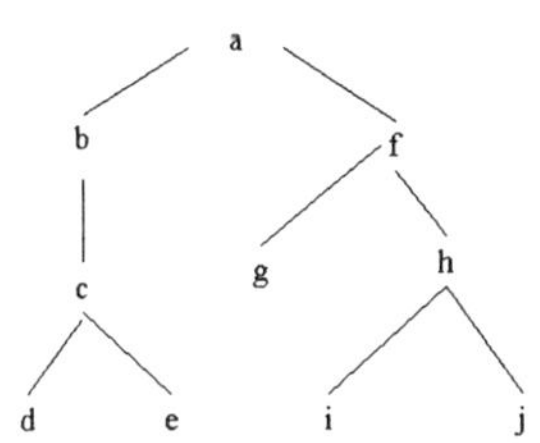

Figure 1. Sample hierarchy

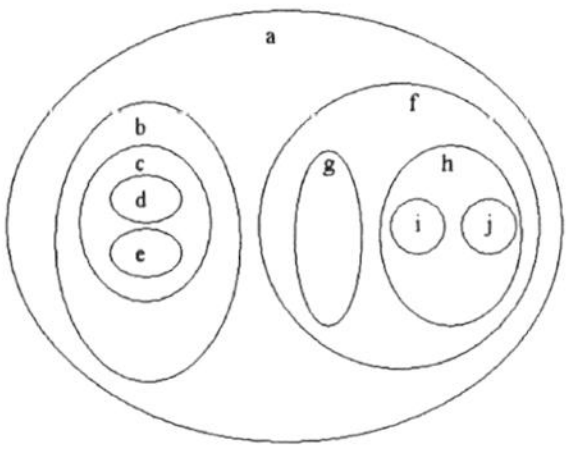

Figure 2. Sample hierarchy using nested sets

The applications we are considering, such as XML and newsgroups the immediate descendants (child nodes) of any node are usually ordered.

The ordering of child nodes is achieved by changing the representation to using sets of closed intervals in an (ordered) space of rational numbers. To obtain these intervals, we assign a rational number to each bracket (see below) so that the ordering of the numbers corresponds to the order of the brackets regardless of whether it is a left or right bracket. Clearly the properties previously stated still apply. Additionally we can see that the children of each node are ordered.

Using Figure 1 as an example we obtain

```
A [ B [ C [ D [   ] E [   ] ] ]
  Q1  Q2   Q3   Q4  Q5     Q6 Q7 Q8 Q9
F [ G [    ] H [ I [   ] J [    ]  ]  ]  ]
 Q10  Q11  Q12   Q13  Q14 Q15   Q16 Q17 Q18 Q19 Q20
```

WITH $Q_I \in \mathbb{Q}$ AND $Q1 < Q2 < \ldots < Q20$.

In this example we can see that the children of `h` are in the order `(i,j)`.

In this paper we adopt the convention of calling the lower bound (the left bracket) of a set the `left bound` and the upper bound (the right bracket) of the set the `right bound`. For example, the left bound and right bound of node f are 10 and 19 respectively. It can easily be seen that the left and right bounds correspond to the preorder and post order of the nodes in the hierarchy [3].

By using consecutive integer numbers instead of ordered rational numbers we further obtain the property that the number of descendants of any given node can be calculated with the formula (RIGHT BOUND – LEFT BOUND-1)/2. As a corollary the leaf nodes have (RIGHT BOUND-LEFT BOUND-1)/2 = 0.

We adopt the convention of assigning the left bound of the root of the hierarchy to 1. The right bound of the root will be equal to the total number of nodes in the hierarchy times two.

Returning to our example we would have the following boundaries:

```
A [ B [ C [ D [ ] E  [ ] ] ]
  1   2   3    4 5      6  7  8  9
F [ G [ ] H [ I [ ] J [  ]  ]  ]  ]
  10   11 12     13  14 15  16   17  18  19  20
```

We can see that the number of children of node F is (19-10-1)/2 = 4.

Figure 3 describes an alternative way to visualize the sample hierarchy.

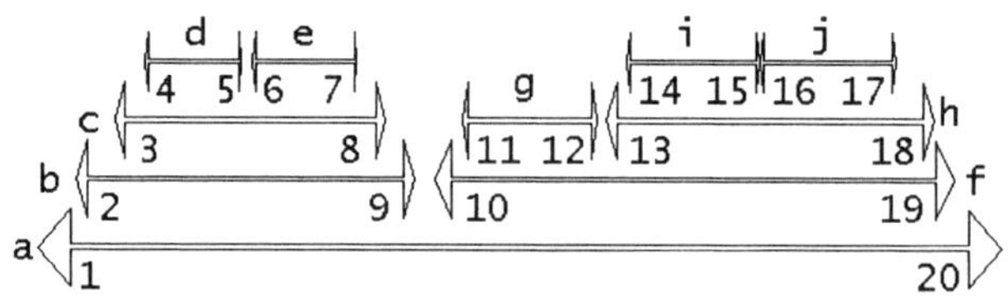

Figure 3. Sample hierarchy depicted as nested line segments

3.2 Representation in SQL

We now continue with an example in SQL. To store hierarchy information we create the following table:

```
CREATE TABLE TREETAB
 (NODE_ID CHAR(1) PRIMARY KEY,
  LEFT_BOUND INT NOT NULL,
  RIGHT_BOUND INT NOT NULL)
```

The sample hierarchy of Figure 1 is represented by the values in Table 1.

Table 1. Values for sample hierarchy

Node Id	Left bound	Right Bound
a	1	20
b	2	9
c	3	8
d	4	5
e	6	7
f	10	19
g	11	12
h	13	18
i	14	15
j	16	17

To ensure that in all cases the value of the table is consistent with our abstract representation we need the following integrity constraints:

1. For all nodes, left bound < right bound
2. No overlapping intervals
3. Existence and uniqueness of root node
4. Left bound of root equal to 1
5. Consecutive numbering

3.3 Sample queries

Using the properties of the representation as stated in section 3.1 we can now formulate some sample queries.

The descendants of the node f can be listed in preorder with the following query:

```
SELECT NODE_ID,
       LEFT_BOUND
  FROM TREETAB
  WHERE LEFT_BOUND > 10 AND
        LEFT_BOUND < 19
  ORDER BY LEFT_BOUND;
```

The ancestors of 'j' can be listed from root downwards with the following query:

```
SELECT NODE_ID,
       LEFT_BOUND
  FROM TREETAB
  WHERE LEFT_BOUND  < 16 AND
        RIGHT_BOUND > 17
  ORDER BY LEFT_BOUND;
```

4. Enhancing the Nested Set Representation

The original representation clearly suffers from problems when new nodes are added to a large hierarchy. Because of the dense numbering adding a new node to the hierarchy results in updating on the average half of all rows. Our first enhancement is to introduce sparse numbering.

The second enhancement is related to the level of the node in the hierarchy.

We proceed by proposing the following improvements to the nested set representation:

1. using sparse numbering to speed up inserts
2. adding the level of each node to the representation

4.1 Speeding up Inserts

One problem with the nested set representation as described in [2] is that when inserting a new node into the tree on average half the nodes need to be updated. This is because the node numbering is dense. For example inserting a new child under 'b' would result in updating the numbering of the nodes 'f', 'g', 'h', 'i', 'j' and 'a'.

With small trees this is not a great problem, but as the amount of nodes increases the amount of updates increases linearly.

The obvious solution to this is to introduce a sparse numbering scheme so that most of the times the insertion of a new node can be done without updating other nodes. The property that is lost is the simple calculation to obtain the number of nodes in a subtree (including the "is leaf" -property).

Naturally there comes a time when a new node does not fit into its designated place and the nodes of the tree have to be renumbered. This renumbering can be done during idle time or on demand. Further, the renumbering can be done locally for "quick relief" or globally for a more lasting effect.

We introduce the simple algorithms ADD_NODE and RENUMBER_ALL. ADD_NODE is optimized for newsgroup type trees where new nodes are always inserted as next (or first) child of parent. RENUMBER_ALL renumbers the bounds so as to use the available space as sparsely as possible.

```
ALGORITHM RENUMBER_ALL

  CONST MAX_INTEGER = 2**31-1

  NODE_COUNT :=
     SELECT COUNT(*) FROM TREE;

  RUNNING_BOUND  := 0;
  BOUND_INTERVAL :=
   MAX_INTEGER/(NODE_COUNT*2)-1;

  FOR EACH
     SELECT NODE_ID,
            LEFT_BOUND AS BOUND_VALUE,
            'LEFT' AS BOUND_TYPE
     FROM TREE
     UNION ALL
     SELECT NODE_ID,
            RIGHT_BOUND AS BOUND_VALUE,
            'RIGHT' AS BOUND_TYPE
     FROM TREE
     ORDER BY BOUND_VALUE

   IF BOUND_TYPE = 'LEFT' BEGIN
     RUNNING_BOUND = RUNNING_BOUND + 1
     UPDATE TREE
       SET LEFT_BOUND = :RUNNING_BOUND
       WHERE NODE_ID = :NODE_ID;
     END
   ELSE BEGIN
     RUNNING_BOUND :=
        RUNNING_BOUND + BOUND_INTERVAL
     UPDATE TREE
       SET RIGHT_BOUND = :RUNNING_BOUND
       WHERE NODE_ID = :NODE_ID;
     END
  END
```

```
ALGORITHM ADD_NODE ( PARENT_NODE_ID,
                     NODE_ID )

  -- GET PARENT NODE INFO

  SELECT LEFT_BOUND,
         RIGHT_BOUND,
         NODE_LEVEL
  FROM TREE
  WHERE NODE_ID =
      :PARENT_NODE_ID;

  -- GET RIGHT-MOST CHILD INFO
   SELECT MAX(RIGHT_BOUND) AS
          MAX_RIGHT_BOUND
   FROM TREE
   WHERE RIGHT_BOUND <
        :RIGHT_BOUND AND
         RIGHT_BOUND >
           :LEFT_BOUND
   IF NOT_FOUND THEN
     MAX_RIGHT_BOUND := LEFT_BOUND

   -- IS THERE ROOM? IF NOT,
    -- RENUMBER AND REFETCH
   -- BOUND-INFORMATION

   IF RIGHT_BOUND-LEFT_BOUND <=2
     BEGIN
       EXECUTE RENUMBER_ALL
       SELECT MAX(RIGHT_BOUND) AS
              MAX_RIGHT_BOUND
         FROM TREE
         WHERE RIGHT_BOUND <
              :RIGHT_BOUND AND
               RIGHT_BOUND >
              :LEFT_BOUND
       IF NOT_FOUND THEN
          MAX_RIGHT_BOUND :=
                    LEFT_BOUND
     END

  NEW_LEFT_BOUND  :=
     MAX_RIGHT_BOUND + 1;
  NEW_RIGHT_BOUND :=
     MAX_RIGHT_BOUND +
      FLOOR( (1/3) *
    (RIGHT_BOUND-NEW_LEFT_BOUND) );
  NEW_LEVEL := NODE_LEVEL + 1;

  INSERT INTO TREE
       (NODE_ID, LEFT_BOUND,
        RIGHT_BOUND, NODE_LEVEL)
  VALUES
     (:NODE_ID, :NEW_LEFT_BOUND,
     :NEW_RIGHT_BOUND, :NEW_LEVEL)
```

4.2 Adding the Hierarchy Level to Each Node

The level of each node can be derived from the number of ancestors it has. We adopt the convention that the level of the node is the number of proper ancestors + 1, i.e. the root has level 1, the children of the root have level 2 etc…

The level of each node can be calculated by joining the tree table with itself and counting the ancestors:

```
SELECT TREE1.NODE_ID,
       COUNT(*) AS NODE_LEVEL
  FROM TREETAB TREE1,
```

```
        TREETAB TREE2
    WHERE TREE1.LEFT_BOUND  >= TREE2.LEFT_BOUND AND
          TREE1.RIGHT_BOUND <= TREE2.RIGHT_BOUND
    GROUP BY TREE1.NODE_ID
```

This is clearly an inefficient and cumbersome way to produce the level information.

We propose that a new attribute, 'level', be added to each node. The level can be easily calculated when the new node is inserted (see algorithm ADD_NODE). We have found no performance penalties for doing this, apart from a small increase in storage space.

Further on we shall see in section 6 that this new attribute is of use when we have to query efficiently for ancestors of a given node in a large tree.

The create statement for the new table is

```
CREATE TABLE TREETABL
 (NODE_ID CHAR(1) PRIMARY KEY,
  LEFT_BOUND INT NOT NULL,
  RIGHT_BOUND INT NOT NULL,
  NODE_LEVEL INT NOT NULL)
```

5. Typical Use Cases and their Access Paths

We will now present some typical queries and describe the access paths using normal B-tree indexes available in any SQL-database. Our aim is to display the best access paths (i.e. the access paths that result in the fastest execution) that are obtainable in each case.

Table 2 contains the B-tree indexes that are used in the queries.

Table 2. Indexes for typical use cases

Name	Unique	Columns
Xtree_node	X	node_id
Xtree_left	X	left_bound
Xtree_leftd	X	left_bound desc, right_bound
Xtree_right	X	right_bound
Xtree_lev_left	X	node_level, left_bound
Xtree_lev_leftd	X	node_level, left_bound desc

5.1 Subtree in Pre- and Post-Order

We obtain a subtree in preorder with the following query:

```
SELECT TREE2.NODE_ID,
       TREE2.NODE_LEVEL,
       TREE2.LEFT_BOUND
  FROM TREETABL TREE1,
       TREETABL TREE2
  WHERE TREE1.NODE_ID = <SUBTREE ROOT> AND
        TREE2.LEFT_BOUND >= TREE1.LEFT_BOUND AND
        TREE2.LEFT_BOUND <= TREE1.RIGHT_BOUND
  ORDER BY LEFT_BOUND
```

Using the indexes XTREE_NODE and XTREE_LEFT we are clearly obtaining an optimal access path: first a lookup on the root id of the sub tree and then a range scan on index xtree_left with no need for sorting. The access path can be portrayed as in figure 4.

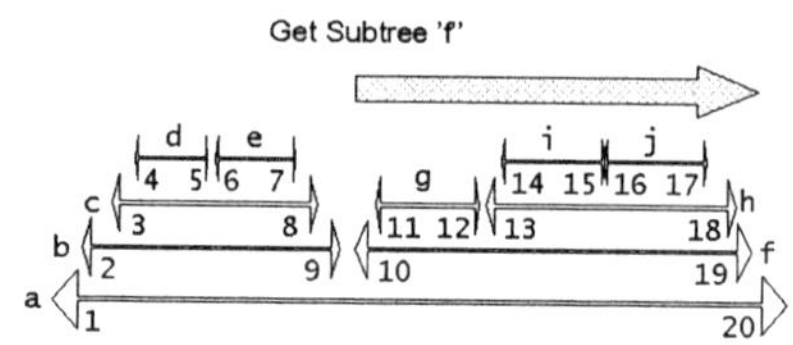

Figure 4. Get subtree of node F

The subtree in postorder is obtainable simply by replacing LEFT_BOUND with RIGHT_BOUND in the order by clause. The optimizer can then choose the index XTREE_RIGHT instead.

5.2 Get Children of a Given Node

```
SELECT TREE2.NODE_ID,
       TREE2.NODE_LEVEL,
       TREE2.LEFT_BOUND
  FROM TREETABL TREE1,
       TREETABL TREE2
 WHERE TREE1.NODE_ID = <PARENT_ID> AND
       TREE2.LEFT_BOUND > TREE1.LEFT_BOUND AND
       TREE2.LEFT_BOUND < TREE1.RIGHT_BOUND AND
       TREE1.NODE_LEVEL+1 = TREE2.NODE_LEVEL
 ORDER BY LEFT_BOUND
```

The query is essentially the same as in 5.1 except for the restriction on NODE_LEVEL. Using indexes XTREE_NODE and XTREE_LEV_LEFT an optimal access path is obtained.

5.3 Ancestors of a Given Node

While the actual query is easy to state, obtaining an acceptable access path is not so trivial. We will start with the obvious solution:

```
SELECT TREE2.NODE_ID,
       TREE2.NODE_LEVEL,
       TREE2.LEFT_BOUND
  FROM TREETABL TREE1,
       TREETABL TREE2
 WHERE TREE1.NODE_ID = <NODE_ID> AND
       TREE2.LEFT_BOUND  < TREE1.LEFT_BOUND AND
       TREE2.RIGHT_BOUND > TREE1.RIGHT_BOUND
 ORDER BY LEFT_BOUND
```

Using the indexes XTREE_NODE and XTREE_LEFTD we can see that the amount of nodes that have to be examined depends on the position of the node. On the average, half of all nodes of the whole tree must be traversed. Hence the time to obtain the ancestors of a given node rises linearly with the number of nodes in the tree. As can be seen in Figure 5, finding the ancestors of 'j' requires searching through nearly the whole tree.

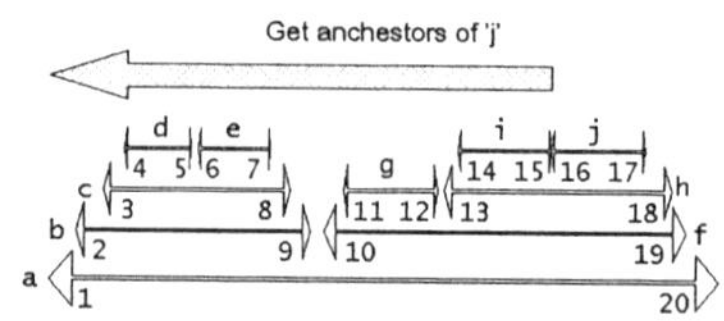

Figure 5.

We will return to this performance problem and present an elegant solution in section 6.

5.4 The Transitive Closure

Due to limitation of the data structure to represent only strict hierarchical trees (rather than arbitrary graphs), the transitive closure is effectively equivalent to the selection of a subtree. However, we would like to produce the output in ordinary relational form, compatible with the output from recursive queries.

The closure of the tree can be obtained by the following query:

```
SELECT TREE1.NODE_ID,
       TREE1.NODE_LEVEL,
       TREE1.LEFT_BOUND,
       TREE2.NODE_LEVEL,
       TREE2.NODE_LEVEL -   TREE1.NODE_LEVEL AS DISTANCE,
       TREE2.LEFT_BOUND
  FROM TREETABL TREE1,
       TREETABL TREE2
  WHERE TREE1.LEFT_BOUND <= TREE2.LEFT_BOUND AND
        TREE1.RIGHT_BOUND > TREE2.LEFT_BOUND
```

Remarks for this query are essentially the same as for the 'Get subtree'- query stated previously.

6. The 'get ancestors' Problem Revisited

The solution to the performance problem for 'get ancestors' query relies on an observation that a node has at most one ancestor on each level. For each of the levels it is possible to obtain that ancestor efficiently. It is the node with the largest left_bound that is smaller than the left_bound of the given node and the level less than the level of the given node.

E.g. node 'j' has one ancestor on each of the levels 1, 2 and 3 (a, f and g respectively). Hence f has the largest left_bound, that is smaller than the left_bound of 'j', than any other node with the same level (which is 2).

This is a query that is efficiently expressible in SQL using standard B-trees and a modern cost based optimizer:

```
SELECT TREE2.NODE_ID,
       TREE2.NODE_LEVEL,
       TREE2.LEFT_BOUND
  FROM TREETABL TREE1,
       TREETABL TREE2
  WHERE TREE1.NODE_ID = <NODE_ID> AND
        TREE2.LEFT_BOUND =
         ( SELECT MAX(LEFT_BOUND)
             FROM TREETABL TREE3
             WHERE TREE3.NODE_LEVEL = 2 AND
             TREE3.LEFT_BOUND <=   TREE1.LEFT_BOUND )
```

Using the indexes XTREE_NODE and XTREE_LEV_LEFTD (see Table 2) the access path will be nested matching index scans on both indexes.

To expand the query to obtain all ancestors we introduce an auxiliary table [5] for enumerating the levels:

```
CREATE TABLE TABLENUM
 ( N INTEGER PRIMARY KEY );
```

and insert the values 1..x into it (x being the maximum level we presume the tree to have).

We now modify the query slightly to obtain each of the ancestors separately for each level that is less than the level of the given node:

```
SELECT TREE2.NODE_ID,
       TREE2.NODE_LEVEL
  FROM TREETABL TREE1,
       TREETABL TREE2,
       TABLENUM NUM
 WHERE TREE1.NODE_ID = <NODE_ID> AND
        NUM.N < TREE1.NODE_LEVEL AND
        TREE2.LEFT_BOUND =
         (SELECT MAX(LEFT_BOUND)
           FROM TREETABL TREE3
           WHERE TREE3.NODE_LEVEL =  NUM.N AND
                 TREE3.LEFT_BOUND <= TREE1.LEFT_BOUND )
```

In spite of the complex nature of the query the optimizers of both DB2 V8.1 and SQLServer 2000 will handle the query as expected. We will examine the performance of this query with the previous version in section 7.2.

7. Comparison and Evaluation

The comparison is made using the enhanced nested set representation, proposed in this work, with the "traditional" adjacency list representation and the original nested set representation with dense numbering. For performance evaluation purposes synthetic data sets were generated. The iterative generator adds the child nodes to random parents. In the sparse numbering scheme we used a space of 1 to 2^{31}-1. The generated trees and their properties can be seen in Table 1.

Table 3. Generated hierarchies

Tree	Number of nodes	Height of tree
1	1000	17
2	2000	18
3	5000	20
4	10000	22
5	20000	22
6	50000	25
7	100000	26
8	200000	27

Performance tests were done using IBM DB2 UDB V8.1 on a computer with a 500MHz Pentium P3 processor and Microsoft Windows 2000 Professional operating system. All measurements are in milliseconds.

7.1 Building the Tree: the Performance of Inserts

The first performance tests concern the insert speed of nodes using either dense or sparse numbering. As can be expected, the sparse numbering scheme outperforms the dense alternative on inserts. Actually, the insertion time grows linearly with the tree size for dense numbering and remains almost constant for sparse numbering.

Our tests include 10 runs for both sparse and dense numbering. In Chart 1 we can see how renumbering requests are done periodically for the sparse tree, on average three times per test run. Other than for the renumbering, the insert speed for the nodes remains constant using the sparse numbering, at about 3,5 ms per node added.

Naturally in some situations, such as loading XML-documents into a database [7], the bounds of the nodes can be precalculated in which case dense numbering will perform well. It is mainly in random inserts that the sparse numbering will bring benefits.

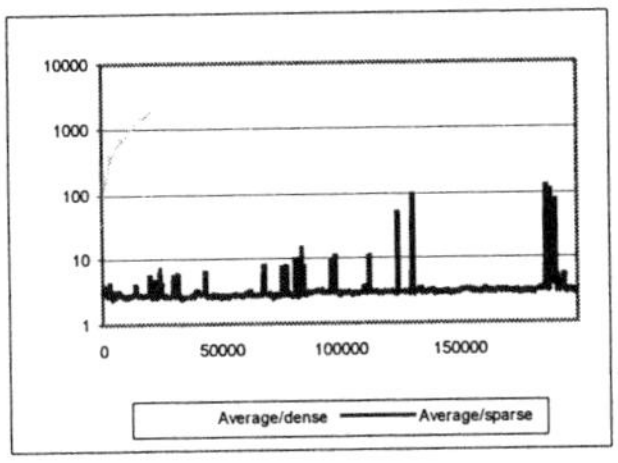

Chart 1. Inserting nodes(ms/size of tree)

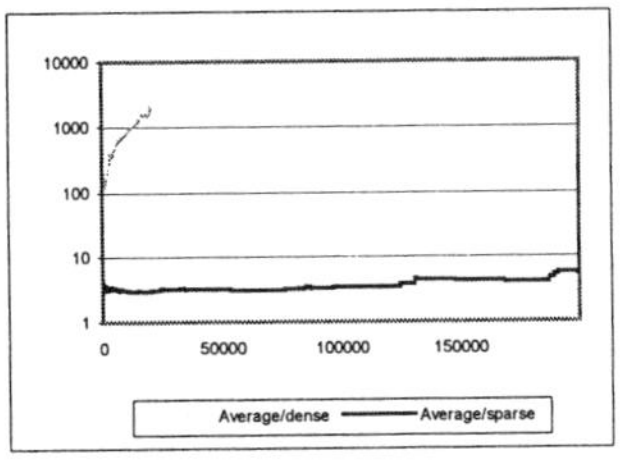

Chart 2. Inserting nodes, averaged (ms/size of tree)

Chart 2 is averaged over the whole space and we can see the total average time slowly rising.

The problem, of course, with sparse numbering is that occasionally a renumbering is required and if it is done for the whole tree it creates a random delay that may not be acceptable. In Chart 3 we see how the node count affects the time needed to renumber the hierarchy.

A practical solution to this problem would be to introduce local renumbering algorithms resulting in less frequent global reorganizations. However, this enhancement has not been done in this research.

Another solution is to do the renumbering periodically during idle time. A larger numbering space (e.g. 64 bit integer) will also lessen the need for renumbering.

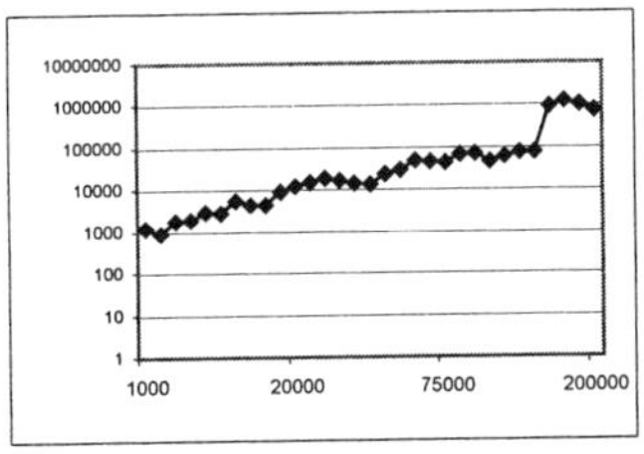

Chart 3. Time needed to renumber hierarchy (ms/size of tree)

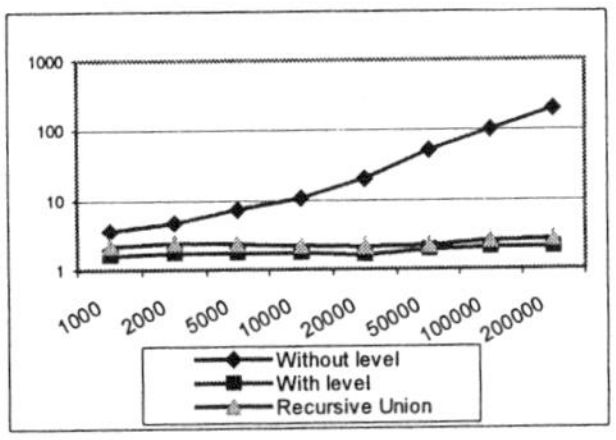

Chart 4. Get ancestors (ms/size of tree)

7.2 Query Performance

To test the usefulness of adding the 'node_level' attribute we compared the performance of three queries:

1. Get ancestors
2. Get siblings
3. Get subtree nodes

Getting ancestors using the original query (see section 5.3) has to scan on average half of all nodes, hence performance degrades as the size of the hierarchy increases regardless of the depth of the tree. As can be seen in Chart 4 the get ancestor-query using the SQL-statement introduced in section 6 clearly outperforms the non-optimized query and its performance does not depend on the size of the tree. What is surprising is that this enhanced query performs about the same or better than the recursive union version.

In the siblings query comparison in Chart 5 we can observe that the query using node levels performs about an order of magnitude faster than the query without node levels materialized.

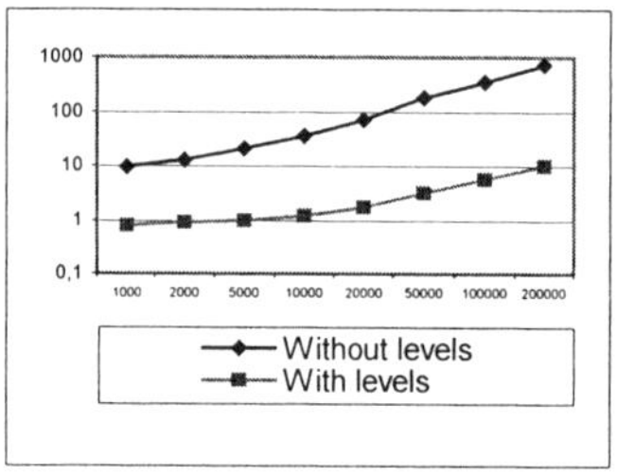

Chart 5. Get siblings (ms/size of tree)

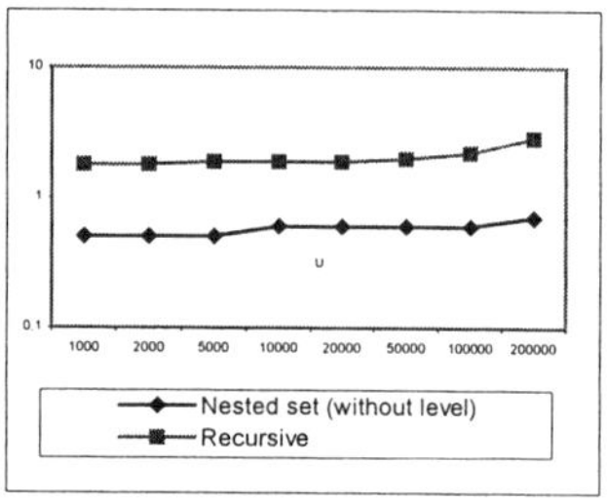

Chart 6. Get subtree in preorder (ms/size of tree)

The subtree query (see section 5.1) is faster than the recursive union version even without the proposed enhancements, as can be seen in Chart 6.

8. Conclusions

We have shown that the nested set representation is useful in situations where strict ordered hierarchies have to be processed in preorder and/or postorder. By adopting the modifications proposed in this paper we can keep the advantages of the nested set representation and still remain competitive with the new SQL language constructs, such as recursive union.

The clearest advantage is in the fact that all queries against the nested set representation can be expressed in standard SQL available on any SQL-based database. The results of the expressions can be further joined to fact tables or to other hierarchies. Because of the numbering scheme we are not relying on any implicit output order for preordering or postordering such as with Oracle's Connect By extension, but can reorder and regroup the nodes at will without losing structure information.

The limits of the original representation as described in [2] and [6] can be overcome by adding sparse numbering and level information to the nodes. We can now get equivalent or better performance compared to the traditional adjacency list representation in queries such as 'get ancestors' and operations such as 'insert node'. It also adds to the types of queries that are possible to express.

Further work is needed to optimize the renumbering algorithm so that the renumbering time and cost can be contained within given limits.

References

[1] Donald E. Knuth. The Art of Computer Programming. Volume 1 pages 312, 317.

[2] Joe Celko. SQL for Smarties (2ed) Morgan Kaufmann Publishers, 1999.

[3] Aho, Alfred V., John E. Hopcroft, and Jeffrey D. Ullman. *The design and analysis of computer algorithms.* AddisonWesley: Reading, MA, 1974.

[4] Ballew, Duncan, Blasingame. Relational Data Structures for Implementing Thesauri http://www.mip.berkeley.edu/mip/related/thesaurus.html

[5] Hugh Darwen. A constant friend. In Relational Database Writings 1985-1989 by C. J. Date, pages 493-500. Prentice Hall, 1990.

[6] Leif Morten Kofoed and Håkon Erdal. Trees and SQL. Presentation at DB2 G.U.I.D.E . April 1995, Lahti Finland.
[7] Torsten Grust. Accelerating XPATH location steps. In. *Proc. ACM SIGMOD 2002.* P. 109-130.
[8] Chun Zhang, Jeffrey Naughton, David DeWitt, Qiong Luo, and Guy Lohman. On Supporting Containment Queries in Relational Database anagement Systems. In Proc. of the ACM SIGMOD Int'l Conference on Management of Data, pages 425-436, Santa Barbara, California, May 2001. ACM Press.

[9] Quanzhong Li and Bongki Moon. Indexing and Querying XML Data for Regular Path Expressions. In Proc. of the 27th Int'l Conference on Very Large Data Bases (VLDB), pages 361-370, Rome, Italy, September 2001.
[10] Srini Venigalla. Expanding Recursive Opportunities with SQL UDFs in DB2 v7.2. http://www7b.boulder.ibm.com/dmdd/library/techarticle/0203venigalla/0203venigalla.html
[11] Ullman, Jeffrey. Principles of database and knowledge base systems. Vol. 2.. Academic Press, 1989.
[12] Torsten Grust, Maurice van Keulen, Jens Teubner. Staircase Join: Teach A Relational DBMS to watch it's (Axis) steps. In Proc. of the 29th Int'l Conference on Very Large Data Bases (VLDB), Berlin, Germany, September 2003.
[13] Vadim Tropashko: Trees in SQL: Nested Sets and Materialized path. http://www.dbazine.com/tropashko4.shtml

Databases and Information Systems
J. Barzdins and A. Caplinskas (Eds.)
IOS Press, 2005

Concurrent Video: Operational Extensions

Oleg PROSKURNIN[1]
University of St.Petersburg, Russia
olegpro@acm.org

Abstract. Today, non-linear video editing technology became ubiquitous, having spread from the film industry to almost every home computer. Dozens of multimedia authoring applications offer a wide range of functional capabilities, yet there is an evident lack of support for collaborative activities among them. To change this situation, a novel data model called concurrent video has been proposed recently. Since it was substantially based on formal aspects of cooperative transactions, it has provided a solid foundation for consistent exchange and sharing of information between co-workers in collaborative environments. The main objective of this work is to extend the concurrent video model by introducing advanced editing operations, which should permit more flexibility within the authoring process and enable a higher level of concurrency among users' actions. Besides, this extension is going to be performed in such a way that major transactional properties of the original proposal will be preserved.

1. Introduction

It is a well known fact that the possibility to collaborate in multimedia authoring environments is going to bring considerable advantages into the process of content creation.

Indeed, post production tasks typically require efforts of various experts – editors, graphics artists and sound engineers, to name a few. In this context, cooperation could emphasize the strengths of different specialists and moreover, could encourage each participant to use the right tool for his particular job, promising in general advanced workflow with minimized time and resources. Hence, support for joint activities would noticeably improve post production technology in most related industries [15].

Recently, a technique for collaborative authoring of media streams, such as video or audio, has been presented in the form of a concurrent video data model [11]. Based on the cooperative activity framework CoAct [13, 18, 9], this model has enabled consistent sharing of information in the face of concurrent modifications, which is often regarded as the key task of multi-user applications design.

The background is that co-authors' actions in the CoAct framework are represented via activity histories, which are basically sequences of editing operations constituting individual working processes. As a result, consistency of information exchange is achieved by means of the history merging mechanism [18] that takes into account compatibility relations of participants' actions in order to detect conflicts in their joint efforts.

To avoid unnecessary conflicts, editing operations offered within the concurrent video model were especially designed to have good commutativity properties. In fact, according to the concepts of timeline-based video authoring [4], these operations allowed participants to work concurrently on different clips, which in the end formed the desired video material.

[1] This work is partially supported by the Russian Foundation for Basic Research under grant No. 04-01-00173

The goal of this paper is to improve the original data model by providing support for non-conflicting cooperative activities within the same clips. However, most transactional aspects of the previous proposal are going to be kept intact.

The rest of this paper is organized as follows. After giving a brief overview of the related work, the improved version of the data model is introduced. Next, transactional properties of concurrent video are discussed and finally, conclusions and further work are presented.

2. Related Work

This section first gives a brief review of the CoAct framework and then presents a summary of previous work on video modeling and collaborative multimedia authoring.

2.1 The Cooperative Activity Model

The cooperative activity framework CoAct [13] defines a generic transactional model that enables consistent sharing and exchange of information in various collaborative scenarios, like multi-user document authoring environments [14].

To support joint activities, the CoAct framework assigns a personal isolated workspace containing copies of shared data to each user involved in the cooperative effort. Also, a separate common workspace is established to represent the current state of team work. Thus, participants are able to manipulate their private versions of shared objects independently from each other, while the common database stores at first initial data, then intermediate and lastly final results of the joint effort.

At the user level, cooperation is achieved by explicit exchange of the contents of workspaces, which can be performed either directly or through the common database by means of dedicated operations. But conceptually, these processes are controlled by the history merging mechanism [18, 9], that takes into consideration semantics of the application domain, ensuring correctness of information exchange.

The point is that participant's actions in CoAct are modeled by an activity history, which is basically a sequence of operations carried out by a certain user in his workspace. Similar to conventional concurrency control [20], a semantic conflict test is defined for each possible pair of operation invocations. Such compatibility predicate is often symmetric and is typically specified in terms of state-independent commutativity relations. Since the order in which commuting operations are executed is irrelevant, the compatibility property is used in CoAct as a basis for merging activity histories. In other words, two operation invocations coming from different users are both allowed to be present in the merge only if they are compatible. Otherwise, the controlling user has to resolve the conflict by discarding unnecessary actions.

Consequently, the merging mechanism of the CoAct model enables semantically consistent incorporation of individual activity histories into a single one that is implicitly formed in the common database and represents the results of cooperative work.

2.2 Collaborative Video Authoring

Although a variety of multimedia authoring tools are available in the market, e.g. [2, 16], none of them supports concurrent video editing and the related research proposals are also very few.

Some of these works investigate cooperative activities based on different locking mechanisms [3, 17], while other approaches consider collaboration in real-time groupware systems only [22]. In contrast to the concurrent video model [11], these proposals do not cover multimedia modeling issues and do not provide consistency guarantees in a transactional sense.

However, there exists a wealth of research works examining various aspects of video data management and some of them have common features with the present paper.

In particular, operations on media segments in video and stream algebra [21, 10], as well as updating mechanisms in other proposals [1, 6] partially resemble editing facilities of the model described below. Moreover, previously introduced hierarchical structures [1, 21] and video objects [12] also have certain correlations with the presented approach.

Nevertheless, the major distinction between concurrent video and former models lies in the suitability of the new proposal for the basic needs of both video authoring systems and cooperative environments, which has not been achieved ever before.

The point is that in the considered authoring systems the movie is created by altering and assembling source clips which are treated as independent units of raw material. These clips along with other media objects are placed on the timeline, which naturally represents the time flow. After the inclusion, objects can be edited, special effects and transitions can be applied and the resulting material can be arranged into a final video production. In addition, since such editing implies just referencing the original files, which are not actually modified, a special process called rendering is required to preview the produced movie.

Concurrent video takes into account these basic concepts of timeline-based authoring and provides a tree-based abstraction for referencing a sequence of media clips. It also presents a set of high-level non-destructive editing operations allowing users to insert and delete, alter and temporally combine video material to form the final production.

To support joint activities, operations on different clips are considered to be commutative, i.e. different users can work concurrently on different clips without any conflicts. Moreover, hierarchical structure that references video data also stores activity histories, providing elegant support for cooperation mechanisms of CoAct.

The above features clearly distinguish concurrent video model [11] from other research proposals in the area of cooperative authoring. In particular, this investigation significantly differs from recent works [8] and [5], which present tree-based data structures for synchronous collaborative document editing.

3. Enhanced Concurrent Video Model

This section outlines improved concurrent video model that provides advanced intra-clip editing operations and is entirely based on its already presented counterpart [11].

3.1 Video Segments

In the considered model video segments represent an abstraction over independent units of video data, which are used as building blocks within the authoring process.

Basically, each video segment references a contiguous part of raw material via a frame sequence and has its own set of attribute-value pairs which describe the proper interpretation of the underlying media at the presentation or rendering level. Frame sequences reflect the stream-like nature of video data, while attributes support implementation of non-destructive editing operations as well as specifications of desired transitions and special effects for the segment.

Actually, both the definition and the purpose of video segments in this work stay the same as in the original proposal [11], however, we quote some facts here for convenience:

Definition 1 (frame sequence). A frame sequence is a finite non-empty sequence $(f_1, \ldots, f_N)$ of video frames f_i, referring to a contiguous block of video data. N is called the length of a video sequence F and is denoted by $Length(F)$.

Definition 2 (video segment). A video segment is a pair (F, A), where F is a frame sequence and A is a possibly empty set of attribute-value pairs $\{a_i:v_i\}$, storing various additional information about the containing segment.

Besides, two trivial operations – concatenation and extraction – can be defined for frame sequences [11], however they won't be referred in this work explicitly and thus are omitted.

3.2 Video Activity Tree

Concurrent video employs a single hierarchical data structure, called video activity tree, for modeling both underlying media and cooperative transactions.

Basically, the structure of an activity tree in this work remains the same as before: leaf nodes, called valid, are associated with video clips that authors place on a timeline, dead nodes stand for previously deleted clips and the rest nodes hold related parts of the activity history. Actually, there exists a one-to-one correspondence between valid nodes of the tree and video clips visible to the users, and moreover, the total order imposed on leaf nodes by the tree structure reflects the order in which mentioned clips appear on the timeline. Also, since we are concerned with collaborative work, activity history elements are marked whether they are private or shared to indicate what part of the tree is present in the common database and what part exists only in the considered local workspace, as illustrated in figure 1.

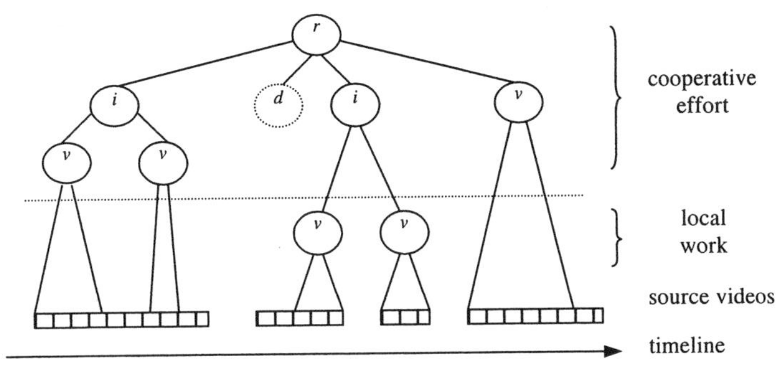

Figure 1. A sample video activity tree in a user's workspace.
Characters *v*, *d*, *i* and *r* denote valid, dead, intermediate nodes and the root node of the tree respectively.

However, unique identifiers assigned to video clips, which have enabled the development of editing operations with state-independent commutativity relations in [11], are not sufficient for the purposes of this work.

In order to provide intra-clip operations with similar commutativity properties, unique values should be assigned to all frames constituting the video stream. Fortunately, such task can be performed by combining clips' identifiers with frames' indices within these clips. This approach is actively exploited by novel editing operations and is reflected below in the redefinitions of activity tree's node structures.

Definition 3 (video activity tree). A video activity tree T is a tuple (*Root*, *Nodes*, *Valid*, *Dead*), where *Root* is the root node of T, *Nodes* is a set of intermediate nodes of T, and *Valid* and *Dead* are disjoint sets of so-called valid and dead leaf nodes respectively.

Additionally, there exists a total order $<_t$ defined over the set of leaf nodes of T, that corresponds to the order in which associated video clips are (or were for dead nodes) located on the timeline.

Definition 4 (valid node). A valid node V of a video activity tree is a tuple (*NID*, *Range*, *Segment*, *History*), where *NID-Range* pair uniquely identifies V's associated video data, *Segment* is the video segment representing this data and *History* is an ordered set of operation instances related to the given node V. *Range* is represented as a pair of bounding values [*Low*, *High*].

Definition 5 (intermediate node and dead node). An intermediate node as well as a dead node of a video activity tree is merely a tuple (*NID*, *Range*, *History*), whose elements are intended for the same purposes as their counterparts in valid nodes (except *NID-Range* pair identifies no data).

Operation instances, mentioned above, can be treated as records describing the fact of execution of a certain editing operation by some user. They act as elementary entities for building activity histories and modeling the authoring process, being the same as in the original work [11]:

Definition 6 (operation instance). An operation instance is a tuple (*Status*, *OID*, *Name*, *Input*, *Output*), where *Status* $\in$ {*private*, *shared*} indicates whether this instance is present only in the current workspace or not, *OID* is a unique identifier of this instance (required for tracing identical instances during the merging process) and *Name* is the name of the executed operation, whose input and output parameters are reflected in *Input* and *Output* sets.

3.3 Editing Operations

The main challenge in the design of editing operations is to provide a solid support for modeling authors' actions in a collaborative environment. Thus, it is worth to remember that the most common thing cooperating authors are likely to do is altering and assembling video clips on a timeline, and what is more, they are likely to do it concurrently.

In contrast with the original proposal [11], which has provided only segment-level editing operations, in this work most attention is focused on frame-level manipulations and achieving commutativity of semantically non-conflicting actions performed within the same video segment, e.g. altering of different regions of frames of the same clip by different users.

First of all, an initialization routine intended for fixing an initial state of the cooperative activity in the common database is provided.

Definition 7 (initialization algorithm). The initialization algorithm takes a sequence of N video clips, denoted by a collection of N video segments (vs_1, …, vs_N) as input and forms a corresponding video activity tree with an empty history:

1. Create N valid nodes V_i: $\forall\ i,\ 1 \le i \le N$: $V_i := (NewID(), [1, Length(vs_i.F)], vs_i, \varnothing)$, where *NewID*() is a function generating a new unique identifier, see [11] for how it may work.
2. Construct a video activity tree T: $T := (Root, \varnothing, \{V_i \mid 1 \le i \le N\}, \varnothing)$, such that: $\forall\ V_i: Parent(V_i) = Root \wedge \forall\ i,j: 1 \le i < j \le N \Leftrightarrow V_i <_t V_j$.

Next, an operation for editing video clips as a whole or for modifying their parts is presented:

Definition 8 (editing algorithm). The editing algorithm takes as input an identifier *id* of the clip and a range *r* of frames within it, which have to be replaced with a new video segment *vs* representing the results of performed modifications. The algorithm splits the affected node of the video activity tree *T* (if necessary) and accordingly updates its activity history:

1. Find a node $V \in T.Valid$: $V.NID = id \wedge r \subset V.Range$,
   Report failure if such node does not exist.
2. *If* $r = V.Range$, *Then* modify V:
   $V := (V.NID, [1, Length(vs.F)], vs, \{O \in V.History \mid O.Status = Shared\})$,
   Append an instance $(private, NewID(), Edit, \{id, r, vs\}, \varnothing)$ to $V.History$,
   Report success.
   Else construct an intermediate node $V' := (V.NID, V.Range, V.History)$,
   And go to step 3.
3. Construct valid nodes V_i (at least two of them with non-empty frame sequences):
   $V_1 = (V.NID, [V.Range.Low, r.Low - 1], LeftSegment, \varnothing)$,
   $V_2 = (NewID(), [1, Length(vs.F)], vs, \varnothing)$,
   $V_3 = (V.NID, [r.High + 1, V.Range.High], RightSegment, \varnothing)$,
   where the contents of video segments *LeftSegment* and *RightSegment* is illustrated in figure 2.
4. Adjust T: $T := (Root, T.Nodes \cup \{V'\}, T.Valid \cup \{V_i\} \setminus \{V\}, T.Dead)$, such that: $Parent(V') = Parent(V) \wedge \forall V_i: Parent(V_i) = V' \wedge \forall i, j: i < j \Leftrightarrow V_i <_t V_j$.
5. Append instance $(private, NewID(), Edit, \{id, r, vs\}, \varnothing)$ to $V_2.History$.

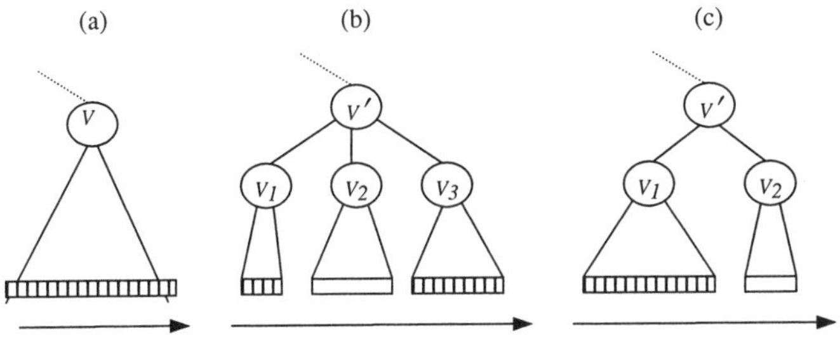

Figure 2. Node splitting during intra-clip editing: (a) – an original valid node with the associated clip, (b) – split node after editing inside of the clip, (c) – split node after right-edge editing.

Evidently, the above algorithm provides an efficient means of modeling frame level manipulations within a particular video clip and at the same time supports operations applied to the whole segment, as already proposed in [11]. Since intra-segment editing is based on the node splitting technique, the affected valid node is replaced with a subtree, whose leaves are arranged to correspond to the resulting video material, as illustrated in figure 2. Moreover, propagation of the original node identifier to the node's respective sub-segments formed after splitting enables concurrent intra-segment manipulations in non-overlapping clip regions.

At this point it is worth to mention that editing algorithms make no modifications in the source video material – only references to the raw data are actually manipulated. Implicitly, such approach assumes the possibility of random access to the frames contained in the underlying video clips. However, this may be unfeasible (at least directly) when clips are coded with compression algorithms exploiting motion estimation, i.e. dependencies between adjacent frames. To overcome this difficulty it can be assumed that video material is accessed from an appropriate starting point, such as the nearest previous key frame, if necessary.

Next, operations for including new material into the video production and removing unnecessary parts from it are described. The insertion algorithm is left almost unchanged [11], while the deletion method now supports intra-clip removals.

Definition 9 (insertion algorithm). This algorithm takes as input a new video segment *vs* and its desired location on the timeline, specified by a destination point *pos* within a clip having given identifier *id*. The affected node is split and the activity history is accordingly updated:

1. Find a node $V \in T.Valid$:
   $V.NID = id \wedge (pos \in V.Range \vee (pos = 0 \wedge V.Range.Low = 1))$,
   Report failure if such node does not exist.
2. Construct valid nodes V_i (at least two of them with non-empty frame sequences):
   $V_1 = (V.NID, [V.Range.Low, pos], LeftSegment, \varnothing)$,
   $V_2 = (NewID(), [1, Length(vs.F)], vs, \varnothing)$,
   $V_3 = (V.NID, [pos + 1, V.Range.High], RightSegment, \varnothing)$,
   and split the node V in a way similar to splitting in the editing algorithm.
3. Append instance $(private, NewID(), Insert, \{vs, id, pos\}, \varnothing)$ to $V_2.History$.

Note, that in contrast to [11], insertion operation instance is stored in the newly created node and not in the split one. Along with propagating of the original segment identifier to its remainders, this enables commutativity of insertions occurred at different positions in the same clip.

Definition 10 (deletion algorithm). This algorithm takes as input an identifier *id* of the clip and a range *r* of frames within it, which need to be removed. The algorithm splits the affected node (if necessary), forms a new dead node and accordingly updates its activity history:

1. Find a node $V \in T.Valid$: $V.NID = id \wedge r \subset V.Range$,
   Report failure if such node does not exist;
   Coordinate with previous deletions and undo local insertions, like in [11].
2. Construct valid nodes V_i (at least two of them with non-empty frame sequences):
   $V_1 = (V.NID, [V.Range.Low, r.Low - 1], LeftSegment, \varnothing)$,
   $V_2 = (V.NID, r, \varnothing)$, this should be actually a dead node,
   $V_3 = (V.NID, [r.High + 1, V.Range.High], RightSegment, \varnothing)$,
   and split V in case its part is removed, otherwise V becomes a dead node itself.
3. Append an instance $(private, NewID(), Delete, \{id, r\}, \varnothing)$ to $V_2.History$.

Similar to [11], the deletion algorithm can be designed to support commutativity between any two removals, even overlapping within the same clip. Also, it can act as an inverse for some local insertions, which are not shared by other users and are not affected by subsequent operations. The latter can be seen as an example of history reduction at the user's side.

In addition to the presented editing operations it turns out to be useful to provide a dedicated moving algorithm, which would enable moving of selected contiguous blocks of the media stream to the specified locations. Though at first sight it may seem that such method is unnecessary because it looks the same as an appropriate sequence of deletion and insertion, in fact it is required since delete-insert pairs are in conflict with concurrent editing of the moved block. Moreover, moving and editing activities over a single segment can be regarded as compatible from a semantic point of view and thus the following algorithm has to be provided.

Definition 11 (moving algorithm) This algorithm takes as input an identifier *mid* of the clip and a range *r* of frames within it, which has to be moved. Their target location is specified by *pos* and *id* parameters, which are similar to their counterparts in the insertion algorithm. The algorithm is especially designed in a way to ensure compatibility of moving operation with concurrent manipulations within the moved segment, such as editing or insertion activities. To achieve this, it is insufficient to move a single valid node like in [11], but it is necessary to move a proper valid (if it exists) or the nearest intermediate node with a certain part of its subtree, which stands for concurrent manipulations mentioned above (if any). Throughout this process, the identifier of the moved clip should be kept intact, as outlined below:

1. Find a node $M \in T.Valid \cup T.Nodes$: $M.NID = mid \wedge r \subset M.Range$; or fail.
2. Apply steps 2, 3 of deletion algorithm to *M*, using *r* and *Move* operation instance.
3. Apply insertion algorithm with *M.Segment* (if *M* is valid node), *r*, *id*, and *pos* as input, such that $V_2 = (mid, r, M.Segment, \varnothing)$, and use *Move* operation instance.
4. Re-execute operation instances which were located in the nodes below *M* (if any);

4. Achieving Cooperation

According to the principles of the CoAct transaction model, two distinct types of commutativity, namely forward and backward, are utilized for manipulating activity histories [9, 19].

In particular, backward commutativity is intended for determining dependencies between user's actions within a single history. The point is that the behavior of editing operations may be influenced by previously executed methods, for example, any modifications of a particular video clip depend on the preceding insertion of this clip into the media stream. Consequently, no operation instance can be exchanged between any workspaces without its relevant predecessors. And backward commutativity allows us to identify closed subhistories [9] having no external dependencies and thus representing consistent units of work, which can be exchanged between cooperating users separately from each other.

Table 1. Forward commutativity relations (symmetric)

Operations	Edit(ID, R, VS)	Insert(VS, ID, POS)	Delete(ID, R)	Move(MID, R, ID, POS)
Edit(id, r, vs)	$id \neq ID \vee r \cap R = \varnothing$			
Insert(vs, id, pos)	$id \neq ID \vee pos \notin R$ *	$id \neq ID \vee pos \neq POS$		
Delete(id, r)	$id \neq ID \vee r \cap R = \varnothing$	$id \neq ID \vee POS \notin r$ *	*true*	
Move(mid, r, id, pos)	$(mid \neq ID \vee r \supset R \vee r \cap R = \varnothing) \wedge (id \neq ID \vee pos \notin R$ *$)$	$(mid \neq ID \vee (r.Low - 1 \neq POS \wedge r.High \neq POS)) \wedge (id \neq ID \vee pos \neq POS)$	$(mid \neq ID \vee r \cap R = \varnothing) \wedge (id \neq ID \vee pos \notin R$ *$)$	$\{mid, id\} \cap \{MID, ID\} = \varnothing$

* Actually, it is also reasonable to treat these operations as conflicting when $pos + 1 \in r$ in order to completely avoid commutativity in boundary cases.

Actually, such dependencies are naturally reflected by the hierarchical structure of the original as well as the extended concurrent video model – dependent operation instances can be found "above" the given one within the tree. Hence, already developed theoretical results and an algorithm for subhistory extraction [11], which is almost as simple as subtree selection, remain valid for the given proposal, providing an efficient way for identifying consistent units of work.

After extraction of consistent units of work from users' workspaces, another type of commutativity – forward – is exploited for detecting conflicts between operations coming from different users, serving as a basis for semantically correct merging of the results of individual activities. Forward commutativity properties for the methods of the data model presented in this paper are summarized in table 1.

Relations in the table demonstrate that editing operations of the expanded concurrent video model provide visible concurrency benefits comparing to the original proposal. It is now clear that users can freely work on the same segments until frame regions they are manipulating do not overlap and what is more, video extracts moved along the timeline can be simultaneously modified by other participants.

5. Conclusions and Further Work

In this paper, the concurrent video data model has been extended by means of the development of intra-segment editing operations.

Such extension has enabled a higher level of concurrency among cooperating users, at the same time providing more accurate support for modeling their activities in collaborative video authoring environments. Additionally, major transactional aspects of the original proposal [11] were preserved in a way to ensure efficient information sharing and exchange.

Moreover, the results of this work are general enough to be applicable to stream data in general. For instance, by renaming of frame sequences and video segments to sample sequences and audio segments, the concurrent video model can be transformed into concurrent audio, which may be found useful in the similar authoring environments.

As a possible direction of further research, investigation of versioning support for stream data on the basis of this work is considered.

Besides, development of a prototype system demonstrating feasibility of the presented approach is regarded as an essential part of the future work.

For this purpose, there is an intention to utilize the recently developed Advanced Authoring Format [7, 15] and its open source SDK. Basically, the AAF file format is an industry-driven standard especially designed for interchange of compositional meta-data between various multimedia authoring tools. At the moment of writing this paper, several video editing applications have already claimed a certain level of compliance with AAF. In particular, Adobe Premiere Pro [2] is able to import and export its projects in this novel form.

Acknowledgements

Author is grateful for encouragement and valuable discussions on the topic provided by Boris Novikov.

References

[1] Adali, S., Candan, K.S., Chen, S., Erol, K., Subrahmanian, V.: The advanced video information system: Data structures and query processing. Multimedia Systems, Vol.4, pages 172-186, 1996

[2] Adobe, Adobe Premiere Pro 1.5: http://www.adobe.com/products/premiere, 2004

[3] Borghoff, U.M., Teege, G.: Structure management in the collaborative multimedia editing system IRIS. In Proc. of the International Conf. on Multi-Media Modeling, pages 159-173, Singapore, 1993

[4] Bulterman, D.C.A., Hardman, L.: Multimedia authoring tools: State of the art and research challenges. In Lecture Notes in Computer Science #1000, Springer-Verlag, 1995

[5] Davis, A.H., Sun, C., Lu, J.: Generalizing operational transformation to the standard general markup language. In Proc. of the 2002 ACM conference on Computer Supported Cooperative Work, pages 58-67, New Orleans, USA, November 2002

[6] Dumas, M., Lozano, R., Fauvet, M.-C., Martin, H., Scholl, P.-C.: A sequence-based object-oriented model for video databases. Multimedia Tools and Applications, Vol.18, Issue 3, pages 249-277, 2002

[7] Gilmer, B.: AAF – The Advanced Authoring Format. http://www.aafassociation.org/html/pr/291_gilmer1_0702.pdf

[8] Ionescu, M., Marsic, I.: Tree-Based Concurrency Control in Distributed Groupware. Computer Supported Cooperative Work, Vol.12, No.3, pages 329-350, 2003

[9] Klingemann, J., Tesch, T., Wasch, J.: Semantics-based transaction management for cooperative applications. In Proc. of the International Workshop on Advanced Transaction Models and Architectures, pages 234-252, Goa, India, August - September 1996

[10] Mackay, W.E., Beaudouin-Lafon, M.: DIVA: exploratory data analysis with multimedia streams. Conf. proc. on Human factors in computing systems, pages 416-423, Los Angeles, USA, April 1998

[11] Novikov, B., Proskurnin, O.: Towards collaborative video authoring. In Proc. of the 7th East-European Conference on Advances in Databases and Information Systems, pages 370-384, Dresden, Germany, September 2003

[12] Oomoto, E., Tanaka, K.: OVID: Design and implementation of a video-object database system. IEEE Transactions on Knowledge and Data Engineering, Vol.5, No.4, August 1993

[13] Rusinkiewicz, M., Klas, W., Tesch, T., Wasch, J., Muth, P.: Towards a cooperative transaction model - The cooperative activity model. In Proc. of the 21st International Conference on Very Large Databases, pages 194-205, Zurich, Switzerland, September 1995

[14] Tesch, T., Wasch, J.: Transaction support for cooperative hypermedia document authoring – a study on requirements. In Proc. of the 8th ERCIM Database Research Group Workshop on Database Issues and Infrastructure in Cooperative Information Systems, pages 31-42, Trondheim, Norway, August 1995

[15] Turner, B.: A new take on teamwork. Video Systems, December 2002. http://videosystems.com/ar/video_new_teamwork

[16] Ulead, MediaStudio Pro 7.0: http://www.ulead.com/msp, 2004

[17] Wang, K.: The design of extending individual multimedia authoring to cooperative multimedia authoring. In Proc. of the Fourth International Conference for Young Computer Scientists (ICYCS'95), Beijing, China, July 1995

[18] Wasch, J., Klas, W.: History merging as a mechanism for concurrency control in cooperative environments. In Proc. of RIDE-Interoperability of Nontraditional Database Systems, pages 76-85, New Orleans, USA, February 1996

[19] Weihl, W.E.: Commutativity-based concurrency control for abstract data types. IEEE Transactions on Computers, Vol.37, No.12, pages 1488-1505, December 1988

[20] Weikum, G.: Principles and realization strategies of multilevel transaction management. ACM Transactions on Database Systems, Vol.16, No.1, pages 132-180, March 1991

[21] Weiss, R., Duda, A., Gifford, D.: Composition and search with a video algebra. IEEE Multimedia, Vol.2, No.1, pages 12-25, 1995

[22] Xiao, B.: Collaborative multimedia authoring: scenarios and consistency maintenance. In Proc. of the Fourth International Workshop on Collaborative Editing Systems, New Orleans, USA, November 2002

Data Mining and Data Warehousing

Databases and Information Systems
J. Barzdins and A. Caplinskas (Eds.)
IOS Press, 2005

From Temporal Data Mining and Web Mining to Temporal Web Mining

Mireille SAMIA*, Stefan CONRAD
Institute of Computer Science
Databases and Information Systems
Heinrich-Heine-University Düsseldorf
D-40225 Düsseldorf, Germany
{samia|conrad}@cs.uni-duesseldorf.de

Abstract. Treating data with temporal information as temporal data, and not as static, is becoming the concern of different domains. Temporal Web mining is a new approach, which deals with real time data. Temporal Web mining extends temporal data mining and Web mining. It concerns the Web mining of data with significant temporal information. It can be used in different areas, such as finance, engineering, environmental sciences, earth sciences, and medicine. Its main goal is to analyze local and Web data in real time in order to discover relevant temporal information. This paper presents the new concept Temporal Web Mining (TWM), focuses on its architecture, and discusses some of its application scenarios, such as volcanism and seismology, and flooding.

1. Introduction

The ability to obtain real time data from multiple different sources, to share it, and to analyze it is of great importance for the development of the information society. Temporal Web mining aims at treating data with temporal information in real time over the Web. It extends temporal data mining and Web mining. Its main goal is to query local and Web data in real time, analyze these temporal sequences in order to discover previously unknown important temporal knowledge. Another purpose is to introduce prediction as a main issue in Web mining, specifically Web content mining. Furthermore, it uses Web data with temporal information in the temporal data mining process. Temporal Web mining can be used in different domains, such as finance, engineering, environmental sciences, medicine, and earth sciences. One of its application scenarios is flooding. Flooding can threaten the public safety and potentially damage properties [14].

Combining and analyzing data about flooding, such as historical flood damage records and flooding data hazards, flooding relationships to other hazards (such as earthquakes, landslides) as well as auxiliary information (such as meteorological data), in real time locally and/or over the Web, can enhance the capability of predicting a flooding.

Another instance is volcanoes and earthquakes. Data about volcanic and seismic phenomena, such as historical volcanic data, historical seismic data, data about any current volcanic or seismic activity, are received from different sources (such as stations), and combined with other data, such as meteorological data, in order to be queried in real time, locally and/or over the Web. Hence, the obtained data is analyzed in order to extract the

* This work is supported by the German Academic Exchange Service (Deutscher Akademischer Austausch Dienst DAAD).

needed temporal data which can be a way to forecast to a certain extent earthquakes and volcanic activities, and a possibility to enhance the safety measures.

Another goal of temporal Web mining is to make good use of the vast amount of data that has been accumulated over the past decades. It encourages and aims at contributing in building a common reference data warehouse which improves data record, data exchange, data manipulation, data standardization and data analysis between different institutions and communities, such as researchers.

In this paper, section 2 gives a brief overview of temporal data mining and Web mining, and discusses related work. In section 3, we define temporal Web mining after describing its process. In section 4, we present the temporal Web mining architecture, and analyze it. Then, we discuss some of its application scenarios in flooding, and earth sciences. Section 5 concludes this paper and points out our directions for future work.

2. Overview and Related Work

This section gives a brief overview of temporal data mining and Web mining before discussing further related work.

2.1 Overview

2.1.1 Temporal Data Mining

Data mining is the process of discovering and extracting relevant hidden relationships and previously unknown interesting patterns that exist in information repositories, such as databases or data warehouses [8].

Temporal data mining mines data collected over time. It concerns the analysis of temporal data and the search for important hidden relationships and patterns between data with temporal information. In other words, its goals are to discover and analyze previously unknown hidden relations between temporal data, and to predict temporal sequences. Depending on the type of temporal data, the approaches to solve the mining problem could differ [1]. A temporal sequence is defined as a sequence of nominal symbols from a particular alphabet, and a time series as a sequence composed of continuous real valued elements.

The discovery of relations between temporal data starts with the representation and modeling of the temporal data. The data can be represented into time series data in either continuous or discrete, linear or non-linear models, stationary or nonstationary models and distribution models (e.g., time-domain representation or time series model representation) [12]. It can also be represented into time series data using a continuous or a discrete transformation (e.g., discrete Fourier transform, discrete wavelet transform and discretization transformation). To find subsequences of events, a sliding window of size w can be used, and placed at every possible position of the sequence [1].

After representing the temporal data, similarity measures between sequences of events are defined. For example, the distance between temporal characteristics in a continuous time domain can be measured using the Euclidean squared distance function. Measuring similarities in discrete spaces can be done using the Shape Definition Language [1].

After finding similarities between sequences of events, temporal data mining techniques are applied, such as the discovery of association rules, classification, and clustering [1, 12].

2.1.2 Web Mining

The Web is a large distributed repository of data, which provides the opportunity to use data mining techniques to analyze Web data. The application of data mining techniques to the Web data is called Web mining.

Web mining is defined as the application of data mining techniques to automatically discover and extract useful information within the Web data and services [6].

Web mining is divided into the following subtasks [6, 10]:

1. Resource Finding: Intended Web documents are retrieved.
2. Information Selection and Pre-processing: Specific information is automatically selected and pre-processed from the retrieved Web resources.
3. Generalization: At individual Web sites and across multiple sites, general patterns are automatically discovered.
4. Analysis: The mined patterns are analyzed.

Web mining can be classified into three major categories [10]:

- *Web Content Mining*: Web document content patterns are automatically discovered.
- *Web Usage Mining* or *Web Log Mining*: Web server access patterns are automatically found.
- *Web Structure Mining*: Hypertext and linking structure patterns are automatically discovered.

2.2 Further Related Work

Roddick and Spiliopoulou [19] discuss two directions of temporal data mining. One concerns the discovery of causal relationships among temporal events that may be causally related. The other concerns the identification of similar patterns within the same time sequence or among different time sequences. Following Antunes and Oliveira [1], the data should be viewed as a sequence of events, and not as an unordered collection of events. Lin et al. [12] discuss two kinds of fundamental problems in temporal data mining. One is the similarity search problem which is defined as the problem of seeking for sequences that are similar to a given query sequence. The other is the periodical problem which is the problem of finding periodic patterns or cyclicity occurring in time-related databases.

The World Wide Web is growing rapidly. Its large amount of data is widely distributed and lacks a unifying structure. Web mining assists different researches in intelligently searching and extracting relevant data. Craven et al. and Madria et al. [5, 13] focus on Web mining in order to discover new knowledge from the information available on the Web. Madria et al. [13] suggest a Warehouse of Web data, called Whoweda, containing Web data to support its research issues of Web mining. Following Leung et al. [11], Web Archeology's approach is to explore the content of the Web and locate relevant information sources. It studies pages collected from millions of Web sites. However, the lack of structure limits the discovery of important information and the data retrieval performance. Following Kosala et al. [10], the problem of personalization of Web information is often associated with the type and presentation of data. For example, although the data can contain temporal knowledge, it is treated as static.

The creation of Web archives is receiving a substantial interest. The Internet Archive project [3, 17] aims at building an Internet library with the purpose of preserving Web information for a long-term and of offering free access to historical digital collections. It receives its data archive from third parties. The main contributor is Alexa Internet which is a Web navigation service [4]. The Austrian On-Line Archive (AOLA) [2, 17] is a project which creates a permanent archive documenting the evolution of the Austrian Web space.

Its aim is to build an archive of the national Austrian Web space. In the Austrian On-Line Archive Processing (AOLAP) project [18], Web data obtained from crawls of the Austrian Web space as part of AOLA and information collected from additional sources, such as the WHOIS database, are combined. The resulting data warehouse facilitates data analysis and the exploration of the selected Web pages.

Temporal Web Mining (TWM) is an extension of temporal data mining and Web mining [21]. We propose to combine temporal data mining and Web mining in order to deal with real time data and multiple sequences. Real time data received by one or multiple sources is analyzed using temporal data mining techniques in order to extract new temporal information. The new temporal information is sent over the Internet and is combined with auxiliary data from other sources. Then, it is used in the Web mining process in order to discover new relevant temporal data in real time and to predict previously unknown temporal information. TWM supports the temporal aspect of Web data by mining Web data with temporal information as temporal data, and not as static data. Its purpose is to introduce prediction as a main issue in Web mining, specifically Web content mining. In other words, TWM aims at predicting temporal data from the content of the Web data. Furthermore, TWM uses Web data, such as temporal data from the Web, in the temporal data mining process.

The Web seems to be too huge and lacks of structure for effective data mining and data warehousing. For this reason, our goal is to encourage building a common reference data warehouse. The essential difference between the existing data warehouses (such as [3, 13]) and our common reference data warehouse is that the analysis of the data derived from the application of temporal data mining techniques and Web mining techniques is stored in addition to the data received from different sources, such as conventional data warehouses, and the World Wide Web. This can enhance the communication, coordination and integration of different societies and communities, such as researchers. It improves data sharing, data exchange and data manipulation between these different communities and societies. Since the analysis of the data obtained by applying the previous mentioned mining techniques is also saved in our data warehouse, TWM improves different mining techniques. New previously unknown crucial relationships and patterns can be extracted. This contributes to discover, for instance, standard patterns and relationships that can be used in different mining techniques, such as classification. It can also be a clue for representing data, defining similarities between sequences, and finding periodic patterns. Interpreting data in real time and improving data analysis enhance data prediction.

3. Temporal Web Mining Definition

In this section, we define temporal Web mining after describing its process.

3.1 Description of the Temporal Web Mining Process

In Figure 1, $Source_A$ receives data with temporal information ($m_1..m_p$) from one or different sources in real time. The combined data T_A is analyzed using temporal data mining techniques. This analyzed temporal data $T_{A_{analyzed}}$ is forwarded to $Source_B$ in real time over the Internet, and added to data obtained from other sources ($n_1..n_q$). This temporal data T_B is analyzed using Web mining techniques. The derived previously unknown temporal data $T_{B_{analyzed}}$ can reveal new valuable information which can support the data analysis of $Source_A$. In this case, $T_{B_{analyzed}}$ is forwarded to $Source_A$, and is used in the data analysis of this source by the application of temporal data mining techniques.

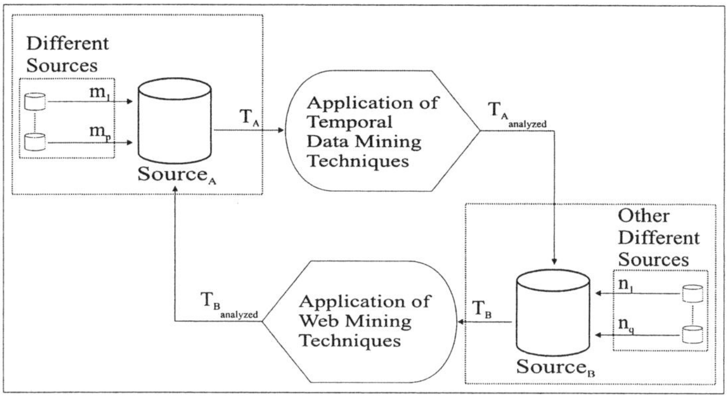

Figure 1. An Overview of the Temporal Web Mining Process

In the TWM process, data with temporal information from the Web is considered as temporal data, and not as static. This temporal data is used in the Web mining process in order to predict new temporal knowledge from the content of the Web data. Then, TWM extends Web mining. TWM aims at using data with temporal information from the Web as temporal data, and at introducing prediction as a main aspect in Web mining, specifically Web content mining. Moreover, by sending the analysis of $Source_B$ to $Source_A$, TWM uses data from the Web, such as temporal Web data, in the temporal data mining process.

3.2 Definition of Temporal Web Mining

Temporal data discovered by the application of temporal data mining techniques is used in the Web mining process in order to retrieve useful data with temporal information in real time over the Web. The derived useful data with temporal information discovered by the application of Web mining techniques is used again in the temporal data mining process.

We define Temporal Web Mining (TWM) as the process of discovering, extracting, analyzing, and predicting data with significant temporal information from the temporal information discovered by the application of temporal data mining techniques, and applying Web mining techniques to this data in real time over the Web (cf. Figure 1) [21].

4. Temporal Web Mining Architecture

In this section, we present the architecture of temporal Web mining, and analyze it. Then, we discuss some of its application scenarios in volcanism and seismology, and in flooding.

4.1 Description of the TWM Architecture

The architecture of temporal Web mining consists of three main components: original source component, destination source component and global source component (cf. Figure 2).

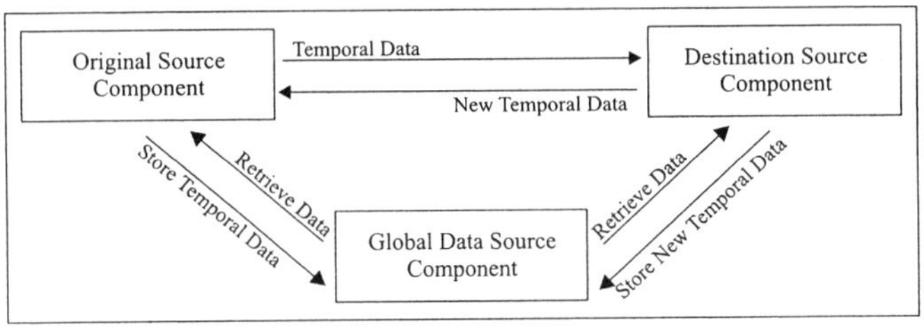

Figure 2. Main Components of TWM Architecture

The aims of our new architecture are to discover, extract, query, analyze, exchange, record, use, and predict (if possible) data between these main components in real time and over the Web.

Figure 3 represents the architecture of TWM. It describes the three TWM main components. The connections between these components depict the exchange of data in the indicated directions. *Temporal data* represents the data derived from the application of temporal data mining techniques. *New temporal data* is the data obtained from the application of Web mining techniques.

4.1.1 Original Source Component

In Figure 3, *multiple different sources* interact, in real time, with each other to retrieve, combine, and analyze data.

One or more of these *multiple different sources* receive real time data. The obtained temporal data is collected in order to be selected and explored. Applying temporal data mining techniques, this temporal data is analyzed. From the *temporal data* database, it is sent to the global source component, where it is saved. Furthermore, in order to reveal more important information, this temporal data may be combined with other data retrieved from the global source data component. It can also be added to other data received from the destination source component. In this case, it is sent over the Internet to the destination source component.

4.1.2 Destination Source Component

The destination source component provides an environment for decision support using mining techniques. The *new temporal data* database receives the analyzed temporal data from the original source component in real time and over the Web (cf. Figure 3). It is examined and, if needed, combined with auxiliary data acquired from *other multiple different sources* or from the global source component. Then, it is analyzed in order to extract derived temporal data. The new temporal data is sent back, in real time over the Internet, from the *new temporal data* database to one or more sources of the original source component. It is also saved in the global source component.

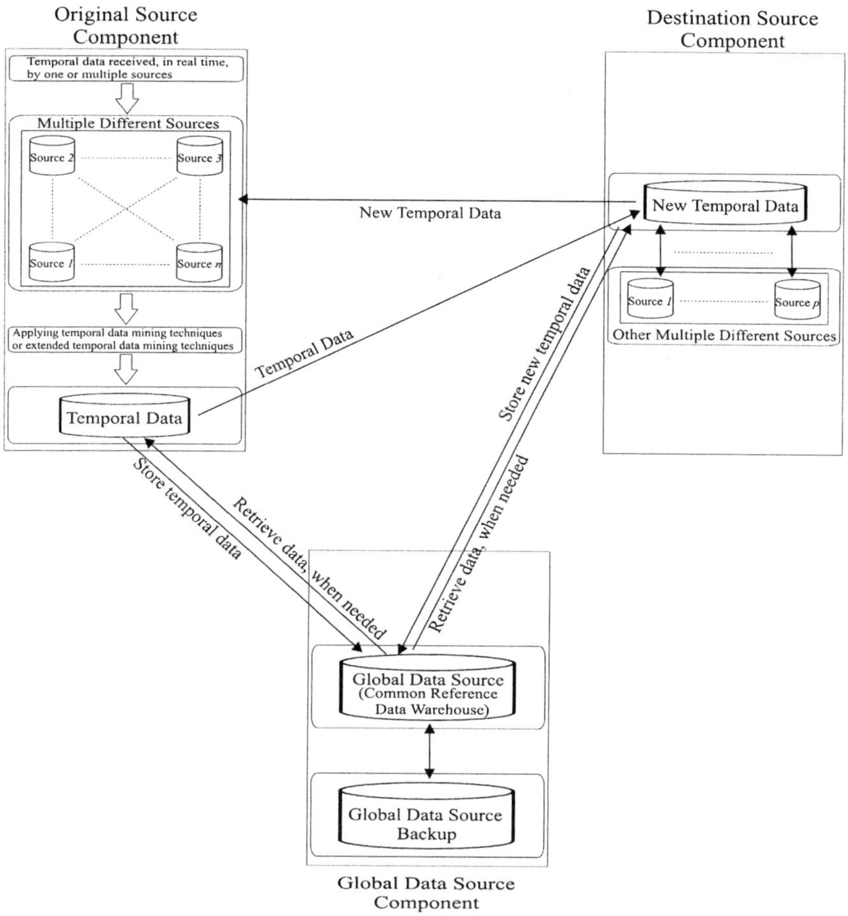

Figure 3. Temporal Web Mining Architecture

4.1.3 Global Source Component

The global source component consists of the *global data source database* (also called common reference data warehouse) and the *global data source backup* database. The *global data source database* receives data from the original source component and the destination source component. It backups this data in the *global data source backup* database.

Whenever needed, the *global data source component* forwards the requested data to the *original source component* or to the *destination source component*. Copies of the same data are sometimes requested simultaneously from different sites. Instead of sending that data to just one site at a given time, it can be sent to more than one site at more or less the same time by using our multicast network (cf. Figure 4) [20]. Hence, our multicast network is responsible for handling concurrent data requests of multiple sites. It is provided with additional functions, such as backup and error recovery. It consists of a source node connected to a number of networks. Each of these networks is connected to an intermediate site (i.e. a router). Each intermediate site is connected to several networks which lead to a number of destination sites. The source node is the *global data source component* and the destination sites are those that requested concurrently the same data.

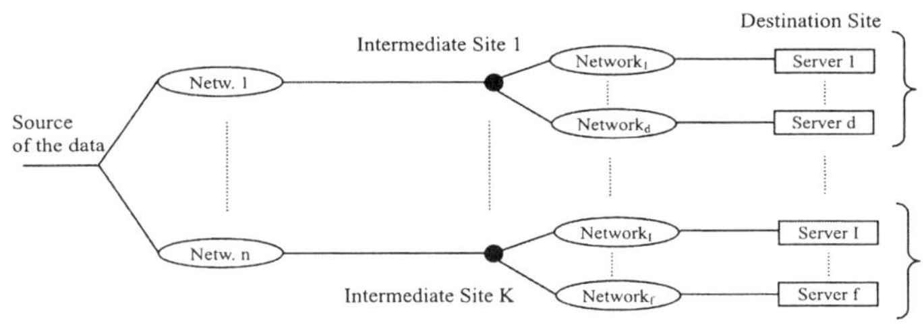

Figure 4. Multicast Network

4.2 Analysis of the TWM Architecture

TWM collects, exchanges, interprets, and predicts (if possible) data between different sources, in real time and over the Web.

Causal relationships among temporal related events have to be discovered by using mining techniques, in order to find the cause of an event and predict its result, if necessary.

The quality of data is one of the most crucial factors to be taken into account [9, 16]. Data is collected from different sources in real time in order to be queried and analyzed. Then, relevant data is to be distinguished from the noise, or any data error. In other words, the ability to deal with any abnormality in data is to be taken into account in order to be reduced.

Since temporal Web mining deals with data changing rapidly over the time, delay and interruptions in data transfer should be minimized. This can be done by optimizing the query execution, by finding a correct representation of the temporal knowledge in question, and by selecting the right path to retrieve the required data.

Because it is difficult to locate and track appropriate data from the huge number of semi-structured documents, our aim is to contribute in building logical structured information sources, such as a common reference data warehouse, that combine data from multiple different sources. This minimizes the cost of data access time, as well as delay, and reduces data redundancy. The structure can be optimized by analyzing the data, the Web logs and the user's behavior. For instance, the user's behavior can be studied by examining Web data, such as the page requested, the time and frequency of request of pages. In other words, Web mining techniques can be applied to interpret and predict user's behavior.

Having a common reference data warehouse, temporal Web mining improves data sharing, data exchange and data manipulation between different societies, institutions and communities. In Figure 3, the *global data source component* establishes a data exchange system that may be used by different communities, such as research communities and different institutions, in order to record, exchange, use, process, and analyze data, assembled from different sources, such as the World Wide Web or conventional data warehouses. It also acts as a backup of the data exchange communities, because the data is regularly collected and stored in the *global data source component* from databases worldwide. It is a possibility to unify the knowledge all over the world for the rapid advance of many fields.

The *global data source component* sends requested data to one or more sites, when needed. Multiple different sites may make concurrent requests of the same data. Instead of delivering this data to just one of these sites at a given time, it is sent to the sites in question at more or less the same time. Our multicast network supports concurrent data requests of multiple sites. In Figure 4, whenever data is needed to be multicasted to different destinations, data is forwarded from the *source of the data* (i.e. *global data source component*) along a distribution tree to each receiver (i.e. *destination site*)[20]. The destination sites (such as *original source component* or *destination source component*) do not communicate directly with the *global data source component* (cf. Figure 3). However, they communicate with their corresponding intermediate site. Each intermediate site supports backup and error recovery functions [20]. Consequently, whenever an intermediate site receives a corrupted piece of data or does not receive the requested piece of data, a message is sent back to the *source of data* requesting the retransmission of the needed data. Moreover, if a destination site does not receive the correct data, a message is sent back to the *intermediate site* asking it to retransmit the corresponding data. Then, a *destination site* does not communicate directly with the *global data source component*. The load on the *global data source component* is reduced which helps to reduce its response time. More clearly, the workload is distributed among the *global data source component* and the *intermediate sites*. This helps to minimize delay and interruptions in data transfer, as well as data loss and data errors.

In Figure 3, our *common reference data warehouse* creates a data exchange standard by, for instance, finding a standard representation for shared data, and by creating common patterns from this data. This can also minimize the cost of data access time, and reduce redundancy. For instance, the *global data source component* gives the possibility for seismologists to use the global seismological data and its analysis, collected in the *global data source component* from many different institutions worldwide, in their data analysis. This makes processing earthquake data, in real time over the Web, easier, faster and more accurate, and helps to extract and predict more relevant important knowledge.

4.3 Application Scenarios for TWM

Collecting and exchanging data between different sources, in real time and over the Web, provide an opportunity for various societies and communities to share information instantly and analyze it in order to contribute to comprehension of different issues that have far-reaching implications for the humankind and environment.

In TWM, different application scenarios can be applied, such as:

- Volcanoes and Earthquakes

  Volcanoes have played an important role in forming and modifying the Earth. More than 80% of the Earth's surface above and below the sea level is of volcanic origin [7]. The keeping of a detailed history of the changes in a volcano and its surroundings helps to characterize the behavior of the corresponding volcano. Visible changes are any observable and often measurable features by researchers that might reflect a change in the state of a volcano. For instance, scientists collect the eruptive products and gases for laboratory analysis, make temperature measurements of lava and gas, and so on. Invisible changes are any changes, which are not visible to the human eye, but measurable by precise and sophisticated instruments and sensors, such as variations in gas compositions.

  An earthquake generates seismic waves which can be sensed and recorded over a wide range of frequencies and seismic amplitudes. The severity of an earthquake depends on many factors, such as the measure of the amplitude of the seismic waves (i.e. the magnitude of the earthquake), the building design and other structures, the geologic conditions, the density of the population and constructions in the area affected by the earthquake.

  Volcanism and seismic activities are often closely related, responding to the same dynamic Earth forces. Real time volcanic data and seismic data can be obtained from, for instance, sensing devices located at different stations. Combining, in real time, from different sources over the Web, the recording and analysis of volcanic phenomena and seismograms from many earthquakes helps to better understand and analyze the Earth's deep interior.

Historical volcanic data, historical seismic data, and historical meteorological data are examples of historical data. Instances of any current data related to a volcanic event are eruption date and time, volcano location, volcano frequency of occurrence, volcano magnitude, temperature of the lava, seismic data, and meteorological data.

Historical data and any current data related to a volcanic event can be in real time exchanged and analyzed between different institutions over the Web in order to better understand, determine and analyze the relations between volcanism and seismic activities. More specifically, volcano data can be date and time of eruption, magnitude, volcano surface temperature, variations in gas compositions, historical volcanic eruption, type of volcanic eruption, volcano elevation, and frequency of occurrence. Seismic data can be magnitude, location, starting and ending date and time of the earthquake, frequency of occurrence, temperature measurements, and barometric pressure.

In Figure 3, whenever a volcanic activity or a seismic activity occurs, volcano data and seismic data are analyzed instantly, if required with other data (such as wind speed data, wind direction data) from *multiple different sources* (such as a station where a wind sensor is located, a seismic station) or from the *global data source*. The obtained temporal data is sent, in real time over the Internet, from the *temporal data* database to the *new temporal data* database, and is also saved in the *global data source*. In the *new temporal data* database, it is combined, if necessary, with new information (such as meteorological data, historical seismic data) from the *new temporal data* database, from *other multiple different sources* (such as meteorological station, seismic station) or from the *global data source*. Then, it is analyzed in order to extract new temporal data that can reveal crucial information about volcanic or seismic activities. The new derived temporal data is sent back to the original source and saved in the *global data source*. Analyzing this data connects different stations and provides mutual support to increase the possibility of searching for any warning sign that can forecast future volcanic or seismic activities.

In this application scenario, temporal Web mining helps to create a better image of the Earth's deep interior, figure out any new idea about volcanoes or earthquakes, and improve the capability for predicting future volcanic or seismic activities. Furthermore, precise determination of every active or inactive volcano location, the detailed analysis of their criteria, as well as the study of damages caused by volcanoes and earthquakes reduce their losses and hazards by enhancing the safety measures and emergencies. They also help researchers, architects, structural engineers, or those who are working in the building or construction domain in their researches and the structure's building and design [15].

- Flooding

The assembling of flood data and flood causes data between different stations, in real time and over the Internet, creates a more complete documentation and understanding of historical flood data (such as flood events, flood causes, flood damages). Hence, flood data can be better analyzed, which helps to enhance the capability of forecasting a flooding, to warn about a flooding, and to present an effective disaster planning to reduce flood losses and hazards and improve the public safety measures, as well as the emergency management.

5. Conclusion

Temporal Web Mining (TWM) is a new concept which combines and extends temporal data mining and Web mining. Its purpose is to introduce prediction as a main issue in Web mining, specifically Web content mining, and to use Web data, such as temporal data from the Web, in the temporal data mining process. It can be used in different domains, such as finance, engineering, medicine, environmental sciences, and earth sciences. Its main goal is to deal with temporal data in real time over the Web, in order to discover, in real time, patterns and relationships among time ordered events that may be causally related. Consequently, some disasters, such as an earthquake, an eruption of a volcano, or a flooding can be to a certain extent forecasted, and effective disaster planning can be enhanced. It also encourages and aims at providing assistance in building a common reference data warehouse which gives better opportunities for data record, data sharing, data manipulation, data standardization and data analysis between different societies, communities, and institutions.

In this paper, we give a brief overview of temporal data mining and Web mining, and discuss related work. We describe the temporal Web mining process, and define temporal Web mining. Then, we present the architecture of TWM, analyze it, and apply application scenarios in volcanism and seismology, and flooding.

Future directions to our research include the improvement of the quality and delivery of information, as well as data analysis, by finding an appropriate data representation. Analyzing the users' behavior in the Web can help to determine an effective query execution plan. Estimating the cost of a query execution and reducing the cost of selecting the adequate path to retrieve the required data can improve the data extraction and data analysis capability. Another work can be to develop an accurate system for data analysis which preserves security and prohibits the access to unneeded confidential information in data records.

References

[1] Claudia Antunes and Arlindo Oliveira. Temporal Data Mining: an Overview. In *KDD Workshop on Temporal Data Mining*, pages 1–13, San Francisco, 2001.

[2] Austrian On-Line Archive. Website. http://www.ifs.tuwien.ac.at/~aola.

[3] The Internet Archive. Website. http://www.archive.org.

[4] Alexa: The Web Information Company. Website. http://www.alexa.com.

[5] Mark Craven, Dan DiPasquo, Dayne Freitag, Andrew K. McCallum, Tom M. Mitchell, Kamal Nigam, and Seán Slattery. Learning to Extract Symbolic Knowledge from the World Wide Web. In *Proceedings of AAAI-98, 15th Conference of the American Association for Artificial Intelligence*, pages 509–516, Madison, US, 1998. AAAI Press, Menlo Park, US.

[6] Oren Etzioni. The World-Wide Web: Quagmire or Gold Mine? *Communications of the ACM*, 39(11):65–68, 1996.

[7] The Federal Emergency Management Agency (FEMA). (2003). Hazards: Volcanoes. Retrieved October 13, 2003, from http://www.fema.gov/hazards/volcanoes/volcano.shtm.

[8] William J. Frawley and Gregory Piatetsky-Shapiro. Knowledge Discovery in Databases. AAAI Press/The MIT Press, 1991.

[9] Jochen Hipp, Ulrich Güntzer, and Udo Grimmer. Data Quality Mining - Making a Virtue of Necessity. In *Proceedings of the 6th ACM SIGMOD Workshop on Research Issues in Data Mining and Knowledge Discovery (DMKD 2001)*, pages 52–57, Santa Barbara, California, May 20 2001.

[10] Raymond Kosala and Hendrik Blockeel. Web Mining Research: A Survey. *ACM SIGKDD Explorations*, 2(1):1–15, July 2000.

[11] Shun-Tak A. Leung, Sharon E. Perl, Raymie Stata, and Janet L. Wiener. (2001). Towards Web-Scale Web Archeology. Research Report 174, Compaq Systems Research Center, Palo Alto, California. Retrieved December 10, 2003 from http://gatekeeper.dec.com/pub/DEC/SRC/research-reports/SRC-174.pdf.

[12] Weiqiang Lin, Mehmet A. Orgun, and Graham. J. Williams. An Overview of Temporal Data Mining. In *Proceedings of the 1st Australian Data Mining Workshop (ADM02)*, Canberra, Australia, December 2002.

[13] Sanjay Kumar Madria, Sourav S. Bhowmick, Wee Keong Ng, and Ee-Peng Lim. Research Issues in Web Data Mining. In *Proceedings of the 1st International Conference on Data Warehousing and Knowledge Discovery (DAWAK99), LNCS 1676*, pages 303–312, Florence, Italy, 1999.

[14] NASA Earth Observatory. (2002). Flooding in Germany. Retrieved June 2003, from http://earthobservatory.nasa.gov/Newsroom/NewImages/images.php3?img_id=10301.

[15] The Association of Bay Area Governments (ABAG). (2003). Impacts of California Earthquakes on Buildings from Shaken Awake! Retrieved December 10, 2003, from http://www.abag.ca.gov/bayarea/eqmaps/shelpop/bldg.html.

[16] Barbara Pernici and Monica Scannapieco. Data Quality in Web Information Systems. In *ER 2002, 21st International Conference on Conceptual Modeling, LNCS 2503*, pages 397–413, 2002.

[17] Andreas Rauber and Andreas Aschenbrenner. (2001). Part of our Culture is Born Digital - On Efforts to Preserve it for Future Generations. Internet-Zeitschrift für Kulturwissenschaften (TRANS. On-line Journal for Cultural Studies). Retrieved November 2003, from http://www.inst.at/trans/10Nr/rauber10.htm.

[18] Andreas Rauber, Oliver Witvoet, Andreas Aschenbrenner, and Robert Bruckner. Putting the World Wide Web into a Data Warehouse: A DWH-based Approach to Web Analysis. In *Proceedings of the DEXA Workshop on Very Large Data Warehouses*, pages 822–826, Aix en Provence, France, September 2002. IEEE.

[19] John F. Roddick and Myra Spiliopoulou. A Bibliography of Temporal, Spatial and Spatio-Temporal Data Mining Research. *ACM SIGKDD Explorations*, 1(1):34–38, June 1999.

[20] Mireille Samia. Optimization of IP Multicast Networks Performance. Master's Thesis, Faculty of Natural and Applied Sciences, Department of Computer Science, Notre Dame University, Lebanon, June 2001.

[21] Mireille Samia. Temporal Web Mining. In *15. GI-Workshop über Grundlagen von Datenbanken (15th GI-Workshop on the Foundations of Databases)*, pages 27–31, Tangermünde, Germany, June 2003.

Databases and Information Systems
J. Barzdins and A. Caplinskas (Eds.)
IOS Press, 2005

Association Rule Mining Meets Functional Dependencies: The AP-FD Algorithm

Jürgen M. JANAS
Fakultät für Wirtschafts- und Organisationswissenschaften
Universität der Bundeswehr München
Werner-Heisenberg-Weg 39, D-85577 Neubiberg, Germany
juergen.janas@unibw-muenchen.de

Abstract. Association rule mining has mainly been studied in terms of transactional data such as the market basket example; comparatively little attention has been paid to its generalization to arbitrary relational data. Although this is justified to some extent because the methods applied to transaction data may easily be adapted to relational data, there are certain peculiarities of relational data that cannot be taken advantage of by these adapted methods. In this paper, we examine the role which functional dependencies – a concept which has been studied rigorously in the area of relational database design – may play in mining relational data for association rules. We will show how the knowledge of functional dependencies may be used to improve the performance of the A Priori algorithm which is the most popular algorithm for finding sets of frequently co-occurring attribute values.

1. Introduction

Data mining is the analysis of typically very large sets of data in order to discover yet unknown relationships among the data and aggregate them in ways which are novel, useful, and understandable to the users of the data. Data mining is a still young and due to its different ancestors, namely statistics, databases and artificial intelligence, heterogeneous discipline. Among its sub-disciplines, association rule mining is probably the one which is most independent from these ancestors.

Association rule mining (originally introduced in [1]) is usually explained in the context of the market-basket problem which is the task of identifying sets of items which frequently occur together in supermarket transaction data. Technically spoken, the rules to be mined in this scenario are single-dimensional association rules, i.e. both sides of these rules are sets of values which come from the same domain, in this case the items that may be purchased in the supermarket. Single-dimensional association rules may be generalized quite naturally to multidimensional association rules which are better tailored to data according to the relational model of data.

Association rule mining is usually done in two phases. During the first phase, one of the numerous variants of an algorithm which is known as "A Priori" is used to determine sets of attribute values which frequently occur together by making a number of sequential scans of the data to be mined. During the second phase, the sets of attribute values found in the first phase are split into the left hand side and the right hand side of one or more association rules according to some simple statistical criteria; it is assumed that these rules might be interesting to the user and therefore they are presented to him for inspection.

In this paper, we will propose an improvement to the first phase of association rule mining that uses knowledge about the functional dependencies that are contained in the data to be mined. Functional dependencies are patterns in data according to the relational model of data that may be observed due to corresponding regularities among the real world objects which are to be modeled by the data; they are commonly employed in the process of database design.

We will show that the knowledge of functional dependencies in the data to be mined allows us to logically infer that certain sets of attribute values occur frequently together without counting their occurrences. This observation is used in a new variant of the A Priori algorithm, namely the AP-FD algorithm (which owes its name to its two main ingredients, i.e. the classical A Priori algorithm and functional dependencies), which is superior to the original algorithm with respect to both runtime and space requirements.

The remainder of this paper is organized as follows. Section 2 serves to introduce the concepts and the notation used within this paper; this comprises concepts both from relational database theory and from data mining; in particular, an adaptation of the A Priori algorithm to multidimensional association rules will be presented. The similarities and the differences of functional dependencies on the one hand and association rules on the other as well as the interactions between these two concepts will be investigated in section 3; moreover, we will use this section for a brief discussion of the question to what extent it is reasonable to expect that functional dependencies will be present in the data to be mined for multidimensional association rules. The fourth section introduces the AP-FD algorithm by showing how it may be developed from the classical A Priori algorithm in two steps; moreover, it contains a discussion of its performance. The final section is used to summarize the results of the paper and also to briefly discuss how the enhancements of the A Priori algorithm proposed in this paper may be combined with other attempts to improve its efficiency that have been reported in the literature.

2. Basic Concepts and Notation

2.1 Relations and Functional Dependencies

We assume that the reader is familiar with the basic concepts of the relational model of data (e.g. on the basis of [2]) and restrict ourselves to informally introducing the concepts we are using in the rest of this paper. When talking about a **relation** we distinguish between the relation schema (denoted by R) which describes the structure of the relation and the relation instance (denoted by r) which contains the stored data and conforms to the structure given by the relation schema.

A **relation schema** R comprises the set of **attributes** (denoted by atts(R)) on which R is defined and a set of **integrity constraints** which have to be satisfied by a relation instance in order to be conforming to R. We will use letters X and Y for sets of attributes and letters A, B, C, ... for individual attributes.

A **relation instance** may be thought of as a set of **rows** each of which consists of a set of values such that for each of the attributes in atts(R) there is exactly one corresponding value; individual such values will be referred to as **attribute values** and be denoted by A:a where a stands for the value corresponding to the attribute A. A **set of co-occurring attribute values** (abbreviated **scav**) consists of one attribute value per attribute from some attribute set $X \subseteq$ atts(R) and will be designated by X:x. We shall not make a distinction between an attribute value and the scav which contains only this attribute value.

The notation $\sigma_{X:x}(r)$ stands for the relation instance which is obtained by selecting those rows from r that contain the scav X:x. Finally, card(r) is used to designate the number of rows contained in the relation instance r.

As far as the integrity constraints of a relation schema are concerned, we restrict our considerations to functional dependencies within this paper. A **functional dependency** is an expression of the form $X \rightarrow A$. We say **$X \rightarrow A$ is satisfied by a relation instance** r if and only if every two rows from r which agree with respect to their values for X, also agree with respect to their values for A; moreover, we say that **$X \rightarrow A$ holds in a relation schema** R if and only if $X \rightarrow A$ is satisfied by every relation instance conforming to R.

We should mention that functional dependencies are usually defined in such a way that also their right hand sides may contain more than one attribute. However, since it can be shown that each functional dependency $X \rightarrow Y$ holds in a relation schema if and only if all of the functional dependencies $X \rightarrow A$ with $A \in Y$ hold in that relation schema, it is legitimate to restrict our considerations to functional dependencies with singleton right hand sides.

A functional dependency $X \rightarrow A$ is called a **trivial functional dependency** if and only if $A \in X$. Obviously, all trivial functional dependencies hold in the respective relation schema.

With regard to the functional dependencies which hold in a relation schema R we distinguish between the set Γ of functional dependencies given as part of the relation schema and the set Γ^+ of all functional dependencies that may be inferred from Γ. Γ^+ always contains all functional dependencies from Γ and all trivial functional dependencies, but it may contain additional functional dependencies which may be derived from Γ according to transitivity and other rules (for details see [2]).

The **closure of a set of attributes** X with respect to Γ is the set of all attributes A such that $X \rightarrow A$ is contained in Γ^+; it will be denoted by X_Γ^+. It is important for the algorithm proposed in this paper that the closure of a set of attributes can be computed in linear time; a corresponding algorithm may be found in [3].

A **key** of a relation schema R is a minimal set $X \subseteq atts(R)$ such that X_Γ^+ is equal to atts(R).

2.2 Association Rules

If R is a relation schema and $X, Y \subseteq atts(R)$, then an expression of the form

$$(X{:}x \Rightarrow Y{:}y,\ s,\ c)$$

where s and c are real numbers from the interval between 0 and 1 is called a **multidimensional association rule**; X:x is called the left hand side and Y:y the right hand side of the rule. If r is a relation instance conforming to R, we say that the multidimensional association rule $(X{:}x \Rightarrow Y{:}y, s, c)$ is satisfied by r if

$$card(\sigma_{Y:y}(\sigma_{X:x}(r))) / card(r) \geq s$$

and

$$card(\sigma_{Y:y}(\sigma_{X:x}(r))) / card(\sigma_{X:x}(r)) \geq c$$

Both s and c are measures for the interestingness of a multidimensional association rule; s is called the **support threshold** and prescribes a lower bound for the fraction of the rows of r that have to contain both the left hand side and the right hand side of the rule for the rule to be satisfied by r.

The **confidence level** c prescribes at least what fraction of the rows of r that contain the left hand side of the rule have to contain the right hand side as well for the rule to be satisfied by r.

Multidimensional association rules are particularly appropriate for mining data according to the relational model of data. For reasons of brevity, multidimensional association rules will be referred to as association rules throughout the rest of this paper.

2.3 The Classical A Priori Algorithm

Mining of association rules is usually done in two phases: During the first phase, the **frequent scavs** are determined, i.e., those scavs which occur in at least that fraction of the rows that is specified by the support threshold. During the second phase, the frequent scavs are split into a left hand side X:x and a right hand side Y:y according to additional criteria such as a lower bound for the confidence level of the resulting rule.

The most popular algorithm for the first phase is the so-called A Priori algorithm which was originally introduced in [4]. The A Priori algorithm is based on the observation that if a scav appears in a fraction s of the rows of a relation instance, then any subset of this scav appears in at least fraction s of the rows, or – to put it the other way round – a scav may appear in a fraction s of the rows only if each of its subsets does so. This observation is sometimes referred to as the **A Priori trick**.

In the relevant literature, the A Priori algorithm is usually explained in the context of the market-basket problem which implies a restriction to single-dimensional association rules. However, as observed in [5], the algorithm may easily be adapted to multidimensional association rules. In Fig. 1, the classical A Priori algorithm is restated on a rather abstract level and in such a way that it applies to multidimensional association rules; thus we make it directly comparable to the algorithms which will be proposed in section 4 of this paper and which are applicable to multidimensional association rules only. The algorithm takes as input a relation instance r and a support threshold s and outputs the set L of all frequent scavs.

The algorithm proceeds levelwise and requires one pass per level through the given relation instance. On level k, the set $\mathbf{L_k}$ of all frequent scavs of size k is determined; this is done in two steps: During the first step, a set $\mathbf{C_k}$ of candidate scavs is constructed from $\mathbf{L_{k-1}}$ by applying the A Priori trick. In the second step, the occurrences of the candidate scavs in the relation instance are counted and thus the actually frequent ones are identified. For further details and a discussion of how to compute the candidate sets $\mathbf{C_k}$ efficiently, the reader is referred to [5].

```
begin
   C₁ := {{A:a}| r contains an occurrence of A:a};
   L₁ := {{A:a}| {A:a} ∈ C₁ ∧ card(σA:a(r))≥s};
   L := ∅;
   k := 1;
   while Lk ≠ ∅ do
      begin
      L := L ∪ Lk;
      k := k + 1;
      Ck  :=  {C|  C  =  {A1:a1,...,Ak:ak}  ∧  Ai≠Aj  for  1≤i<j≤k
                         ∧ C\{Ai:ai} ∈ Lk-1 for 1≤i≤k};
      Lk := {X:x| X:x ∈ Ck ∧ card(σX:x(r))≥s}
      end
end.
```

Figure 1. The classical A Priori algorithm

3. A Common View on Association Rules and Functional Dependencies

3.1 Technical Interactions

At first glance, association rules and functional dependencies seem to show a good deal of similarities: As a matter of fact, both of them are meant to characterize situations in which the values with respect to one or more attributes of a relation determine the values with regard to some other attribute of that relation. However, there are three aspects in which association rules and functional dependencies differ from each other considerably.

First of all, functional dependencies have a coarser granularity than association rules. This means, while a functional dependency $X \rightarrow A$ is a statement which refers to the entirety of the values with regard to X and to the entirety of the values with regard to A in a relation instance, an association rule $(X{:}x \Rightarrow A{:}a, s, c)$ is a statement which relates to the combination of only one particular combination of values with regard to X and one particular value with regard to A.

Secondly, functional dependencies are more stringent than association rules; that means, while a functional dependency enforces that all of the rows which contain the same values with regard to X have to agree with respect to their value for A, an association rule requires only that a certain percentage (expressed by means of the confidence level c) of the rows which contain the same values with regard to X have to have the same value with regard to A.

Finally, functional dependencies relate to the relation schema, whereas association rules relate to a specific relation instance. This implies that a functional dependency will be satisfied by every relation instance which conforms to the respective relation schema; thus the validity of a functional dependency cannot be affected by changes to the relation instance. An association rule, by way of contrast, is a statement about one particular relation instance only and therefore such a statement may become invalid due to a change to the respective relation instance.

It is an immediate consequence of the latter of these differences between association rules and functional dependencies that the existence of one or more association rules will by no means have any effect on the set of functional dependencies which hold in the respective relation schema.

On the other hand, a functional dependency $X \rightarrow A$ which holds in a relation schema implies the validity of all association rules $(X{:}x \Rightarrow A{:}a, s, c)$ for those values with regard to X and to A which occur in combination in the respective relation instance; this implication is independent of the respective relation instance and the confidence level c of each of the implied association rules is equal to 1. Of course, this kind of relationship is owed to the different granularity of functional dependencies and association rules and it is of a rather trivial nature.

The following lemma will show, however, that there is another, more useful kind of interaction between functional dependencies and association rules. As a matter of fact, this interaction is not expressed in terms of association rules, but rather in terms of frequent scavs from which the association rules that are satisfied by a particular relation instance may be derived.

Lemma 1. Let R be a relation schema, r a relation instance conforming to R, and $Y \rightarrow A$ a functional dependency with $Y \subseteq X \subseteq \text{atts}(R)$ that holds in R; if X:x is a frequent scav then $X{:}x \cup A{:}a$ is also a frequent scav with A:a being the (uniquely determined) attribute value which is contained in all rows of r which contain X: x.

Proof: obvious. □

3.2 Functional Dependencies in the Data to Be Mined

Before we will turn to the implications the above lemma has for finding frequent scavs, we have to address a much more general question: Functional dependencies are a concept which is commonly used in the design of relational databases; therefore, it is legitimate to ask whether we may reasonably expect that functional dependencies will be present at all in sets of data to be mined for association rules. It may at first glance seem as if the answer to this question is completely dependant on the respective data set, however, a closer look reveals that certain answers of a more general nature can be given.

Functional dependencies are used in relational database design to define certain normal forms of relation schemes that are considered desirable because they avoid redundancy in the conforming relation instances. The best normal form which may be achieved – as long as only functional dependencies are regarded – is the so-called Boyce/Codd normal form (BCNF); according to the definition of BCNF, if $X \rightarrow A$ is a non-trivial functional dependency that holds in a relation schema which is in BCNF then X has to contain a key of that relation schema.

It is easy to see from this definition that any functional dependency $X \rightarrow A$ which holds in a BCNF relation schema will be of hardly any help for mining frequent scavs because any scav which comprises all of the attributes from X will occur at most once in a conforming relation instance and therefore will hardly ever be considered frequent. As a consequence, functional dependencies hardly ever will be useful for mining associations between multiple attribute values which are contained in the same relation, particularly if the relation schema is in BCNF.

The situation is different, however, as soon as associations between attribute values from multiple relations are considered. In this case, the relations which contain the attribute values have to be joined as a prerequisite to counting the co-occurrences of the attribute values. For reasons of brevity we restrict ourselves to the case where only two relation schemes R_1 and R_2 have to be joined.

The attribute(s) on which the join of R_1 and R_2 is performed must be a key in at least one of the relation schemes R_1 and R_2 because otherwise the join would be "lossy", i.e., the relation resulting from the join would not comprise the same information as R_1 and R_2 (cf. [2] for a thorough discussion of that matter). On the other hand, if the join attribute(s) constitute a key in R_1 it is very unlikely that they are a key of R_2 as well because in that case there would have been no need to separate the information contained in the two relations during database design.

As a consequence, the left hand side of any functional dependency which holds in R_1 will not contain a key of the joined relation even if it contains a key of R_1 and therefore, a scav with attributes from $X \subseteq \text{atts}(R_1)$ may be contained in the joined relation an arbitrary number of times and, in particular, may be frequent. So all the functional dependencies that hold in R_1 may be useful for mining multidimensional association rules even if their left hand sides contain a key of R_1.

It is obvious that the number of functional dependencies in the data to be mined increases the number of relations which result from the normalization process and therefore increases the number of joins that are required to build the relation instance in which the scavs may be counted.

Apart from the above observation, we would like to point out that the data to be mined may also contain functional dependencies for other reasons; we only mention dimension tables in data warehouses which are deliberately denormalized in order to increase query performance (cf. [6]).

4. Developing the AP-FD Algorithm

In this section, we introduce the AP-FD algorithm in two steps. In the first subsection, we develop a simple enhancement of the A Priori algorithm originally introduced in [7] that reduces the size of the candidate sets generated by the algorithm and thereby makes the algorithm applicable to larger data sets than the classical Apriori algorithm. In the second subsection, the AP-FD algorithm, which as an additional benefit may require fewer scans of the data to be mined, is introduced.

4.1 A Simple Enhancement of the Classical A Priori Algorithm

Lemma 1 has direct implications for generating frequent scavs according to the A Priori algorithm: Suppose, at some time during the execution of the A Priori algorithm we know that X:x is contained in the set $\mathbf{L}_{k-1}$ and we also know that there is a functional dependency $Y \rightarrow B$ with $Y \subseteq X$ and $B \notin X$ that holds in R; then we may conclude that $X{:}x \cup B{:}b$ is an element of $\mathbf{L}_k$ for some appropriate value b.

The important point about that is the fact that we can make this conclusion without verifying that all subsets of size k-1 of the set $X{:}x \cup B{:}b$ are elements of $\mathbf{L}_{k-1}$ and without counting the occurrences of $X{:}x \cup B{:}b$ in r either. Instead, we may immediately assign such a scav to $\mathbf{L}_k$ without making it a member of the candidate set $\mathbf{C}_k$.

However, there is still another way in which we can benefit from this situation: Once we have concluded that $X{:}x \cup B{:}b$ is contained in $\mathbf{L}_k$ due to the existence of some functional dependency as described above, it is apparent that any scav $X{:}x \cup B{:}b'$ where b' is different from b cannot be an element of $\mathbf{L}_k$ even if all subsets of size $k-1$ of $X{:}x \cup B{:}b'$ were elements of $\mathbf{L}_{k-1}$; and again we may make this inference without counting the occurrences of $X{:}x \cup B{:}b'$ in r. This implies a further reduction of the size of the candidate sets $\mathbf{C}_k$. In order to illustrate these effects the following example is provided.

Example 1. Let R be a relation schema with atts(R) = {A, B, C} and let $A \rightarrow B$ be a functional dependency which holds in R. It is easy to verify that the relation instance r given in Table 1 is conforming to R.

Table 1. Example relation instance

A	B	C
a_1	b_1	c_1
a_1	b_1	c_2
a_2	b_2	c_1
a_2	b_2	c_3
a_3	b_3	c_4
a_4	b_3	c_5

Given a support threshold s = 0,3 the first set of frequent scavs found by the A Priori algorithm is

$$\mathbf{L}_1 := \{\{A{:}a_1\}, \{A{:}a_2\}, \{B{:}b_1\}, \{B{:}b_2\}, \{B{:}b_3\}, \{C{:}c_1\}\}$$

Because of the functional dependency $A \rightarrow B$ and the fact that a_i and b_i occur together in some row of r for $1 \le i \le 2$, we may conclude that

$$\{A{:}a_1, B{:}b_1\}, \{A{:}a_2, B{:}b_2\} \in \mathbf{L}_2$$

and furthermore that

$$\{A{:}a_1, B{:}b_2\}, \{A{:}a_1, B{:}b_3\}, \{A{:}a_2, B{:}b_1\}, \{A{:}a_2, B{:}b_3\} \notin \mathbf{L}_2$$

The remaining five combinations of two frequent attribute values, namely

$$\{A{:}a_1, C{:}c_1\}, \{A{:}a_2, C{:}c_1\}, \{B{:}b_1, C{:}c_1\}, \{B{:}b_2, C{:}c_1\}, \{B{:}b_3, C{:}c_1\}$$

would become elements of $\mathbf{C}_2$. Note that according to the classical A Priori algorithm all eleven possible pairs of frequent attribute values would be included in the candidate set $\mathbf{C}_2$ and their occurrences in r be counted just to finally end up with the same result. □

Given a frequent scav X:x from $\mathbf{L}_{k\text{-}1}$ and a functional dependency $Y \rightarrow B$ with $Y \subseteq X$ there seems to remain the problem of finding the attribute value B:b which corresponds to X:x according to the functional dependency. Of course, this attribute value may be found in any row which contains X:x, but we certainly would not like to make an additional scan of the relation instance in order to find such a row. It turns out, however, that this is not really a problem: Since B:b is functionally dependent on some of the attribute values in X:x, it is guaranteed that there is only one such value.

Hence, whenever during subsequent traversals of the loop of the A Priori algorithm we have to count the occurrences of a scav which contains one or more such unknown attribute values, we may increment the respective count without checking the value of these particular attributes. Obviously, such missing values may be added either during the next counting of the occurrences of the candidate scavs or at the end of the algorithm after all frequent scavs have been found.

Altogether, this leads to the enhanced version of the A Priori algorithm shown in Fig. 2 in which the unknown values are taken into account by using the symbol "*" which may be understood as some kind of wildcard for an unknown, yet uniquely determined value.

```
begin
    C1 := {{A:a}| r contains an occurrence of A:a};
    L1 := {{A:a}| {A:a} ∈ C1 ∧ card(σA:a(r))≥s};
    L := ∅;
    k := 1;
    while Lk ≠ ∅ do
        begin
        L := L ∪ Lk;
        k := k + 1;
        Ck := {C| C = {A1:a1,...,Ak:ak} ∧ Ai≠Aj for 1≤i<j≤k
                      ∧ C\{Ai:ai} ∈ Lk-1 for 1≤i≤k
                      ∧ Ai ∉ ({A1,...,Ak}\{Ai})r+ for 1≤i≤k};
        Lk := {X:x| X:x ∈ Ck ∧ card(σX:x(r))≥s}
              ∪ {C| ∃l ∈ {1,...,k} C = {A1:a1,...,Al:*,...,Ak:ak}
                      ∧ Ai≠Aj for 1≤i<j≤k
                      ∧ C\{Al:al} ∈ Lk-1
                      ∧ Al ∈ ({A1,...,Ak}\{Al})r+}
    end
end.
```

Figure 2. Enhancement of the classical A Priori algorithm

It should be mentioned that the computation of the closures of sets of attributes – contrary to what the above algorithm may suggest at first sight – need not be done anew for each possible element of $\mathbf{C}_k$ or $\mathbf{L}_k$, respectively. As closures are defined in terms of attributes rather than in terms of attribute values, it is sufficient to compute all the closures which are needed during a traversal of the while-loop only once when the loop is entered. Moreover, one can easily see from the above algorithm that the closure of an attribute set X is required for the computation of $\mathbf{C}_k$ and $\mathbf{L}_k$ only if $\mathbf{L}_{k\text{-}1}$ contains some element X:x.

When comparing the enhanced and the classical A Priori algorithm with regard to runtime there is not much difference. The enhanced A Priori algorithm will occasionally require one scan less of the relation instance r (namely, whenever one attribute value of the largest scavs in **L** is functionally dependent on the remaining attribute values), but the gains during counting due to smaller candidate sets will be moderate at best.

Actually, the main advantage of the enhanced A Priori algorithm is the fact that it avoids to include candidates into the sets $\mathbf{C_k}$ unnecessarily. This is very important because, according to [8], the size of the candidate sets $\mathbf{C_k}$ (particularly $\mathbf{C_2}$) is the bottleneck of the A Priori algorithm and "the number of candidates must be sufficiently small that a count for each can fit in main memory". For that reason, enhanced A Priori will be able to cope with larger sets of data than classical A Priori.

4.2 The AP-FD Algorithm

The reason why the enhanced A Priori algorithm proposed in the previous subsection is not faster than the classical A Priori algorithm is owed to the fact that the attribute values which are included into the frequent scavs due to functional dependencies are inserted one at a time only during each scan of the data. The following generalization of lemma 1 suggests that we might do better than that.

Lemma 2. Let R be a relation schema, r a relation instance conforming to R, and $Y \subseteq X \subseteq$ atts(R); if X:x is a frequent scav then $X{:}x \cup Y_\Gamma^+{:}y'$ is also a frequent scav with $Y_\Gamma^+{:}y'$ being the (uniquely determined) attribute values which are contained in all rows of r which contain X: x.

Proof: obvious. □

It is easy to see that lemma 2 may be employed in the A Priori algorithm to infer the existence of a frequent scav of size k+m from a frequent scav of size k in a single step even if m > 1. Thus it is possible that the largest scavs will be found after a fewer number of scans of the data.

However, the price that has to be paid for that is the fact that the frequent scavs found within a single scan of the data are no longer all of equal size. Moreover, it is not at all obvious how these larger scavs which have already been found should be taken account of when generating the next set of candidate scavs. The AP-FD algorithm described in the following deals with these problems essentially by splitting each frequent scav that is generated into two subsets one of which is of fixed size, whereas the other may contain any number of attribute values, particularly zero.

Like the classical and the enhanced A Priori algorithm, the AP-FD algorithm alternately computes a set $\mathbf{C_k}$ of candidates and a set $\mathbf{L_k}$ of frequent scavs. However, unlike the other two algorithms the scavs contained in $\mathbf{L_k}$ may be of different size; more precisely, a scav contained in $\mathbf{L_k}$ is at least of size k and consists of a "counting-part" and an "fd-part". The **counting-part** of a scav in $\mathbf{L_k}$ is a scav of fixed size k and the **fd-part** of a scav in $\mathbf{L_k}$ is the set of attribute values which are functionally dependent on the attribute values in the counting-part of the same scav.

The separation of counting-part and fd-part of a scav is made in order to be able to apply the classical A Priori algorithm to the counting-parts which are of equal size within a single set $\mathbf{L_k}$. The fd-parts are used to avoid the generating of candidates for which one may infer that they are frequent because of the given functional dependencies. For that reason a scav C will become a candidate only if none of its attribute values is contained in the fd-parts of a subset of C.

The AP-FD algorithm is given in Fig. 3; like the enhanced A Priori algorithm, it requires – compared to the classical A Priori algorithm – a set of functional dependencies as an additional input. Note that the fd-parts of the scavs in $\mathbf{L_k}$ are not given explicitly, rather, only their attributes are mentioned because the actual values are uniquely determined by the counting-part of the respective scav anyway. Where we need to refer to such unknown values (namely in the construction of the output set **L**) in the AP-FD algorithm, we do so once again by using the symbol "*".

```
begin
   C1 := {{A:a}| r contains an occurrence of A:a};
   L1 := {({A:a}, AΓ+\{A})| {A:a} ∈ C1 ∧ card(σA:a(r))≥s};
   L := ∅;
   k := 1;
   while Lk ≠ ∅ do
      begin
      L := L ∪ {(X:x ∪ Y:*)| (X:x, Y) ∈ Lk};
      k := k + 1;
      Ck := {C| C = {A1:a1,...,Ak:ak} ∧ Ai≠Aj for 1≤i<j≤k
                   ∧ ∀i ∈ {1,...,k} ∃(X:x, Y) ∈ Lk-1
                            (X:x = C\{Ai:ai} ∧ Ai ∉ Y)};
      Lk := {(X:x, XΓ+\X)| X:x∈Ck ∧ card(σX:x(r))≥s}
      end
end.
```

Figure 3. The AP-FD algorithm

It is easy to see from the way $\mathbf{C_k}$ is built that there will never be any functional dependency among the attribute values which are contained in a candidate scav and as a consequence there will also never be any functional dependency among the attribute values which are contained in the counting-part of a frequent scav. Thus, the AP-FD algorithm generates as few candidates as possible when building a candidate set $\mathbf{C_k}$; in this respect the AP-FD algorithm is equal to the enhanced A Priori algorithm described in the previous subsection.

However, the AP-FD algorithm is superior to both the classical A Priori algorithm and the enhanced A Priori algorithm with regard to the number of scans of the data that are required to find a frequent scav of size k. Both the classical A Priori algorithm and the enhanced A Priori algorithm require exactly k scans under any circumstances, whereas the number of scans that are required by the AP-FD algorithm is equal to the number of attributes contained in the counting-part of the scav. Thus the overall number of scans required by the AP-FD algorithm will be reduced whenever the largest ones of the frequent scavs have functional dependencies among their attribute values.

Example 2. Let R be a relation schema with atts(R) = {A, B, C, D, E} and let $\Gamma = \{A \rightarrow B, BC \rightarrow D, E \rightarrow C\}$ be the set of functional dependencies which hold in R. It is easy to verify that the relation instance r given in Table 1 is conforming to R.

Table 2. Example relation instance

A	B	C	D	E
a_1	b_1	c_1	d_1	e_1
a_1	b_1	c_2	d_2	e_2
a_2	b_1	c_1	d_1	e_3
a_3	b_2	c_2	d_1	e_2
a_3	b_2	c_2	d_1	e_2

We compare the performance of the classical A Priori algorithm and the AP-FD algorithm by applying both of them to r with s = 0.4 as support threshold. It is easy to see from looking at the two bottom rows of r that the classical A Priori algorithm will require five scans of r until it discovers the largest frequent scav, namely $\{A{:}a_3, B{:}b_2, C{:}c_2, D{:}d_1, E{:}e_2\}$.

The AP-FD algorithm, on the other hand, generates the following sets one after the other:

$$\mathbf{C}_1 = \{ \{A{:}a_1\}, \{A{:}a_2\}, \{A{:}a_3\}, \{B{:}b_1\}, \{B{:}b_2\}, \{C{:}c_1\}, \{C{:}c_2\}, \{D{:}d_1\}, \{D{:}d_2\}, \{E{:}e_1\}, \{E{:}e_2\}, \{E{:}e_3\} \}$$

$$\mathbf{L}_1 = \{ (\{A{:}a_1\}, \{B\}), (\{A{:}a_3\}, \{B\}), (\{B{:}b_1\}, \varnothing), (\{B{:}b_2\}, \varnothing), (\{C{:}c_1\}, \varnothing), (\{C{:}c_2\}, \varnothing), (\{D{:}d_1\}, \varnothing), (\{E{:}e_2\}, \{C\}) \}$$

$$\mathbf{C}_2 = \{ \{A{:}a_1, C{:}c_1\}, \{A{:}a_1, C{:}c_2\}, \{A{:}a_1, D{:}d_1\}, \{A{:}a_1, E{:}e_2\}, \{A{:}a_3, C{:}c_1\}, \{A{:}a_3, C{:}c_2\}, \{A{:}a_3, D{:}d_1\}, \{A{:}a_3, E{:}e_2\}, \{B{:}b_1, C{:}c_1\}, \{B{:}b_1, C{:}c_2\}, \{B{:}b_1, D{:}d_1\}, \{B{:}b_1, E{:}e_2\}, \{B{:}b_2, C{:}c_1\}, \{B{:}b_2, C{:}c_2\}, \{B{:}b_2, D{:}d_1\}, \{B{:}b_2, E{:}e_2\}, \{C{:}c_1, D{:}d_1\}, \{C{:}c_2, D{:}d_1\}, \{D{:}d_1, E{:}e_2\} \}$$

$$\mathbf{L}_2 = \{ (\{A{:}a_3, C{:}c_2\}, \{B, D\}), (\{A{:}a_3, D{:}d_1\}, \{B\}), (\{A{:}a_3, E{:}e_2\}, \{B, C, D\}), (\{B{:}b_1, C{:}c_1\}, \{D\}), (\{B{:}b_1, D{:}d_1\}, \varnothing), (\{B{:}b_2, C{:}c_2\}, \{D\}), (\{B{:}b_2, D{:}d_1\}, \varnothing), (\{B{:}b_2, E{:}e_2\}, \{C, D\}), (\{C{:}c_1, D{:}d_1\}, \varnothing), ((C{:}c_2, D{:}d_1\}, \varnothing), (\{D{:}d_1, E{:}e_2\}, \varnothing) \}$$

Note that during the third traversal of the while-loop the largest frequent scav in r, namely $\{A{:}a_3, B{:}b_2, C{:}c_2, D{:}d_1, E{:}e_2\}$ is already inserted into the set L; moreover, it turns out that $\mathbf{C}_3$ is the empty set and therefore the AP-FD algorithm terminates. Thus in this case, the AP-FD algorithm requires only two full table scans whereas the classical A Priori algorithm requires five.

A detailed comparison of the size of the sets generated by the two algorithms is shown in Table 3. □

Table 3. Size of the sets generated by the classical A Priori and the AP-FD algorithm

	C_1	L_1	C_2	L_2	C_3	L_3	C_4	L_4	C_5	L_5
A Priori algorithm	12	8	25	14	11	11	5	5	1	1
AP-FD algorithm	12	8	19	11	–	–	–	–	–	–

A thorough comparison of Fig. 1 and Fig. 3 will show, that the result of the classical A Priori algorithm and the AP-FD algorithm are not exactly the same: The classical A Priori algorithm collects definitely all frequent scavs in **L**, whereas the AP-FD algorithm fails to insert certain scavs because it inserts only the union of a counting-part and the corresponding fd-part but not the counting-part by itself. More precisely, the AP-FD algorithm inserts a frequent scav into **L** only, if there is no superset of that scav which is already known to be frequent and therefore can be inserted into **L** at the same time. It is obvious that this discrepancy could be overcome easily if required.

Finally, we would like to point out that the potential benefit of the AP-FD algorithm to reduce the number of candidate sets will obviously grow with the number of frequent scavs X:x which exist for a given X. It is easy to see that in this respect, the above example will be outperformed by real life relations, particularly, if they contain a large number of rows.

5. Conclusion

In this paper, we have investigated the relationship between functional dependencies and multidimensional association rules. It turns out that the knowledge of functional dependencies in the data to be mined may be used in the A Priori algorithm which is commonly used to determine all frequent sets of co-occurring attribute values. For that purpose, we have proposed a new variant of the A Priori algorithm called the AP-FD algorithm. This algorithm is superior to its classical counterpart with respect to both runtime and the number of candidate sets that are generated. In particular, this latter point is of great importance as the number of candidate sets generated is the bottleneck for the applicability of the A Priori algorithm to very large data sets.

There have been numerous proposals in the literature that focus on improving the efficiency of the classical A Priori algorithm. It seems worth mentioning that the AP-FD algorithm described in this paper has not to be understood as an alternative to such proposals, but rather may be combined with many of these proposals thus resulting in a further gain in performance. We only mention the most important ones of these variations of the classical A Priori algorithm.

Hash-based techniques (such as [9]) use one or more hash tables the entries of which are counters. While computing $\mathbf{L}_k$ from $\mathbf{C}_k$, these algorithms map all scavs of size k+1 to the buckets of the hash table and increment the corresponding counter; a scav of size k+1 then has to be inserted into $\mathbf{C}_{k+1}$ only if the corresponding counter exceeds the support threshold. It is obvious that this procedure is equally applicable if the AP-FD algorithm is employed instead of the classical A Priori algorithm.

Transaction reduction methods (as described in [4]) make use of the observation that a row which does not contain a frequent scav of size k cannot contain a frequent scav of size k+1. Therefore all such rows are ruled out from the scan of the data during which $\mathbf{L}_{k+1}$ is determined by counting. Clearly this may be done as well if the AP-FD algorithm is used instead of the classical A Priori algorithm.

The **partitioning technique** (cf. [10]) applies the classical A Priori algorithm to partitions of the data to be mined that are sufficiently small in order to fit into main memory. The union of all scavs which are frequent in at least one partition is then used in a second scan of the data to find out which ones are frequent with respect to the entire data set. Clearly the AP-FD algorithm may be used to speed up the first phase of this procedure.

Sampling techniques (see [11] for an example) mine a subset of the given data for frequent scavs and then verify those scavs with respect to the entire data set by means of one or more full scans of the data. Obviously, the classical A Priori algorithm may be replaced by the AP-FD algorithm when mining the samples.

As a matter of fact, there are other methods for mining frequent scavs that do not directly build upon A Priori; we only mention DIC (for "dynamic itemset counting", described in [12]) and CARMA (for "continuous association rule mining algorithm", cf. [13]). Although one cannot incorporate the AP-FD algorithm into these methods, they may well benefit from the knowledge of functional dependencies. This is because both DIC and CARMA – as well as most other research on the subject – "are variations of the 'bottom-up theme' proposed by the A Priori algorithm" as it was called in [14]. "Bottom-up theme" means that the frequent scavs are obtained by starting with singleton scavs and then incrementally generating larger and larger scavs. Without looking at the details of such a procedure it is obvious that the knowledge of functional dependencies will always offer opportunities to overleap certain steps. How this can be done efficiently in detail, however, requires further study.

References

[1] Agrawal R., Imielinski T., Swami T. Mining Association Rules between Sets of Items in Large Databases. Proc. ACM SIGMOD Int. Conf. on Management of Data, 1993. p. 207 – 216

[2] Garcia-Molina H., Ullman J.D., Widom J. Database Systems: The Complete Book. Upper Saddle River, NJ. Prentice Hall, 2002

[3] Beeri C., Bernstein P.A. Computational Problems Related to the Design of Normal Form Relation Schemas. ACM Trans. on Database Systems 1979; 4(1): p. 30 – 59

[4] Agrawal R., Srikant R. Fast Algorithms for Mining Association Rules. Proc. Int. Conf. Very Large Data Bases, 1994. p. 487 – 499

[5] Han J., Kamber M. Data Mining: Concepts and Techniques. San Francisco. Morgan Kaufmann Publishers, 2001

[6] Kimball R. The Data Warehouse Toolkit - Practical Techniques for Building Dimensional Data Warehouses. New York. Wiley, 1996

[7] Janas J.M. An Enhanced A Priori Algorithm for Mining Multidimensional Association Rules. Proc. 25th Int. Conf. Information Technology Interfaces, 2003. p. 193 – 198

[8] Ullman J.D. Data Mining Lecture Notes. http://www-db.stanford.edu/~ullman/mining/mining.html [03/06/2003]

[9] Park J.S., Chen M.S., Yu P.S. An Effective Hash-Based Algorithm for Mining Association Rules. Proc. ACM SIGMOD Int. Conf. on Management of Data, 1995. p. 175 – 186

[10] Savasere A., Omiecinski E., Navathe S. An Efficient Algorithm for Mining Association Rules in Large Databases. Proc. Int. Conf. Very Large Data Bases, 1995. p. 432 - 443

[11] Toivonen H. Sampling Large Databases for Association Rules. Proc. Int. Conf. Very Large Data Bases, 1996. p. 134 – 145

[12] Brin S., Motwani R., Ullman J.D., Tsur S. Dynamic Itemset Counting and Implication Rules for Market Basket Analysis. Proc. ACM SIGMOD Int. Conf. on Management of Data, 1997. p. 255 – 264

[13] Hidber C. Online Association Rule Mining. Proc. ACM SIGMOD Int. Conf. on Management of Data, 1999. p.145 – 156

[14] Aggarwal C.C., Yu P.S. Mining Large Itemsets for Association Rules. Bulletin of the IEEE Computer Society TC on Data Engineering, March 1998. p. 23 – 31

Databases and Information Systems
J. Barzdins and A. Caplinskas (Eds.)
IOS Press, 2005

Data Mining Query Scheduling for Apriori Common Counting

Marek WOJCIECHOWSKI, Maciej ZAKRZEWICZ
Poznan University of Technology, Institute of Computing Science,
ul. Piotrowo 3a, 60-965 Poznan, Poland
{marek,mzakrz}@cs.put.poznan.pl

Abstract. In this paper we consider concurrent execution of multiple data mining queries. If such data mining queries operate on similar parts of the database, then their overall I/O cost can be reduced by integrating their data retrieval operations. The integration requires that many data mining queries are present in memory at the same time. If the memory size is not sufficient to hold all the data mining queries, then the queries must be scheduled into multiple phases of loading and processing. We discuss the problem of data mining query scheduling and propose a heuristic algorithm to efficiently schedule the data mining queries into phases.

1. Introduction

Data mining is a database research field that aims at discovery of trends, patterns and regularities in very large databases. We are currently witnessing the evolution of data mining environments towards their full integration with DBMS functionality. In this context, data mining is considered to be an advanced form of database querying, where users formulate declarative *data mining queries*, which are then optimized and executed by one of data mining algorithms built into the DBMS. One of the most significant issues in data mining query processing is long execution time, ranging from minutes to hours.

One of the most popular pattern types discovered by data mining queries are *frequent itemsets*. Frequent itemsets describe co-occurrences of individual items in sets of items stored in the database. An example of a frequent itemset can be a collection of products that customers typically purchase together during their visits to a supermarket. Such frequent itemsets can be discovered in the database of customer shopping baskets. Frequent itemsets are usually discovered using *level-wise algorithms*, which divide the problem into multiple iterations of database scanning and counting occurrences of candidate itemsets of equal size.

Due to long execution times, data mining queries are often performed in a *batch mode*, where users submit sets of data mining queries to be executed during low database activity time (e.g., night time). It is likely that the batches contain data mining queries that operate on similar parts of the database. If such queries are executed separately, the same parts of the database are retrieved multiple times. We could reduce the overall I/O activity of the batch of data mining queries if we integrated their data retrieval operations on the same portions of the database.

For a system with unlimited memory, the integration of execution of multiple data mining queries consists in *common counting* [13][14] of candidate itemsets for all the queries so that every portion of the database needs to be read only once per iteration. However, if the memory is limited, we are not able to keep all candidate itemsets of all the data mining queries in the memory at the same time. The whole process must then be split into multiple phases of loading and counting the candidates, and therefore the data mining queries must be divided into subsets to be executed in separate phases. We refer to the problem of assigning the data mining queries to execution phases as to *data mining query scheduling*.

In this paper we discuss the problem of data mining query scheduling and we introduce a heuristic algorithm to perform the scheduling for a system with limited memory. The goal of the algorithm is to schedule the data mining queries in such a way that the overall I/O cost for the whole batch is minimized.

1.1 Related Work

The problem of mining association rules was first introduced in [1] and an algorithm called *AIS* was proposed. In [2], two new algorithms were presented, called *Apriori* and *AprioriTid* that are fundamentally different from the previous ones. The algorithms achieved significant improvements over *AIS* and became the core of many new algorithms for mining association rules. *Apriori* and its variants first generate all frequent itemsets (sets of items appearing together in a number of database records meeting the user-specified support threshold) and then use them to generate rules. *Apriori* and its variants rely on the property that an itemset can only be frequent if all of its subsets are frequent. It leads to a level-wise procedure. First, all possible 1-itemsets (itemsets containing 1 item) are counted in the database to determine *frequent 1-itemsets*. Then, frequent 1-itemsets are combined to form potentially frequent 2-itemsets, called *candidate 2-itemsets*. Candidate 2-itemsets are counted in the database to determine *frequent 2-itemsets*. The procedure is continued by combining the frequent 2-itemsets to form *candidate 3-itemsets* and so forth. A disadvantage of the algorithm is that it requires k or $k+1$ passes over the database to discover all frequent itemsets, where k is the size of the greatest frequent itemset found.

In [4], an algorithm called *FUP* (Fast Update Algorithm) was proposed for finding the frequent itemsets in the expanded database using the old frequent itemsets. The major idea of *FUP* algorithm is to reuse the information of the old frequent itemsets and to integrate the support information of the new frequent itemsets in order to reduce the pool of candidate itemsets to be re-examined. Another approach to incremental mining of frequent itemsets was presented in [11]. The algorithm introduced there required only one database pass and was applicable not only for expanded but also for reduced database. Along with the frequent itemsets, a *negative border* [12] was maintained.

In [10] the issue of interactive mining of association rules was addressed and the concept of *knowledge cache* was introduced. The cache was designed to hold frequent itemsets that were discovered while processing other queries. Several cache management schemes were proposed and their integration with the *Apriori* algorithm was analyzed. An important contribution was an algorithm that used itemsets discovered for higher support thresholds in the discovery process for the same task, but with a lower support threshold.

The notion of data mining queries (or *KDD* queries) was introduced in [6]. The need for Knowledge and Data Management Systems (KDDMS) as second-generation data mining tools was expressed. The ideas of application programming interfaces and data mining query optimizers were also mentioned. Several data mining query languages that are extensions of *SQL* were proposed [3][5][7][8][9].

2. Basic Definitions and Problem Formulation

2.1 Frequent Itemsets

Let $L=\{l_1, l_2, ..., l_m\}$ be a set of literals, called items. Let a non-empty set of items T be called an *itemset*. Let D be a set of variable length itemsets, where each itemset $T \subseteq L$. We say that an itemset T *supports* an item $x \in L$ if x is in T. We say that an itemset T *supports* an itemset $X \subseteq L$ if T supports every item in the set X. The *support* of the itemset X is the percentage of T in D that support X. The problem of mining frequent itemsets in D consists in discovering all itemsets whose support is above a user-defined support threshold.

2.2 Apriori Algorithm

Apriori is an example of a level-wise algorithm for association discovery. It makes multiple passes over the input data to determine all frequent itemsets. Let L_k denote the set of frequent itemsets of size k and let C_k denote the set of candidate itemsets of size k. Before making the k-th pass, *Apriori* generates C_k using L_{k-1}. Its candidate generation process ensures that all subsets of size k-*1* of C_k are all members of the set L_{k-1}. In the k-th pass, it then counts the support for all the itemsets in C_k. At the end of the pass all itemsets in C_k with a support greater than or equal to the minimum support form the set of frequent itemsets L_k. Figure 1 provides the pseudocode for the general level-wise algorithm, and its *Apriori* implementation. The *subset*(t, k) function gives all the subsets of size k in the set t.

This method of pruning the C_k set using L_{k-1} results in a much more efficient support counting phase for *Apriori* when compared to the earlier algorithms. In addition, the usage of a hash-tree data structure for storing the candidates provides a very efficient support-counting process.

```
C1 = {all 1-itemsets from D};
for (k=1; Ck ≠ ∅; k++) do
begin
  count(Ck, D);
  Lk = {c ∈ Ck | c.count ≥ minsup};
  Ck+1 = generate_candidates(Lk);
end;
Answer = UkLk;
```

```
L1 = {frequent 1-itemsets};
for (k = 2; Lk-1 ≠ ∅; k++) do
begin
  Ck = generate_candidates(Lk-1);
  forall tuples t ∈ D do
  begin
    Ct=Ck ∩ subset(t, k);
    forall candidates c ∈ Ct do
      c.count++;
  end;
  Lk = {c ∈ Ck | c.count ≥ minsup}
end;
Answer = UkLk;
```

Figure 1. A general level-wise algorithm for association discovery (left) and its Apriori implementation (right)

2.3 Data Mining Query

A *data mining query* is a tuple (R, a, Σ, Φ, β), where R is a database relation, a is an attribute of R, Σ is a selection predicate on R, Φ is a selection predicate on frequent itemsets, β is the minimum support for the frequent itemsets.

Example. Consider the database relation $R_1(attr_1, attr_2)$. The data mining query $dmq_1 = (R_1$, "$attr_2$", "$attr_1 > 5$", "$|itemset|<4$", $3)$ describes the problem of discovering frequent itemsets in the set-valued attribute $attr_2$ of the relation R_1. The frequent itemsets with support above 3 and length less than 4 are discovered in records having $attr_1>5$.

2.4 Multiple Data Mining Query Optimization

Let $DMQ=\{dmq_1, dmq_2, ..., dmq_n\}$ be a set of data mining queries, where $dmq_i=(R, a, \Sigma_i, \Phi_i, \beta_i)$, Σ_i is of the form "$(l^i_{1min}<a<l^i_{1max}) \vee (l^i_{2min}<a<l^i_{2max}) \vee ... \vee (l^i_{kmin}<a<l^i_{kmax})$", and there are at least two data mining queries $dmq_i=(R_i, a, \Sigma_i, \Phi_i, \beta_i)$ and $dmq_j=(R, a, \Sigma_j, \Phi_j, \beta_j)$ such that $\sigma_{\Sigma i}R \cap \sigma_{\Sigma j}R \neq \emptyset$. The problem of *multiple data mining query optimization* is to generate an algorithm to execute *DMQ* with the minimal I/O cost.

2.5 Data Sharing Graph

Let $S=\{s_1, s_2, ..., s_k\}$ be a *set of elementary data selection predicates* for *DMQ*, i.e., selection predicates over the attribute a or the relation R such that for all i, j we have $\sigma_{si}R \cap \sigma_{sj}R=\emptyset$ and for each i there exist integers a, b, ..., m such that $\sigma_{\Sigma i}R=\sigma_{sa}R \cup \sigma_{sb}R \cup ... \cup \sigma_{sm}R$ (example in Fig. 2). A graph $DSG=(V,E)$ is called a *data sharing graph* for the set of data mining queries *DMQ* iff $V=DMQ \cup S$, $E=\{(dmq_i,s_j) | dmq_j \in DMQ, s_j \in S, \sigma_{\Sigma i}R \cap \sigma_{sj}R \neq \emptyset\}$.

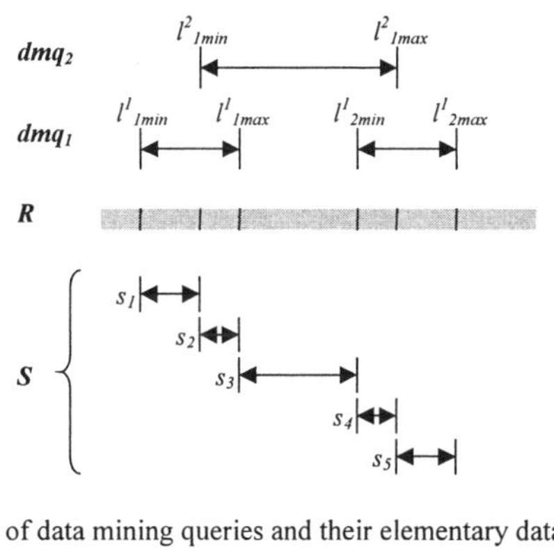

Figure 2. Example set of data mining queries and their elementary data selection predicates

Example. Consider the relation $R_1=(attr_1, attr_2)$ and three data mining queries: $dmq_1=(R_1$, "$attr_2$", "$5<attr_1<20$", $\emptyset$, $3)$, $dmq_2=(R_1$, "$attr_2$", "$10<attr_1<30$", $\emptyset$, $5)$, $dmq_3=(R_1$, "$attr_2$", "$15<attr_1<40$", $\emptyset$, $4)$. The set of elementary data selection predicates is then $S=\{s_1=$"$5<attr_1<10$", $s_2=$"$10<attr_1<15$", $s_3=$"$15<attr_1<20$", $s_4=$"$20<attr_1<30$", $s_5=$"$30<attr_1<40$"$\}$. The data sharing graph for $\{dmq_1, dmq_2, dmq_3\}$ is shown in Fig. 3.

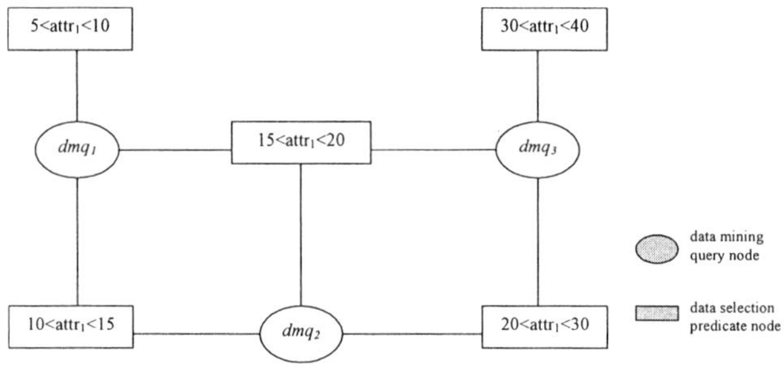

Figure 3. Example data sharing graph

2.6 Apriori Common Counting

A straightforward way to perform multiple data mining query optimization is the *Apriori Common Counting* algorithm. The algorithm proceeds as follows. In the first step, *Apriori Common Counting* constructs separate candidate 1-itemset hash trees (in memory) for each data mining query. Next, all database partitions corresponding to the elementary selection predicates are scanned and the candidate itemsets for the appropriate data mining queries are counted. This process is repeated for each iteration: for candidate 2-itemsets, candidate 3-itemsets, etc. Notice that if a given elementary selection predicate is shared by multiple data mining queries, then the specific part of the database needs to be read only once (per iteration). This property helps reduce the overall I/O cost of batched data mining query execution. The idea of the *Apriori Common Counting* algorithm is depicted in Fig. 4.

for (i=1; i<=n; i++) /* n = *number of data mining queries* */
 C_1^i = {all 1-itemsets from $\sigma_{s1 \cup s2 \cup \ldots \cup sk} R$, $\forall s_j \in S$: $(dmq_i, s_j) \in E$} /* *generate 1-candidates* */
for (k=1; $C_k^1 \cup C_k^2 \cup \ldots \cup C_k^n \neq \varnothing$; k++) **do begin**
  for each $s_j \in S$ **do begin**
   $CC = \bigcup C_k^l$: $(dmq_l, s_j) \in E$; /* *select the candidates to count now* */
   if $CC \neq \varnothing$ **then** *count*(CC, $\sigma_{sj} R$);
  end;
  for (i=1; i<=n; i++) **do begin**
   $L_k^i = \{c \in C_k^i \mid c.count \geq minsup^i\}$; /* *identify frequent itemsets* */
   $C_{k+1}^i = generate\_candidates(L_k^i)$;
  end;
end;
for (i=1; i<=n; i++) **do**
  $Answer^i = \bigcup_k L_k^i$; /* *generate responses* */

Figure 4. Apriori Common Counting

3. Data Mining Query Scheduling

The basic *Apriori Common Counting* described in the previous section assumes unlimited memory for its operation. However, if the memory is limited, then it is not possible to construct candidate hash trees for all the data mining queries. The whole algorithm must then be split into multiple *phases* and every phase must consist in executing a subset of the data mining queries. The key problem is which data mining queries should be performed in the same phase and which of them can be performed in separate phases. The task of dividing the set of data mining queries into subsets is referred to as *data mining query scheduling*.

There are several aspects to consider when designing a data mining query scheduling algorithm. Firstly, it is obvious that system memory size restricts the number of data mining queries that may be processed in the same phase. Memory requirements for the data mining queries are based on sizes of their candidate hash trees, which in turn depend on data characteristics and the specific iteration of the algorithm (typically, sizes of candidate hash trees systematically reduce for iterations 3, 4, etc.). Since the candidate hash tree sizes change in each iteration, the data mining query scheduling algorithm should be used before generating every new tree, not only at the beginning of the data mining query processing. Another aspect is that the goal of *Apriori Common Counting* is to reduce the overall I/O activity. Therefore, similarities between data mining queries should be taken into account when putting data mining queries into the same phase. Data mining queries that operate on separate portions of the database can be processed in separate phases, while data mining queries that operate on highly overlapping database portions should be executed in the same phase. To measure the "overlapping" between data mining queries one can rely on a traditional DBMS query optimizer, which estimates predicate costs based on database statistics.

In order to schedule data mining queries, the sizes of their candidate hash trees must be known. There are two options to derive the hash tree sizes. The first option is to calculate the upper bounds on the candidate hash trees and use the upper bounds in the scheduling algorithm. The upper bounds can be evaluated based on the number of frequent itemsets discovered in the previous iteration. A disadvantage of this approach is that the real candidate hash trees are smaller than the estimates, so the scheduling algorithm is likely to miss the optimal solution. The second option is to generate the candidate hash trees first, measure their sizes, save them in temporary files, perform the scheduling and then retrieve the appropriate trees from the files while performing the phases. The main advantage of this approach is that the scheduling algorithm operates on the exact sizes of the trees, and therefore it is able to find the optimal solution. However, the additional I/O cost is introduced because of the need to temporarily store the candidate hash trees on disk. Nevertheless, when dealing with very large databases (in case of which candidate tree sizes are by several orders of magnitude smaller than the database) that extra cost is going to be compensated by reduction of database reads thanks to *Common Counting*.

Let us consider an example of data mining query scheduling based on our previous set of data mining queries from Fig. 3. Let *cost*(s) be the I/O cost of retrieving database records that satisfy the data selection predicate s. Let *treesize*(dmq,k) be the k-item candidate hash tree size for the data mining query dmq. Sample costs and tree sizes (e.g., for the third *Apriori* iteration) are given in the table below. Let us assume the system memory limit of 10MB, meaning that at most two of the data mining queries can fit in at a time (i.e., in one phase).

s_i	$cost(s_i)$
$5<attr_1<10$	5000
$10<attr_1<15$	7000
$15<attr_1<20$	2000
$20<attr_1<30$	2000
$30<attr_1<40$	1000

dmq_i	$treesize(dmq_i,3)$
dmq_1	4M
dmq_2	5M
dmq_3	3M

There exist four different schedules that satisfy the given constraints. The schedules and the total costs of executing the sample set of data mining queries are given below. The *Schedule A* represents a sequential execution of all the data mining queries. One can notice that the optimal solution is the *Schedule B*, which reduces the overall cost by 30%. This schedule has also been depicted in Fig. 5.

Schedule A

phase	*data mining queries*	*trees size*	*data selection predicates*	*phase cost*
1	dmq_1	4M	s_1, s_2, s_3	14,000
2	dmq_2	5M	s_2, s_3, s_4	11,000
3	dmq_3	3M	s_3, s_4, s_5	5,000
total cost				**30,000**

Schedule B

phase	*data mining queries*	*trees size*	*data selection predicates*	*phase cost*
1	dmq_1, dmq_2	9M	s_1, s_2, s_3, s_4	16,000
2	dmq_3	3M	s_3, s_4, s_5	5,000
total cost				**21,000**

Schedule C

phase	*data mining queries*	*trees size*	*data selection predicates*	*phase cost*
1	dmq_1, dmq_3	7M	s_1, s_2, s_3, s_4, s_5	17,000
2	dmq_2	5M	s_2, s_3, s_4	11,000
total cost				**28,000**

Schedule D

phase	*data mining queries*	*trees size*	*data selection predicates*	*phase cost*
1	dmq_2, dmq_3	8M	s_2, s_3, s_4, s_5	12,000
2	dmq_1	4M	s_1, s_2, s_3	14,000
total cost				**26,000**

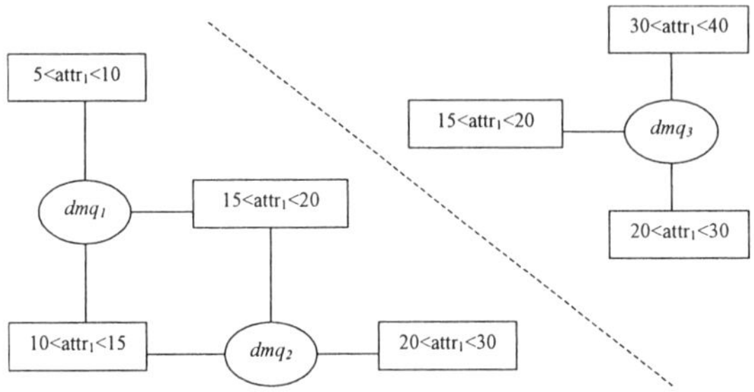

Figure 5. Data sharing graph for the optimal schedule for the sample set of data mining queries

The data mining query scheduling problem can be solved using a combinatorial approach, in which all possible (allowable) schedules are generated first, and then their overall costs are calculated. The combinatorial approach can be suitable for a small number of data mining queries in the set. However, for complex problems, involving large numbers of data mining queries, the overhead of the approach would be unacceptable. For a given number of data mining queries, the number of all possible schedules is determined by the Bell number – e.g., for 13 queries the number of schedules exceeds 4 millions. Therefore we introduce a heuristic algorithm for finding suboptimal schedules for executing a set of data mining queries.

3.1 Heuristic Scheduling Algorithm: CCRecursive

The algorithm iterates over all the elementary selection predicates, sorted in descending order with respect to their I/O costs. For each elementary selection predicate we identify all the data mining queries that include the predicate. If none of the identified queries has been already scheduled, then we create a new phase and we put all the queries into the new phase. Otherwise, we merge the phases to which the scheduled queries belonged and we assign the other queries to this new phase. If the size of the newly created phase exceeds the memory limit, then the phase is split into smaller ones by recursive execution of the algorithm. At the end of the algorithm, we perform *phase compression*, which consists in merging those phases that do not consume all the available memory. The detailed structure of the algorithm is given in Fig. 6. The auxiliary function *treesize*(Q), where Q is a set of data mining queries, represents total memory size required to hold candidate hash trees for all the data mining queries in Q.

```
Phases ← {∅};
sort S = <s_i , s_2 ,..., s_k> in descending order with respect to cost(s_i);
CCRecursive(S, DMQ, Phases):
begin
  ignore in S those predicates that are used by less than two dmqs;
  for each s_i in S do begin
    tmpDMQ ← {dmq_j | dmq_j =(R, a, Σ_j, Φ_j, β_j), s_i ⊆Σ_j, dmq_j ∈ DMQ};
    commonPhases ← {p ∈ Phases | p∩ tmpDMQ ≠ ∅};
    if commonPhases = ∅ then
      newPhase ← tmpDMQ;
    else
      newPhase ← tmpDMQ ∪ Up| p∈ commonPhases;
    end if;
    if treesize(newPhase) ≤ MEMSIZE then
Phases ← Phases \ commonPhases;
Phases ← Phases ∪ newPhase;
    else
    Phases ← CCRecursive(<s_i+1, …, s_k>, newPhase, Phases);
    end if;
  end;
  add phase for each unscheduled query;
  compress Phases containing queries from DMQ;
  return Phases;
end.
```

Figure 6. Heuristic scheduling algorithm: CCRecursive

4. Experimental Evaluation

In order to evaluate our heuristic algorithm *CCRecursive* we performed a series of simulations on a PC with *AMD Duron 1200 MHz* processor and 256 MB of main memory. We focused on the isolated problem of scheduling queries into phases fitting in main memory in a given iteration of *Common Counting*. We compared the amount of data read from the database by our heuristic algorithm and the complete "brute-force" algorithm testing all possible assignments of queries to phases.

We simulated actual batches of frequent set discovery tasks by randomly generating a collection of queries. For each query, the database selection predicate and the size of the candidate tree were randomly generated. Then the total amount of main memory was also randomly chosen in such a way that the number of queries fitting into it ranged from one query to all the queries.

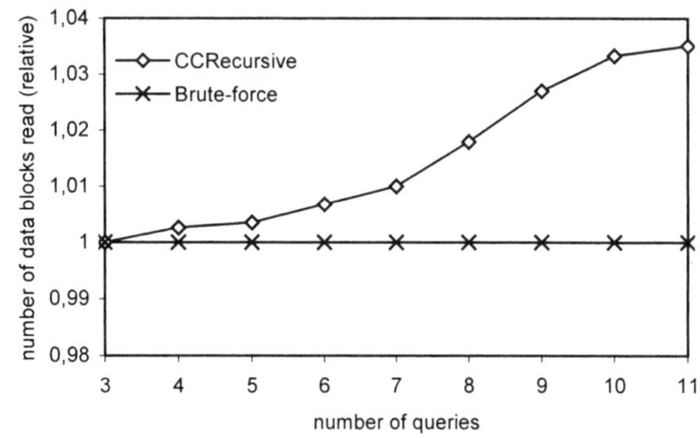

Figure 7. Amounts of data read by CCRecursive schedules and optimal schedules

We performed several series of experiments varying the number of queries. Each of the series consisted of 100 simulations. Figure 7 presents how the accuracy of our heuristic algorithm changes with the number of queries. To assess the accuracy we measured the relative amount of data read from the database by schedules generated by our heuristics compared to the optimal schedules (generated by the complete brute-force scheduling algorithm). For example, in the case of 11 queries, *CCRecursive* generates schedules that read on average about 3.5% more data than the optimal schedules.

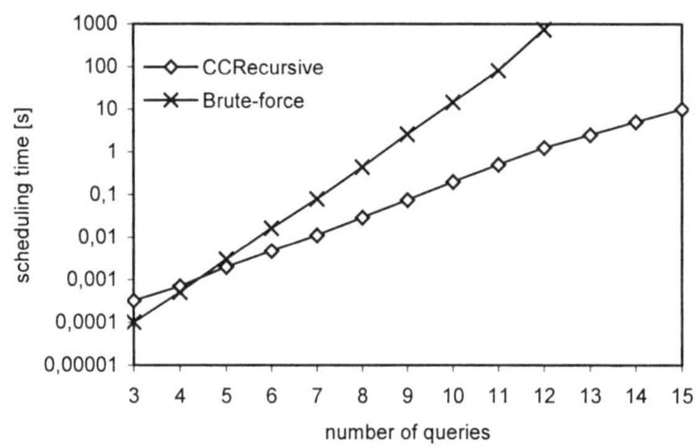

Figure 8. Execution times (logarithmic scale) of CCRecursive and the brute-force scheduling algorithm

Figure 8 presents the execution times (times needed to generate schedules) of *CCRecursive* and the brute-force algorithm. Although *CCRecursive* still scales exponentially with the number of queries, its execution time increases less rapidly than in case of the brute-force solution. For instance, the brute-force algorithm consumes more than 1000 s already for 12 queries, while *CCRecursive* exceeds that threshold in case of 22 queries (the chart presents the times for up to 15 queries).

The results of conducted experiments show that *CCRecursive* significantly outperforms the brute-force solution (with the exception of cases with 3 and 4 queries when execution times of both algorithms are negligible), which makes it applicable for larger batches of data mining queries. We believe that the accuracy of our heuristics (shown in Fig. 7) is acceptable. However, it should be noted that the actual trade-off between extra disk accesses (introduced by the heuristics) and reduction in the scheduling time cannot be assessed without knowing the database size and hardware parameters.

5. Concluding Remarks

In this paper we addressed the problem of *common counting* of candidate itemsets for multiple data mining queries. We have formally defined the problem of *data mining query scheduling*, which consists in splitting the set of data mining queries into subsets (phases) such that the candidate hash trees can fit in limited memory and the overall I/O cost is minimized.

Since the number of possible schedules growth rapidly with the number of queries, we proposed a heuristic scheduling algorithm, called *CCRecursive*. The experiments show that our heuristics generates schedules that are close to optimal and is more efficient than the brute-force solution and thus applicable for much greater number of queries.

References

[1] Agrawal R., Imielinski T., Swami A. Mining Association Rules Between Sets of Items in Large Databases. Proc. of the 1993 ACM SIGMOD Conf. on Management of Data, 1993.

[2] Agrawal R., Srikant R. Fast Algorithms for Mining Association Rules. Proc. of the 20th Int'l Conf. on Very Large Data Bases, 1994.

[3] Ceri S., Meo R., Psaila G. A New SQL-like Operator for Mining Association Rules. Proc. of the 22nd Int'l Conference on Very Large Data Bases, 1996.

[4] Cheung D.W., Han J., Ng V., Wong C.Y. Maintenance of Discovered Association Rules in Large Databases: An Incremental Updating Technique. Proc. of the 12th ICDE, 1996.

[5] Han J., Fu Y., Wang W., Chiang J., Gong W., Koperski K., Li D., Lu Y., Rajan A., Stefanovic N., Xia B., Zaiane O.R. DBMiner: A System for Mining Knowledge in Large Relational Databases. Proc. of the 2nd KDD Conference, 1996.

[6] Imielinski T., Mannila H. A Database Perspective on Knowledge Discovery. Communications of the ACM, Vol. 39, No. 11, 1996.

[7] Imielinski T., Virmani A., Abdulghani A. Datamine: Application programming interface and query language for data mining. Proc. of the 2nd KDD Conference, 1996.

[8] Morzy T., Wojciechowski M., Zakrzewicz M. Data Mining Support in Database Management Systems. Proc. of the 2nd DaWaK Conference, 2000.

[9] Morzy T., Zakrzewicz M. SQL-like Language for Database Mining. ADBIS'97 Symposium, 1997.

[10] Nag B., Deshpande P.M., DeWitt D.J. Using a Knowledge Cache for Interactive Discovery of Association Rules. Proc. of the 5th KDD Conference, 1999.

[11] Thomas S., Bodagala S., Alsabti K., Ranka S. An Efficient Algorithm for the Incremental Updation of Association Rules in Large Databases. Proc. of the 3rd KDD Conference, 1997.

[12] Toivonen H. Sampling Large Databases for Association Rules. Proc. of the 22nd Int'l Conference on Very Large Data Bases, 1996.

[13] Wojciechowski M., Zakrzewicz M. Methods for Batch Processing of Data Mining Queries. Proc. of the 5th International Baltic Conference on Databases and Information Systems, 2002.

[14] Wojciechowski M., Zakrzewicz M. Evaluation of Common Counting Method for Concurrent Data Mining Queries. Proc. of the 7th ADBIS Conference, 2003.

Databases and Information Systems
J. Barzdins and A. Caplinskas (Eds.)
IOS Press, 2005

Controlling Access to Data Warehouse Data within the Database

Laila NIEDRITE, Liga GRUNDMANE
University of Latvia, Department of Computer Science
19 Raina boulevard, Riga, Latvia

Abstract. There are some alternative approaches as to how to control user access to restricted information. The goal of this paper is to provide two models of data access control specific for data warehouses implemented in RDBMS. The models *"Role Based Security Model"* and "*Dynamically Adapting Views Based Data Access Control Model"* are based on the idea of revoking all information, restricting tasks from the application layer, which is important for data warehouses where different user tools may be provided. Our proposed two solutions also support the possibility of deriving data access rights from data sources.

Introduction

The data warehouse is storage of data that is extracted and integrated from different data sources. A data warehouse often contains sensitive data and certain user/s are either prevented or permitted as the case may be from accessing unauthorized data as defined and set out in accordance to the various criteria's within either the *"Role Based Security Model"* or "*Dynamically Adapting Views Based Data Access Control Model"*. Both these models will be discussed later in this paper, however, as any system or theoretical model/s, they must all be set up to serve their respective clients needs and over-all corporate objectives while maintaining system data integrity and security of certain types of data while maximizing the throughput time (such as query time) of the system itself. Both models offer inherent advantages and disadvantages and they must be considered if proper overall data base operation is a necessity, particularly in a setting where there are or could be literally thousands of users on line at one time. One of theses concerns is of particular importance and that relates to certain overlaps of users, which may through employment or other be required to perform multi functions which directly results in their ability to cross over access of certain data bases. The data access security system must be set up to recognize and distinguish and assign the proper access rights in such cases.

In particular how can consistent and simultaneous data access rights be accomplished and secured knowing that one person (as mentioned in the previous paragraph) may be a data source systems user with different data access rights to similar content data sets, e.g. PERSONS in one data source and EMPLOYEES of one particular department in the other data source? If certain data access rights can be derived from data sources, can it be done automatically and do specific data warehouse features create additional needs for data access control?

One of the existing approaches uses data sources users' access rights to derive the data warehouse users' access rights directly [1], [2]. In this approach the data warehouse is treated as a view over data sources, where every data warehouse user has a view over only authorized data from data sources.

Another approach [3] is based on analyzing the user's analytical business functions and the information necessary to fulfill these functions. The former data source data access rights are not taken into account, because of the new nature of the business functions. Data warehouse data can also be treated as new information, which is created in the process of integration of data source data. Some data warehouse specific functionality provided by client side tools e.g. OLAP operations drill-down, roll-up can add new restrictions to the accessibility of data, for example, the user having access to detailed data on one particular department level, can have full access to aggregated data on company level.

Another consideration, which is discussed in this paper, is at what level the data access rights ought to be implemented on, the application or the data base level?

The main criteria which solution is most appropriate in any given or particular case are further discussed in [4]. The most important issues, which need to be considered, are the number of potential users and the number of planned client side tools when developing access rights to a data warehouse.

In Section 1 we describe some specific characteristics of data warehouses and considerations about data access restrictions in data warehouses. Some definitions are given in this section and an example of data warehouse star schema is introduced.

In Section 2 we present basic structure of our first solution called *"Role Based Security Model"* control model in data warehouses. Our second solution *"Dynamically Adapting Views Based Data Access Control Model"* to data access control problem is presented in Section 3.

In Section 4 some tests on Query response time are presented and results are briefly discussed. Finally, Section 5 concludes the paper and sketches our future research work.

1. Data Warehouse Specific Characteristics and Restricted Data Access

If we compare data warehouses to on-line transaction processing systems, we can point out some distinguishing characteristics, which can result in specific requirements for data access control. The following issues for building data access control architecture in a data warehouse were taken into account [5], [6].

- The data warehouse integrates data from different data sources with different data access control mechanisms, if we consider the implementation side, and with different data sensitivity level and with different scope of allowable data for each user on the information side.
- The typical data access rights for users in the data warehouses are read operations.
- The purpose of data usage in a data warehouse can be data analysis, knowledge discovery, and data mining. These purposes can define sets of users different from data source users. In this case the users are managers and data analysts and they are not always the users of data sources. If the reporting purpose is specified too, the users and their data access rights can be similar to data source users.
- There are no limitations on the number and type of client side tools used. The client side applications can vary from reporting and OLAP tools to data mining applications. These tools don't only provide predefined reports and queries, but also the capability to query the database directly. In this case also, the users should have limited access to data. The data from different data sources are consolidated in the same physical data structures in the database. It is possible that the users of one data source operational system can have limited access to other data source data.
- In the on-line transaction systems, which serve as data sources for the data warehouse, the limitations can be implemented on the application side and the server side jointly. For the data warehouse, implementation of the server side limitations leads to a more secure and consistent solution with easier administration.

A question arises however, how a consistent user access rights control system, which would derive access permissions from data sources and minimize the administration can be implemented, if data warehouse specific additional restrictions have to be defined. One possible solution is shown in Figure 1.

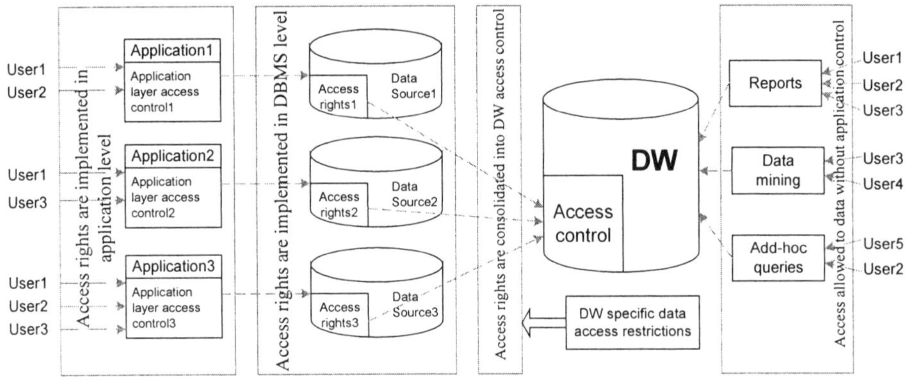

Figure 1. DW data access control architecture with derived access rights

We have to specify some key concepts to describe two alternative models of the data access control architecture implementation.

Two possible ways of controlling data access, further referred to as horizontal restrictions and vertical restrictions:

- The **horizontal restriction** on the database table is a set of rows of the table accessible to the user.
- The **vertical restriction** is a set of columns accessible to the user.

With **data access right** we denote the ability to read data in data warehouse tables according to the horizontal and vertical restriction.

If the user has only access to a subset of all rows of the table, they should believe that the data warehouse table contains only these rows. There are some existing techniques for implementation of horizontal restrictions, e.g. views, dynamic views and virtual private database in database systems [7].

There may be a case, when a user can only access particular columns of the database table, whether the table implements a dimension or a fact table of the star schema. One possible solution for the problem is implementing the restrictions on columns in the user tools, if they have corresponding features. But this solution doesn't satisfy the direct querying needs and in case of many user tools leads to difficult or even impossible data access rights administration. Another solution is the implementation of vertical restrictions with database views or dynamic views.

We will provide a simple example of a data warehouse star schema to illustrate our solutions and show the necessity for these restrictions in Figure 2. The star schema consists of three dimension tables DEPARTMENT, TIME and PERSON and a fact table FACT_TABLE with two fact measures FACT1 and FACT2. For 'User1' the following data access rights are defined – he can see the data in all rows in all tables by departments 'Dept2', 'Dept3' and 'Dept4', he can also access all columns except the columns MARITAL_STATUS and FACT1. Data accessible to 'User1' in Figure 2 are marked dark grey. The light grey colour shows the data accessible only in one direction – vertical or horizontal.

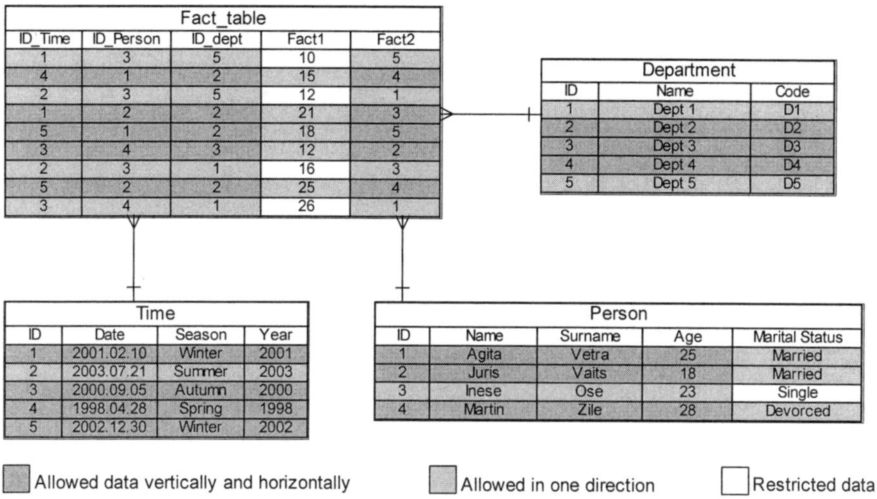

Figure 2. The example star schema

2. Role Based Data Access Control Model Design

Assigning roles is one of the ways that makes implementing data warehouse data access control design possible. The role can be considered as a set of unique data access rights. These rights grant a user access to data in the data warehouse.

If we assume that the data warehouse is a set of all its data, stored in rows and columns of the database table, we can look at the data warehouse as two different data subsets – rows and columns. A role can be defined as a data set ROLE=$\{z|x\in H\&y\in V\&z=x\cap y\}$ where H is used to denote horizontal restriction and V is used for vertical restriction. If roles are defined in this way, all users having access to the same set of rows and columns get the same role. We control user data access rights with roles, because the number of roles is usually smaller than that of data warehouse users.

We will use well-known way for implementing data access restriction. Our model is based on views, where horizontal and vertical restriction is accomplished, and database object privileges, in the way, that makes possible automatic data access control model implementation, administration and maintenance. By using further described approach, it is possible to use derived data access rights from data sources.

Each role needs its own set of views. All data access rights are built in these views. The role's physical implementation can be the traditional database views and materialized views. In the latter case, it is necessary to refresh all role based views after the data warehouse refreshment to keep data access management implementation consistent with the roles' specifications.

Not all tables, especially dimensions, need to hide data from users. For example, no restrictions exist for the time dimension. Other similar dimensions could also be found in data warehouses. Not only horizontal restrictions have exceptions. Some of the columns can be always seen for all users. These are, for example, all primary keys columns. Since primary keys for dimensions are usually automatically generated numbers and a fact table has composite primary key that consists of foreign keys to dimensions, it isn't necessary to hide them from the user. Before defining each restriction, it is important to evaluate the necessity of this restriction. Needless restrictions prolong query runtime and increase the database.

2.1 Implementation with One Restricted Dimension

One simple example of data access rights restriction is the case when the data warehouse star schema is restricted by only one dimension, as it is shown in Figure2.

The **restricted dimension** is a star schemas dimension table with one **restricted column** which values determine the horizontal restriction.

The accessibility of star schemas fact table depends on foreign keys values, which corresponds to restricted dimension primary keys of horizontal restriction. The accessibility of other dimension tables depends on corresponding foreign keys values of the set of allowed records in the fact table.

The dimension that horizontally restricts all data available in the star schema in Figure 2 is the DEPARTMENT dimension. The restricting column in this example is the CODE column. This means that users from one department cannot see the data from other departments. Because companies have quite a complicated hierarchical organizational structure, horizontal restriction cannot be simple either.

Information about roles und users ought to be saved in the data warehouse. One possible solution is shown in Figure 3, which represents two tables: one for users and one for roles. The ROLES table maintains records for all roles and their data access rights on certain rows in the DEPARTMENT dimension and all accessible columns in all dimensions and fact tables.

Column ROWS in ROLES table contains the allowed values of restricted dimension's restricted column. Column COLS contains the allowed tables and columns in the star schema, and it serves as a basis for vertical restriction.

Roles			
ID	Name	Rows	Cols
1	Role1	D%	Time,Department,Person,Fact_table
2	Role2	D2	Time,Department,Fact_table
3	Role3	D2,D3,D4	Time,Department,Fact_table(Fact2), Person(Name,Surname,Age)

Users		
ID	Username	Role
1	User1	2
2	User2	3
3	User3	1
4	User4	3

Figure 3. Access restriction information

The ROLES table in Figure 3 has three different roles: 'Role1' users can see all data without any exceptions. 'Role2' users can see data from 'Dept2', but dimension PERSON is hidden from them. 'Role3' is earlier demonstrated in Figure 2. Here we have to denote that we make no vertical restrictions on primary key columns.

Except for data structures, where information about users and roles is stored, it is necessary to implement the views depending on roles.

General notation for the view that corresponds to the horizontally restricted dimension is represented in the code (1). General notation for unrestricted dimensions is as follows in (2). The third and last type of data warehouse tables that needs views are fact tables. Its views generally look as in (3).

```
Create view <restricted_dimension>_<role> as select <coll>, <col2>,…,<coln> from
<restricted_dimension> where <col_restricted> like <rows_code1> or <col_restricted> like
<rows_code2> or … or <col_restricted> like <rows_coden>
                                                                              (1)

Create view <base_dimension>_<role> as select <coll>, <col2>,…,<coln> from
<base_dimension> where <primary_key> in (select <foreign_key_base_dimension> from
<facts_table>_<role>)                                                         (2)

Create view <facts_table>_<role> as select <coll>, <col2>,…,<coln> from <facts_table>
where <foreign_key_restricted_dimension> in (select <primary_key> from
<restricted_dimension>_<role>)                                                (3)
```

For example, for the situation shown in Figure 2 where data access restriction for 'Role3' is given, four views have to be made in the data warehouse. Create statements according to the previously defined notation are in (4).

```
Create view department_role3 as select id, name, code from department where code like
'd2' or code like 'd3' or code like 'd4'
Create view fact_table_role3 as select id_time, id_person, id_dept, fact2 from
fact_table, where id_dept in (select id from department_role3)
Create view person_role3 as select name, surname, age from person where id in (select
id_person from fact_table_role3)
Create view time_role3 as select id, date, season, year from time where id in (select
id_time from fact_table_role3)
                                                                              (4)
```

To automate these operations certain procedures need to be followed. After the data warehouse refreshment they can accomplish all of the above mentioned statements (1),(2),(3) and a little bit more. Let's look at the basic steps that need to be fulfilled in the following procedure.

```
If role_inserted then insert into TMP_TABLE values (role,'I','');
If role_deleted then insert into TMP_TABLE values (role,'D','');
If role_updated then insert into TMP_TABLE values (role,'U','');
If user_role_add then insert into TMP_TABLe values (User,'I',role);
If user_role_delete then insert into TMP_TABLE values (User,'D',role);
If user_role_change then begin
  Insert into TMP_TABLE values (User,'D',role_old);
  Insert into TMP_TABLE values (User,'I',role_new);
end;
For all_role_changed(TMP_TABLE) do
  If view_op in ('D','U') then drop_view(role);
  If view_op in ('I','U') then
       create_view (role,is_true_rows(role),is_true_cols(role));
End;                                                                          (5)
```

In our solution we propose to make a table TMP_TABLE (<role> or <user>, <operation>, null or <role>) for saving some temporary information. Our pseudo code (5) uses some functions, e.g. function `role_inserted` returns true if a new role is inserted into the ROLES table. Function `user_role_add` returns true if a role is added to user's record.

After the ROLES table refreshment we need to make corrections in our views based data access control system. Some other functions are needed, e.g. function `all_role_changed` gathers all records from TMP_TABLE that describe changes in the ROLES table.

Unfortunately, not all work is done by refreshing views. We have to grant SELECT rights to some users and revoke SELECT rights from other users depending on changes made in the tables ROLES and USERS. It can be accomplished by making the following basic steps:

```
For all_user_role_changed and all_role_changed in tmp_table do
  If User_op='D' then revoke_select_on (role,User);
  If User_op='I' then grant_select_on (role,User);
  If view_op='U' then grant_select_on (View_role,all_view_users(role));
End;                                                                          (6)
```

All changes in view based data access control structure can be made automatically, but the specific features of user side data access tools can sometimes add manual tasks.

2.2 Model with more than One Restricted Dimension

As data warehouse usually contains data from more than one data source, restrictions can be defined on more than one dimension. In previous section all horizontal restrictions were made corresponding to one dimension e.g. DEPARTMENTS dimension. We can also add horizontal restrictions, for example, on time dimension, which will show how old the data that the user is able to see are. For implementing the model with restriction on year, it is possible to modify the ROLES dimension a little. An example is shown in Figure 4 below:

Roles				
ID	Name	Rows_dept	Rows_time	Cols
1	Role1	D%	19%	Time, Department, Person, Fact_tabe
2	Role2	D2	2001,2003	Time, Department, Fact_tabe
3	Role3	D2,D3,D4	2%	Time, Department, Person(Name,Surname,Age), Fact_table(Fact2)

Figure 4. Roles with two restricted dimensions

In such way it is possible to add as many dimensions as necessary. If in situation with one restricted dimension accessible rows were found starting with the restricted dimension, then in this case the restriction process starts from the fact table. So in general the fact table's view structure is:

```
Create view <facts_table>_<role> as select <col1>,…,<coln> from <facts_table> where
<foreign_key_restricted_dimension1> in (select <primary_key> from
<restricted_dimension1> where <code_rd1> like <rows_code1_rd1> or...or <code_rd1> like
<rows_coden_rd1>)
Intersection
select <col1>,…,<coln> from <facts_table> where <foreign_key_restricted_dimension2> in
(select <primary_key> from <restricted_dimension2> where <code_rd2> like
<rows_code1_rd2> or...or code_rd2> like <rows_coden_rd2>)
                                                                                    (7)
```

All dimensions are defined in the same way as previously we defined dimensions without horizontal restriction (see Section 2.1). Pseudo code (5) and (6) describing basic steps for refreshing views remain unchanged as in the previous section.

2.3 Model with Data Access Rights Derived from more than One Data Source

The most complicated solution is necessary when data access rights are derived from more than one data source. We propose that all these data access rights are saved in one table. Our example in Figure 5 is an illustration of this solution.

Data_sources	
ID	Name
1	Data source1
2	Data source2
3	Data source3

Roles	
ID	Name
1	Role1
2	Role2

Users		
ID	Username	Role
1	User1	2
2	User2	1
3	User3	2

Rights				
ID	ID_Role	ID_data_source	Hor_rights	Vert_rights
1	1	1	D5	Person(Surname,Age), Fact_table(Fact2), Time, Department
2	1	2	D2	Person(Name, Surname, Marital_status), Fact_table, Time, Department
3	2	1	D1	Person,Time,Department
4	2	2	D3	Fact_table,Time,Department

Figure 5. Data access restriction from more than one data source

Horizontal restriction remains simple and is almost the same as in the model with more than one restricted dimension. Only this time, instead of intersection we have to use union. Vertical restriction is a more complex issue. Some columns need a function that returns either the cell value, if it is accessible, or an empty cell. Such function is part of dynamically changing views and will be discussed in the Section 3.3.

2.4 What Are the Decision Points?

The Role based data access control design is easy to use, if there are comparatively few roles and each role is associated with many users. If there are many roles, administration of such structure isn't an easy task. If there are changes in the role's rights, corresponding changes have to be made in view definitions. This problem could be solved automatically as described earlier in this paper in the example of the pseudo code (5) and (6), but the complexity of the process can lead to security threats. When the user's role is changed, some manual changes can be done in client side data access tools, for example changing accessible predefined reports.

We have to admit that changing role's rights isn't a good idea as it often means more users will be added. It can lead to an erroneous situation when new rights are suitable for none of the same role users, but one user can see something prohibited and data access control in the data warehouse is breached.

Role based data access control is not an appropriate solution in situations when there are many roles or almost each user has their own role. Consequently, you will build a cumbersome view structure in your data warehouse. In the case when number of users is close to the number of roles, it is better to use usernames as roles. That removes from list the tasks concerning changes in relationship user-role.

Role based data access control is a way to implement a lot of requirements for data warehouse data security by leaving query response time in previous limits, if it is possible to use materialized views for query optimization, but it can take a lot of additional administration work depending on used client side tools as mentioned in Section 2.1.

3. Dynamically Adapting Views Based Data Access Control Model

In the previous section the implementation of the described method has some administration difficulties in the case of many warehouse users and roles. We have developed another method, where there is no need for administration and huge view structures. Each base table has one view and this view contains available data depending on each user's rights.

By **dynamically adapting views** we are denoting views fulfilled with only those data, which are accessible in the base table to a corresponding user.

If the user has no access to any data in the base table, the defined corresponding view is blank. The view can be blank in two situations:

- The user has no access to any rows in the base table;
- The user has no access to any columns in the base table.

For security reasons the blank views have to be hidden from the user. It can be easily made on the database level by removing SELECT rights from a particular user. More problems can arise connected with client side data access tools, when it is possible that blank objects are visible. Nevertheless, there are tools where this problem is solved.

By defining data access control plan for dynamically adapting views, we always have to keep in mind that every restriction affects query performance.

3.1 Dynamically Adapting Views with One Restricted Dimension

To describe this implementation model, we would like to return to the data warehouse example schema in Figure 2 and to added security tables in Figure 3. Only this time, we remove column ROWS from the ROLE table and add it to USERS table. This change has the following goals:

- To separate the horizontal restriction from the vertical restriction. In this approach roles are used only for defining the vertical restriction. We tried to avoid using the vertical restriction where possible, as it impacts on the query performance and makes indexes and materialized views impossible to use.
- In this way the number of roles is much smaller so it is possible to make a materialized view for each role, where the vertical restriction is necessary. It decreases query execution time as no vertical calculation is needed during query execution time.

In the implementation of dynamically adapting views we followed three steps.

Step 1. One of them is creating a function for the horizontal restriction. This function is included in the horizontally restricted dimension view definition in such way that this views returns only user's accessible records. If there is no additional vertical restriction on the horizontally restricted dimension the general form of the corresponding view is (8). In our example in Figure 2 such view definition is only for the DEPARTMENT dimension. It looks as follows in (9).

```
Create view <restricted_dimension>_view as select * from <restricted_dimension> where
F1(<restricted_col_val>,<user>, <rows_col_name>)=1
                                                                                  (8)
Create view DEPARTMENT_VIEW as select * from DEPARTMENT where F1(CODE,user,ROWS)=1;
                                                                                  (9)
```

The main structure of Function F1 is quite simple. The Function has three input parameters, a restricted column value, username and the name of the column in USERS table where accessible rows are stored.

```
Select rows_col_name from USERS where username=user;
If restricted_col_val=substr(rows_col_name) then return 1 else return 0;
                                                                                  (10)
```

Step 2. The next step is creating the vertical restriction. This part is more complicated and consists of many similar functions. The purpose of these functions is either to return columns cell value or a blank cell, if the column isn't accessible. These functions can be divided into two groups:

- *Universal functions* that are used in the dimension views' definitions. As an input parameter they get a username, a column name, a dimension name and records primary key, which in our solution is an automatically generated number in the ID column. Because data types for returned value could be NUMBER, DATE and CHAR, it is impossible to implement them with only one function.
- *Fact table functions*. Each fact table needs its own functions. Their number depends on data types used in the fact table. Almost all fact tables have different primary keys that depend on the related dimension number. For this reason fact table functions as input parameters don't get a table name, but receive the whole composite primary key instead.

The main structure of the vertical restriction on columns can be made in three steps.

```
1.  Select col into cell_value from base_table where primary_key=primary_key_columns;
2.  Select COLS into access_cols from ROLES, USERS where username=user and
    roles.id=users.role;
3.  If col=substr(access_cols) then return cell_value else return null;          (11)
```

If functions for the horizontal and vertical restriction are created, general structure of the fact table's view is as in (12). Dimensions' views are defined as follows in (13).

```
Create view <fatct_table>_view as select <primary_key_columns>,
<func(<user>,<coll>,<primary_key_columns>)>,...,<func(<user>,<coln>,
<primary_key_columns>)> from <fact_table> where <foreign_key_restricted_dimension> in
(select <primary_key> from <restricted_dimension>_view)                           (12

Create view <base_dimension>_view as select <primary_key>,
<func(<Username>,<PK>,<coll>,<base_dimension>)>,...,<func(<Username>,<PK>,
<coln>,<base_dimension>)> from <base_dimension> where <primary_key> in (select
<foreign_key_base_dimension> from <facts_table>_view)                             (13
```

In our example from Figure 2 the views for fact tables and persons are as follows, by assuming that the DEPARTMENT dimension has no vertical restriction but the time dimension is open for all:

```
Create view FACT_TABLE_VIEW as select ID_TIME, ID_PERSON, ID_DEPT,
col_fact_numb(user,'fact1',ID_TIME,ID_PERSON,ID_DEPT) as FACT1,
col_fact_numb(user,'fact2',ID_TIME,ID_PERSON,ID_DEPT) as FACT2, from FACT_TABLE where
ID_DEPT in (select ID from DEPARTMENT_VIEW);

Create view PERSON_VIEW as SELECTID, col_dim_varch(user,ID,'name','person') as NAME,
col_dim_varch(user, ID,'surname','person') as SURNAME,
col_dim_varch(user,ID,'age','person') as AGE,
col_dim_varch(user,ID,'marital_status','person') as MARITAL_STATUS from PERSON where ID i
(select ID_PERSON from FACT_TABLE_VIEW).                                          (14)
```

Step 3. When dynamically adapting views are made, a procedure for grants is also necessary. This procedure controls the existence of blank views for any user. If there are any, the procedure revokes SELECT rights from the corresponding user on the corresponding view. The general steps of this procedure are:

```
For all_users except owner do revoke select on tablename_view from user;
For all_users except owner do
  If has_data(tablename_view,user) then
       grant select on tablename_view to user                                     (15
```

If the data access control model is implemented in this way, there is no need for additional administration. We have to add new users and roles but all further steps are done automatically.

3.2 Dynamically Adapting Views with more than One Restricted Dimension

The idea behind this solution has already been described by and large in Section 2.2. However, some changes will have to be added. Columns with restriction on rows are moved from the ROLES table to the USERS table. The rationale for doing this was described in the previous section. The system of defining views also remains the same. Some changes are made in the fact table view general definition.

```
Create view <facts_table>_view as select <primary_key_columns>,
<func(<username>,<col1>,<primary_key_columns>)>,...,<func(<username>, <coln>,
<primary_key_columns>)>from <facts_table> where <foreign_key_restricted_dimension1> in
(select <primary_key> from <restricted_dimension1> where
F1(<restricted_col_val>,<user>,<rows_col_name1>)=1)
Intersection
Select <primary_key_columns>,
<func(<username>,<col1>,<primary_key_columns>)>,...,<func(<username>,
<coln>,<primary_key_columns>)>from <facts_table> where <foreign_key_restricted_dimension2
in (select <primary_key> from <restricted_dimension2> where
F1(<restricted_col_val>,<user>,<rows_col_name2>)=1)                                  (16)
```

All dimensions are defined in the same way as horizontal restriction free dimensions were defined previously (see Section 3.1).

3.3 Dynamically Adapting Views with more than One Data Source

The system of saving rights form multiple data sources is perfectly illustrated by the model in Figure 5. To allow 'User2' view the information shown in Figure 6 we made some changes in our functions that are responsible for the vertical restriction. First, we define the fact table with a structure as in the previous section. Only this time instead of intersection we use union. Dimensions are also defined as in the previous section. Functions responsible for the vertical restriction return cell value, if this cell is contained in any pair <hor_rights, vert_rights> for this role, if not these functions return blank cell.

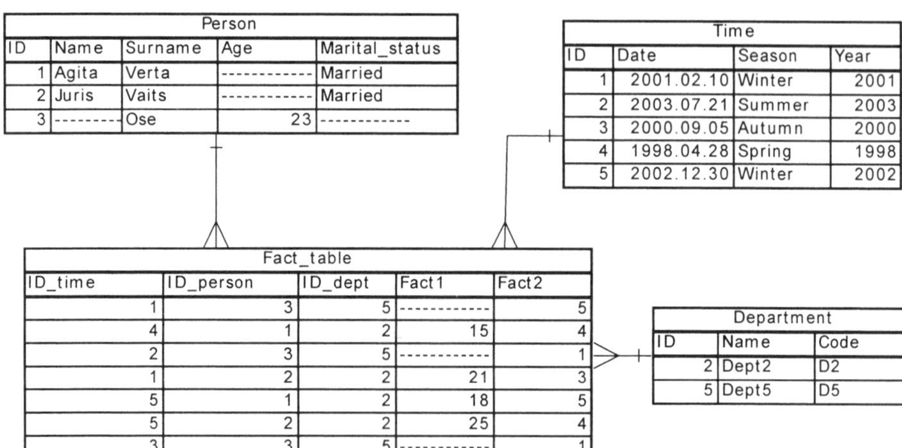

Figure 6. Access restriction from multiple data sources

Basic steps for the function implementing the vertical restriction in this case are as follows:

```
1.      Select col into cell_value from base_table where
primary_key=primary_key_columns;
2.      Corsor C1 is select * from rights, roles, users where rights.id_role=roles.id and
roles.id=users.role and username=user
3.      For each_cursor_record do
If contains(col,primary_key_columns,hor_rights,vert_rights) then return cell_value else
return null;                                                                          (17)
```

4. Model-Performance Testing

After the development of our proposed models, we tested the query response times. We ran them on our data warehouse data mart, which consisted of eight dimensions. The largest of these dimensions are persons/people representing 75,000 records, followed by orders having 60,000 records, education with 11,000 records and addresses with 17,000 records. The "Fact table" consists of 450,000 (four hundred fifty thousand) rows. We took queries from a standard user report, divided the results in two groups, e.g. time with materialized views and time without materialized views. Queries were run on views, described in previous sections. The users were made as users from different departments with different data access rights.

The average results for each query are included in Figure 7.

	Without materialised views			With materialised views		
	Response time			Response time		
	Base tables	Role based DAC	Dynamically adapting views	Base tables	Role based DAC	Dynamically adapting views
Query1	6,11	5,063	19,719	0,141	0,047	0,89
Query2	5,328	4,737	18,937	0,219	0,109	0,328
Query3	5,859	4,718	18,796	0,125	0,125	0,375
Query4	2,093	4,532	68,203	0,203	3,468	11,859
Query5	2,422	3,625	69,719	0,234	2,11	10,125
Query6	1,515	1,5	24,093	0,984	0,141	1,375
Query7	1,281	1,344	22,531	0,859	0,156	0,484
Query8	0,969	1,547	17,641	0,516	0,578	1,141
Query9	2,703	9,39	230,578	----	-----	----
Query10	3,765	8,484	318,079	----	-----	----

Figure 7. Models performance testing

The tests were performed on Oracle 9.0 by using Oracle SET TIMING ON. Each query is presented with time used. The queries used a different structure, which included aggregation functions and some manually written functions such as calculation of age, if birth date is given. Not all the queries we ran were possible to make suitable materialized views. Queries also included additional conditions, sorting, grouping and set operations.

As seen in Figure 7 and is mentioned in this paper, the role based data access control mechanism queries uses considerably less time than dynamically adapting views queries. Future query testing using appropriate indexes and materialized views shall be considered.

5. Conclusions

In this paper we introduced two solutions for implementing data warehouse security policies. The *Role based security model* is easy to understand and can be made in such a way that query response time isn't seriously impacted. From the other side this technique can lead to a huge database and a complicated views structure that needs a lot of administration work. The other solution is *Dynamically adapting views* that have opposite characteristic features from the *Role based security model*. This solution needs almost no administration, but as the views are made dynamically, query response time can increase dramatically, if no additional work is done.

We suggest using the role based security system, if the data warehouse doesn't have a lot of roles and there are resources for administrating all objects that need to be created by using this model. *Dynamically adapting views* are more appropriate if the data warehouse has many users with different access rights and there are no resources for security administration.

A lot of restriction methods are already built in DBMS and data access tools. For this reason it is possible to use the implemented methods. It is a good decision, if the security policy requirements are not complicated, for example, only horizontal restriction on one or more dimensions is required. But if in the security policy vertical restriction is necessary, there will be problems of finding use for the built-in security mechanism.

Our future work will have to deal with making improvements to our methods. The basic guideline is making queries to return data faster. This can be done in several ways:

- by adding rights indicator to each record in each table;
- by using built-in access restriction functions like virtual private database for separating the horizontal restriction from the vertical one;
- by using materialized views.

References

[1] A. Rosenthal, and E. Sciore. How Can Data Sources Specify Their Security Needs to a Data Warehouse? IEEE Workshop on Security in Distributed Data Warehousing, October 2001.

[2] A. Rosenthal, and E. Sciore. View Security as the Basis for Data Warehouse Security. Proceedings of the International Workshop on Design and Management of Data Warehouse DMDW'2000, Sweden, June 2000.

[3] E. Weippl, O. Mangisengi, W. Essmayr, F. Lichtenberger, and W. Winiwarter. An Authorization Model for Data Warehouses and OLAP. Proc. of the Workshop on Security in Distributed Data Warehousing. New Orleans, Louisiana, October 2001.

[4] C. Silbernagel. Data security: Protecting the warehouse from within. DM Direct, June 1999.

[5] W. H. Inmon. Building the Data Warehouse, John Wiley, 1996.

[6] R. Kimball, and M. Ross. The Data Warehouse Toolkit, John Wiley & Sons, 2nd edition, 2002.

[7] K. B. Edwards, and G. Lumpkin. Security and the Data warehouse. Oracle Corporation, http://otn.oracle.com/ow2003/ow_security.html, 2003.

Specifications

Databases and Information Systems
J. Barzdins and A. Caplinskas (Eds.)
IOS Press, 2005

Use of Knowledge Engineering Techniques for Creation and Analysis of Aggregate Specifications

Henrikas PRANEVICIUS, Germanas BUDNIKAS
Business Informatics Department, Kaunas University of Technology
Studentu 56-301, LT-3031 Kaunas, Lithuania
e-mail: hepran@if.ktu.lt

Abstract. The paper presents a technique that applies knowledge engineering techniques for creation of Aggregate specifications. Application knowledge base (KB) is created by using the knowledge acquisition technique joined with a piece-linear aggregate model. The production rules of the application KB are transformed to single hit decision tables, and the static properties of the KB are checked in Prologa system. Further, the application KB is combined with the defined KB of validated properties and validation method, and dynamic properties of application KB are checked in the expert system in CLIPS. A validated application KB is used defining a framework of Aggregate specification using Praxis editor and supplementing Praxis generated framework with the application functional description. The technique is illustrated with an example of a single channel queuing system.

1. Introduction

Formal methods and specification languages are widely used for design of distributed systems. The most popular formal specification languages being used for description of distributed systems are SDL, Lotos, Estelle.

Specification language Estelle/Ag is based on the Aggregate method [1]. There are some differences between Estelle/Ag and Estelle—the piece-linear aggregate model is used in Estelle/Ag. The use of such a model instead of a finite-state automaton, which is the formal background of standard Estelle, enables to create validation and simulation models based on a single specification. This is possible due to the special structure of the piece-linear aggregate. Further in the paper we use notion of Aggregate specification as the one written in Estelle/Ag language.

A construction of specifications in this language is performed in two phases. Specification editor Praxis helps to define the specification framework that describes an interaction of specified system aggregates, their states and conditions of state change. Next, the framework is heuristically supplemented with the knowledge from a conceptual model about behaviour of an analysed system.

There are a lot of works where knowledge-based systems are used for creation of (formal) specifications. A key feature of techniques used in such a works is the analysis of correctness that is performed both during the construction of the initial knowledge description and while a target model or KB has been created.

This paper presents an approach that utilises techniques of knowledge representation as well as knowledge engineering for building up a knowledge base (KB) aimed at transformation to Aggregate specification and validation* of the KB static and dynamic properties. A transformation phase to the specification is described too.

The static properties are characteristics of a KBS that can be evaluated without its execution. Such an evaluation is often referred to as static verification. During static verification, a KB is checked for anomalies. Preece and Shinghal [2] present a classification of the anomalies that may be present in rule-based systems. It is necessary to note the difference between an anomaly and error. The anomaly indicates the existence of a possible error. The dynamic properties are those characteristics of a rule-based system that can be evaluated only by examining how the system operates at a run time.

Knowledge base and Aggregate specification are used for description of the same problem. Many authors (e.g. Bruynooghe *et al.* [3]) believe that the declarative style of description that is used in KBs is more understandable and acceptable than the procedural style (the latter is used in Aggregate specifications). This is because a problem is described at the knowledge level at which the knowledge engineer specifies expertise during knowledge acquisition. Therefore, the knowledge description is presented not using strict mathematic notation but concepts of an application domain that is natural. Due to this reason, we suppose that the creation of Aggregate specifications using KB is more attractive in that sense.

Until now, knowledge-based techniques were not used for the creation of Aggregate specifications except Praxis system. However, this system generates structure of the specification only and does not define the behaviour of an application. Our approach does it by using KB.

Our technique is similar to the one proposed by Fuchs and Schwitter [4] in such a way that they also offer transformation of problem domain description to representation structures and then to an executable language. However, our technique checks general properties during validation and verification while Fuchs and Schwitter [4] technique checks specific invariant properties. Our approach is similar to that of [5] since they both offer the use of declarative languages for a description of the user needs. We use aggregate model concepts for representation of the object structure, while Specht [5] use "object as theory" model. In our approach, in contrast to the compared one, we emphasise validation of the created declarative description. A transformation to target representation is used in both approaches. Our approach is similar to Arentze *et al.* [6] by the use of decision tables (DT) for representation of state-based systems. Our approach as well as many others, example of which is [7], exploit advantages of tabular representation in order to perform verification by transforming certain representation to DTs. However, our approach is specific in the sense that verification is performed on knowledge base that is oriented at creation of Aggregate specifications. In [8], already created Aggregate specifications are validated using first order predicate logic and Prolog. While in our work we use knowledge techniques for analysis of the specifications to be created using knowledge base. Sellini *et al.* [9] and we use intra- and inter- validation for analysis of the acquired knowledge. They also use an intermediate formalised description (application KB in our case) for construction of a knowledge model (specification in our case). While performing static verification of an application KB, we use results of Vanthienen *et al.* [10], whereas when analysing the dynamic properties we use the reachable state method that is similar to the functional validation method suggested by Preece *et al.* [11] in a view of analysis of execution paths.

An applicability of our technique is defined by the applicability of the PLA method and Aggregate specifications. The technique is oriented for creation of Aggregate specifications which primary domain of usage is communication protocols and business systems.

* In the paper, for the sake of briefness, sometimes we refer to validation having in mind both validation and verification because "validation subsumes verification" [11].

Our technique was successfully applied for the creation of Aggregate specifications of the single channel queuing system, alternating bit protocol and a network of queuing systems. The scalability of the proposed technique is limited by software tools that are used in creating the specification.

The paper is structured as follows. The main stages of the proposed scheme of Aggregate specification creation are presented in section 2. Section 3 describes construction of the application KB and analysis of its static and dynamic properties. Section 4 describes procedures for generation of specification structure with Praxis and addition of functional description of the application from the KB to the generated structure. Section 5 presents an example of the proposed technique. Conclusions sum up the proposed approach.

2. The Main Stages of the Proposed Scheme

The developed approach is depicted in Fig.1. The application KB is created using the knowledge acquisition technique that was adapted for the creation of the specific KB, which is referred to as KB_{Ag} too and intended for mapping to Aggregate specification. Knowledge about the problem domain is represented in the KB_{Ag} in the context of the PLA model.

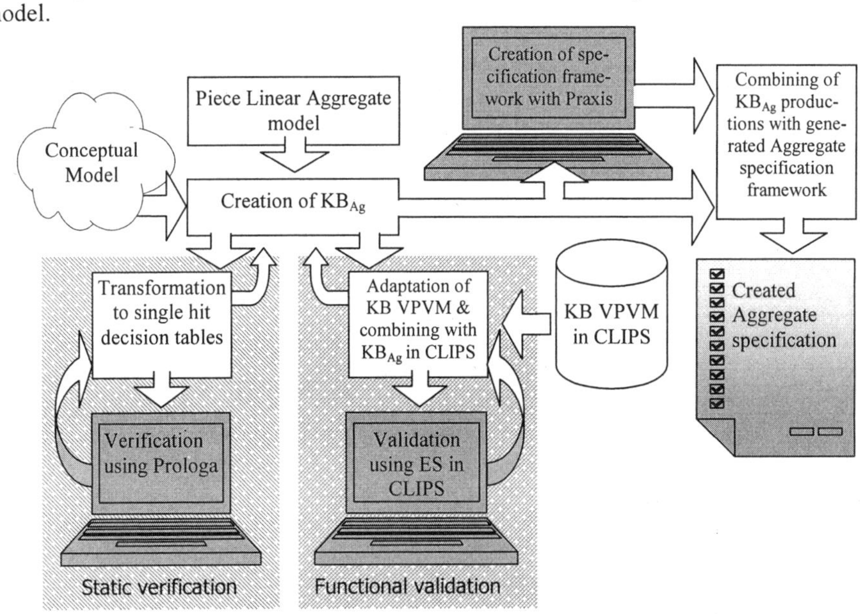

Figure 1. Scheme of the proposed approach for creation of Aggregate specifications

Production rules of the KB_{Ag} are transformed to single hit decision tables in Prologa, and static verification is performed in this system. The Prologa system is an interactive design tool for computer-supported construction and manipulation of decision tables. The system offers design techniques and additional features to enhance the construction and validation of decision tables [12]. The verification in Prologa is implemented using the tabular verification method [10] that belongs to a group of static verification methods. The transformation to Prologa DTs is specific with respect to PLA model and employs some of its concepts—aggregates, internal and external events, input and output signals, discrete state component coordinate, etc. Production rules are transformed to the DTs of certain groups thus enabling to fully exploit advantages of tabular representation to perform static verification. Functional validation is performed using the expert system (ES) in CLIPS. CLIPS (C

Language Integrated Production System) is a tool for productive development and delivery of expert systems [13]. The ES is constructed by combining the KB_{Ag} with the KB of validated properties and validation method (KB VPVM). The functional validation is implemented using the reachable state validation method that can be classified as being a member of the group of formal proof methods. Further, using validated and verified KB_{Ag} one defines Aggregate specification framework during the session with specification editor Praxis. The generated framework is supplemented with knowledge about system behaviour extracted from the KB_{Ag} using defined mappings.

A distinctive feature of our approach is the fact that validation and verification task is performed at the initial stage of Aggregate specification creation. Validated knowledge is used both for creation of the specification framework and for supplementation to the framework. In addition, our approach defines the creation of Aggregate specifications.

3. Creation and Analysis of KB Aimed at Transformation to Aggregate Specifications

KB_{Ag} represents problem domain knowledge using production rules. This representation is chosen due to its similarity to Aggregate specification language constructions describing conditions for state change: `when` *condition* `begin ... end`. Since most of the common verification and validation problems in rule-based systems can be solved using DTs [12] and the tabular verification method is computerised in Prologa system, in our approach static verification will be performed using this method. Thus, in order to perform the static verification of the KB_{Ag}, its production rules have to be transformed to Prologa DTs. Moreover, as stated in [12], a DT is equivalent to a set of production rules, and their transformation to the tables can be performed without too much effort.

Because the KB_{Ag} will be used for the creation of Aggregate specifications, it has to contain knowledge about the PLA model. To acquire this knowledge, we applied the knowledge acquisition technique [14] that was adapted for our needs. The following knowledge about the PLA model is used.

- Succeeding concepts: (a) Aggregates; (b) Input and output signals and their components, coordinates of discrete and continuous state components.
- Relations: (a) Interconnection scheme of aggregates; (b) Changes of state and signal outputs.

In our approach, the predicates and production rules are used for description of the above-mentioned concepts and relations. Applying our approach for specification of an application one has to use the defined predicates and production rules for problem representation. Thus, the KB_{Ag} is created by filling in the knowledge about the application as predicates and production rules of the defined form. Next, we present examples of predicates and productions of KB_{Ag}.

$$(\text{ArrivalOfInputSignal } an_i \; iip_i^j \; x_i^j \; x_j^1 \ldots x_j^{c_{ic}^i}),$$

where an_i – symbolic name of the ith aggregate; iip_i^j – interaction point, $x_j^1, \ldots, x_j^{c_{ic}^j}$ – components of signal x_i^j.

$$(\text{State } an_i \; w_i^1 \ldots w_i^{c_{cc}^i} \; d_i^1 \ldots d_i^{c_{dc}^i}),$$

where $w_i^1,\ldots, w_i^{c_{cc}^i}$ and $d_i^1,\ldots, d_i^{c_{dc}^i}$ are the coordinates of continuous and discrete state components.

The production rule that describes state change and signal output after occurrence of an external event has the following general form:

IF (ArrivalOfInputSignal $an_i\ iip_i^j\ x_i^j\ x_j^1 \ldots x_j^{c_{ic}^i}$)

AND (State $an_i\ w_i^1 \ldots w_i^{c_{cc}^i}\ d_i^1 \ldots d_i^{c_{dc}^i}$)

AND (Aux $an_i\ w_i^1 \ldots w_i^{c_{cc}^i}\ d_i^1 \ldots d_i^{c_{dc}^i}$)

THEN (State $an_i\ w_i^{1*} \ldots w_i^{c_{cc}^i *}\ d_i^{1*} \ldots d_i^{c_{dc}^i *}$)

AND (OutputSignal $an_i\ oip_i^j\ x_i^j\ x_j^1 \ldots x_j^{c_{oc}^j}$))

where (Aux $an_i\ w_i^1 \ldots w_i^{c_{cc}^i}\ d_i^1 \ldots d_i^{c_{dc}^i}$) describes auxiliary conditions $LC_1, \ldots, LC_{p_m}$ that check values of coordinates of state components:

(Aux $an_i\ w_i^1 \ldots w_i^{c_{cc}^i}\ d_i^1 \ldots d_i^{c_{dc}^i}$) $\equiv LC_1$ AND|OR…AND|OR LC_{p_m}, where $i = \overline{1,n}$, p_m – the number of additional logical conditions for the m-th production rule that describes state alteration.

3.1 Static Verification of the KB_{Ag}

Most of the validation problems in rule-based systems like redundant, ambivalent, categorised, cyclic or missing rules, redundant conditions, unused action parts may be resolved using DTs [12].

Our technique uses DT representation for static verification of KB_{Ag} since DTs clearly demonstrate incompleteness and in-consistency of knowledge [7] and software tool Prologa that assists verification process. Anomalies in DTs have direct correspondence to anomalies in production rules, which make the KB_{Ag}.

Static Anomalies in Rule Bases and Decision Tables

The decision table DT is defined [10]:

$$DT: CS_1 \times CS_2 \times \ldots \times CS_m \rightarrow AV_1 \times AV_2 \times \ldots \times AV_n,$$

where CS_i is a set of condition states, AV_j is a set of action values.

The most important criterion when distinguishing tables is the question whether all columns are mutually exclusive (*single hit* versus *multiply hit*). In a single hit table, in contrast to multiply hit table, each possible combination of condition can be found in exactly one and only one column.

The following anomalies are distinguished: *intra*-tabular, which occur in a single DT, and *inter*-tabular anomalies, that originate from interactions between several DTs [15]. A relation between *intra*-tabular anomalies and anomalies in rule bases is depicted in Fig.2. A classification of inter-tabular anomalies can be found in [16].

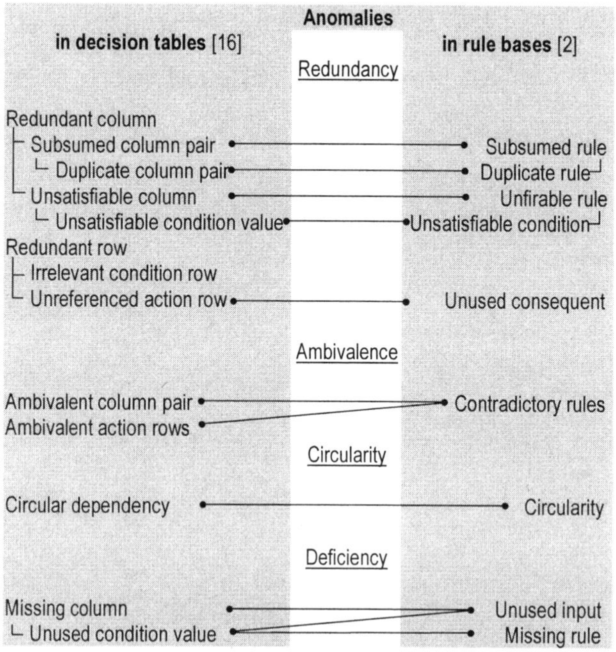

Figure 2. Relation between intra-tabular anomalies in DTs and anomalies in rule bases

Transformation of KB_{Ag} Productions to Single Hit Decision Tables

In our technique, static verification of KB_{Ag} is performed in Prologa system. The system uses single hit DT representation. Next, we present an outline of suggested steps of transformation of KB_{Ag} productions to single hit DTs.

1. Productions that describe certain types of events are transformed to corresponding event tables;
2. Predicates of antecedent (consequent) part of a rule are written in a form of condition (action) subjects in a DT;
3. Decrease of a coordinate of discrete state component by a constant value is marked with not $(d_i^j = d_i^j + 1)$. This notation is used while ambivalence property is being checked.

Technique for Detection of Intra-, Inter- Tabular Anomalies in KB_{Ag}

This technique is mostly based on works of [10], [12] and operating in Prologa system. Below we present a portion of this technique.

A *subsumed column pair* (a case of intra-tabular **redundancy**), which definition is as presented below [15]: a DT contains a *subsumed column pair*, if and only if it includes a pair of columns $(CS_j; AS_j)$, $(CS_k; AS_k)$ $(1 \le j,\ k \le t)$, $j \ne k$ for which: $CS_j \subseteq CS_k$ and $AS_j \supseteq AS_k$,
in a single hit DT is represented by single column that corresponds to several KB_{Ag} productions. Thus, the anomaly of the subsumed column pair is detected if several KB_{Ag} productions correspond to the same DT column.

Full description of the technique for anomaly checking in KB_{Ag} as well as transformation steps of KB_{Ag} productions to single hit DTs can be found in [17].

3.2 Functional Validation of the KB_{Ag}

Functional validation of the application KB is carried out using the expert system in CLIPS. It is built by combining already verified application KB with a KB of validated properties and validation method.

KB of Validated Properties and Validation Method and its Join with KB_{Ag}

In our approach, we define the KB of validated properties and validation method that is implemented in CLIPS. It may be used for various kinds of applications with minor adaptations. The instances of dynamic properties whose validation is implemented in the KB VPVM are the absence of static deadlocks, final state reachability, boundedness, and completeness. The reachable state validation method is realised in the KB VPVM. In a view of analysis of the execution paths, this method is similar to functional validation method suggested by Preece *et al.* [11] that analyses the sequences of rules that must fire to achieve a goal.

When performing the join of KB_{Ag} with KB VPVM, the needed adaptation changes are minor—they are an adaptation of description of validated properties for a specific application. For instance, in order to check the absence of the static deadlock property, the adaptation includes definition of specific continuous state coordinates; in order to check boundedness property, individual bounds on discrete state component coordinates have to be defined. Having combined the KB VPVM with the KB_{Ag}, the ES in CLIPS is built.

Validation of the KB_{Ag} Using the Expert System in CLIPS

The expert system in CLIPS consists of the combined KB VPVM and the KB_{Ag}, and CLIPS inference engine that is based on the forward chaining strategy [18]. In order to perform the functional validation using the ES, the initial and the final states of an analysed application are defined. An application model operates according to the reachable states method. If the validated properties are violated, the expert system generates a corresponding report and a designer corrects the KB accordingly.

4. Forming the Specification

In the first step, specification framework is constructed during a session with the specification editor Praxis [1] using the knowledge extracted from the validated and verified application KB.

In the second step, the generated specification framework is extended by adding with the behavioural description taken from the application KB. The supplementation process consists of the following phases:

- Finding the rules in the KB that correspond to Aggregate specification constructions. These constructions describe either initial state of an aggregate, or condition of state change due to an event. Mapping between the rules in the KB_{Ag} and Aggregate specification constructions is presented in Table 1;
- Transformation of antecedents and consequents of the found rules to Aggregate specification operators. The corresponding mapping has been defined too. A part of it is given in Table 2;
- Supplementation of specification constructions with the operators resulted from the previous phase.

Table 1. Mapping between productions in KB_{Ag} and Aggregate specification constructions

Predicates in antecedent part of KB_{Ag} productions	*Specification constructions*
(Initialize)	`Initialize` `begin … end;`
(EndOfOperation an_i *operation*)	`when eop.`*operation* `begin … end;`
(ArrivalOfInputSignal an_i iip_i^j x_i^j $x_j^1 \ldots x_j^{c_{ic}^i}$)	`when` $iip_i^j . x_i^j$ `begin … end;`

Table 2. Fragment of mapping between predicates in consequent part of productions in KB_{Ag} and Aggregate specification operators

Predicates in production rule consequent	*Specification operators*
(OutputSignal an_i oip_i^j x_i^j $x_j^1 \ldots x_j^{c_{oc}^j}$)	`Output` oip_i^j $(x_i^j, x_j^1, \ldots, x_j^{c_{oc}^j})$;
w_i^j = on	`Start` w_i^j;
w_i^j = off	`Cancel` w_i^j;

5. Example

We illustrate creation of inconsistent and redundant fragment of a KB_{Ag} of the single channel queuing system (QS) with priorities. Conceptual description for the described fragment states that two types of requests may arrive to a queuing system: with high or low priority. The system has queue counters for the corresponding requests. The arrived request to the empty system is started to serve. If service operation is active in the system, the arrived request is placed to the corresponding queue. Inconsistencies and redundancies are detected during the static verification. A portion of a formed Aggregate specification from queuing system KB_{Ag} is presented too.

The Fragment of the KB_{Ag} of QS

r1 "system is idle, low piority request has arrived"
IF (ArrivalOfInputSignal ip2 ReqLP)
  AND (State *service serv_req_pr hpqc lpqc*)
  AND *service* = off
  AND *hpqc* = 0 AND *lpqc* = 0
THEN *service* = on
  AND *serv_req_pr* = 0
  AND (State *service serv_req_pr hpqc lpqc*)

where *service* stands for service operation, *serv_req_pr* - a priority of served request, *hpqc* and *lpqc* - high and low priority queue counters, respectively.

r2 "request is being served, low priority request has arrived"
IF (ArrivalOfInputSignal ip2 ReqLP)
  AND (State *service serv_req_pr hpqc lpqc*)
  AND *service* = on
THEN *lpqc* =*lpqc* + 1
  AND (State *service serv_req_pr hpqc lpqc*)

r2a "low priority request is being served, and low priority request has arrived"

The rule is identical to r2, except an antecedent part that is supplemented with the condition *serv_req_pr* = 0.

r2b "high priority request is being served, and high priority request has arrived"

The rule is identical to r2, except an antecedent part that is supplemented with *serv_req_pr* = 1 condition.

r2c "high priority request is being served, and low priority request has arrived"

The rule is identical to r2b, except a consequent part where *lpqc* = *lpqc* + 1 is changed with *lpqc* = *lpqc* – 1.

Static Verification of KB_{Ag} of QS

While illustrating a detection of intra-tabular anomalies, only a group of some productions were transformed to a single table. This was done in order to show a fewer number of anomalies in a DT at the same time.

Redundancy - *subsumed column pair.* Productions r1, r2, r2a and r2b are used for illustration of this kind of anomaly. r12ab single hit DT is presented in Fig.3.

In this table, production r1 is described by the 2nd column, productions r2, r2a and r2b are described by the 1st column. Regardless of the possible states (0 or 1) of the 6th condition subject, a set of corresponding actions is the same. Due to this reason the condition subject state is marked with "–" (irrelevant) symbol.

Productions r2, r2a and r2b are represented by the same 1st column and the production r2 has the least number of condition subjects. This means, the production r2 subsumes r2a and r2b, and thus these two rules are redundant.

Irrelevant condition row anomaly is illustrated by the r12ab DT that is presented in Fig.3. The 6th condition subject of this DT is irrelevant and redundant. It means the condition of r2a and r2b productions that checks *serv_req_pr* value is redundant.

r12ab						
1. ^(State *service serv_req_pr hpqc lpqc*)	True					False
2. ^(ArrivalOfInputSignal ip2 ReqLP)	True				False	-
3. *service*	on	off			-	-
4. *hpqc*	-	=0		>0	-	-
5. *lpqc*	-	=0	>0	-	-	-
6. *serv_req_pr*	-	-	-	-	-	-
1. *service* = on	.	x	.	.	.	.
2. *serv_req_pr* = 0	.	x	.	.	.	.
3. *lpqc* = *lpqc* + 1	x	.	.	.	.	.
4. (State *service serv_req_pr hpqc lpqc*)	x	x	.	.	.	.
	1	2	3	4	5	6

Figure 3. r12ab single hit DT

Ambivalence - *ambivalent column pair* anomaly. Productions r1, r2b and r2c are used for anomaly illustration. r12bc single hit DT is presented in Fig.4. The 2nd column of this table corresponds to productions r2b and r2c. A cross of the 3rd row of an action subject and the 2nd column of the table is marked by "?" symbol. It means that these two productions define ambivalent actions, namely *lpqc* = *lpqc* + 1 and not (*lpqc* = *lpqc* + 1), while their conditions are overlapping. This anomaly indicates an occurrence of a typing error when defining

r12bc							
1. ^(State *service serv_req_pr hpqc lpqc*)	True						False
2. ^(ArrivalOfInputSignal ip2 ReqLP)	True					False	-
3. *service*	on		off			-	-
4. *hpqc*	-		=0		>0	-	-
5. *lpqc*	-		=0	>0	-	-	-
6. *serv_req_pr*	0	1	-	-	-	-	-
1. *service* = on	.	.	X	.	.	.	.
2. *serv_req_pr* = 0	.	.	X	.	.	.	.
3. *lpqc* = *lpqc* + 1	.	?	.	.	.	.	.
4. (State *service serv_req_pr hpqc lpqc*)	.	X	X	.	.	.	.
	1	2	3	4	5	6	7

Figure 4. r12bc single hit DT

the action — instead of "+" sign "–" sign was specified. After fixing this error, the duplicate column pair anomaly is indicated — productions r2b and r2c, which correspond to the 2nd column of DT in Fig.4, are identical.

A Portion of a Formed Aggregate Specification

A fragment of the created Aggregate specification of the QS example using our technique is presented below. The description of the QS behaviour in case of arrival of a low priority request is included in the productions r1 and r2. These rules have been used to supplement `when ip2.ReqLP` construction of Aggregate specification framework. The Praxis generated code is written in `typewriter` font, the added code—in *italic*.

```
Trans
 when ip2.ReqLP
  begin
    if ((service = off) and (hpqc = 0) and (lpqc = 0)) then
      begin
        serv_req_prior = 0;
        Start service;
    end;
    if (service = on) then
      begin
        lpqc = lpqc + 1;
      end;
  end;
```

6. Conclusions

Summarising the paper we would like to emphasise the following:

- This paper has showed a possibility to perform static verification of Aggregate specification that is expressed in a knowledge base;
- The use of KB defines the creation of Aggregate specifications. The production rules representation technique combined with PLA model is used for description of application structure and behaviour. The description at the knowledge level permits to operate not using strict mathematic notation but the concepts of the application domain;

- KB validation and verification permits to perform analysis of the specification under creation at its initial construction stage.

References

[1] Pranevicius, H., V. Pilkauskas, A. Chmieliauskas. *Aggregate Approach for Specification and Analysis of Computer Network Protocols.* Technologija, Kaunas. 1994: 254 p.

[2] Preece, A., R. Shinghal. Foundation and Application of Knowledge Base Verification. *International Journal of Intelligent Systems*, **9**, 1994: 683-702.

[3] Bruynooghe, M., N. Pelov, M. Denecker. Towards a more declarative language for solving finite domain problems. In *Proc. of the ERCIM/ COMPULOG Workshop on Constraints.* 1999: 1-14.

[4] Fuchs, N.E., R. Schwitter. Attempto Controlled English. In *Proc. of the CLAW 96, The First International Workshop on Controlled Language Applications*, Katholieke Universitet Leuven, 1996.

[5] Specht, G. O!-LOLA - Extending the Deductive Database System LOLA by Object-Oriented Logic Programming. *Informatica*, Vol.9, No.1, 1998: 107-118.

[6] Arentze, T.A., A.W.J. Borgers, H.J.F. Timmermans. The integration of expert knowledge in decision support systems for facility location planning, *Computers Environment and Urban Systems*, 19(4), 1995: 227-247.

[7] Degelder J; M. Steenhuis. A knowledge-based system approach for code-checking of steel structures according to Eurocode 3. *Computers-and-Structures.* 67(5), 1998: 347-355.

[8] Pranevicius, H., R. Miseviciene, V.Miliute. Use of Aggregate Specification and Logic Programming for Knowledge Base Validation and Verification. In *Proc. of International Conference "Modelling and Simulation of Business Systems"*. Technologija, 2003: 203-208.

[9] Sellini, F., C. Vargas, P.-A. Yvars. Considerations about validation of knowledge models in KBE systems. In *Proc. 4th European Symposium on the Validation and Verification of Knowledge Based Systems.* Katholieke Universiteit Leuven, 1997: 83-93.

[10] Vanthienen, J., C. Mues, G. Wets. Inter-tabular Verification in an Interactive Environment. In *Proc. 4th European Symposium on the Validation and Verification of Knowledge Based Systems.* Katholieke Universiteit Leuven, 1997: 155-165.

[11] Preece, A., C. Grossner, T. Radhakrishnan. Validating Dynamic Properties of Rule-Based Systems. *International Journal of Human-Computer Studies*, **44**, 1996: 145-169.

[12] Vanthienen, J. *Prologa v.5 User's manual*, Katholieke Universiteit Leuven, 2000: 121 pp.

[13] Giarratano, J.C. *CLIPS Reference Manual. Basic Programming Guide, CLIPS Version 6.0.* Software Technology Branch, Lyndon B. Johnson Space Center, JSC-25012, 1993: 388 pp.

[14] Russel, S., P. Norvig. *Artificial Intelligence—a Modern Approach.* Prentice Hall, 1995.

[15] Mues C. *On the Use of Decision Tables and Diagrams in Knowledge Modeling and Verification*, PhD dissertation, Katholieke Universitet Leuven, 2002: 223 p.

[16] Vanthienen, J., C. Mues, A. Aerts. An illustration of verification and validation in the modelling phase of KBS development. *Data & Knowledge Engineering* 27, 1998: 337-352.

[17] Pranevicius, H., G. Budnikas. Static Verification of Aggregate Specifications. *Information technology and control*, No.4(29), 2003, 39-44.

[18] Pranevicius, H., G. Budnikas. Creation of Estelle/Ag Specifications Using Knowledge Bases. *Informatica*, Vol.14, No.1, 2003: 63-74.

Databases and Information Systems
J. Barzdins and A. Caplinskas (Eds.)
IOS Press, 2005

An Approach to Evaluate Quality in Use of IS Specification Language

Albertas CAPLINSKAS, Jelena GASPEROVIC
Institute of Mathematics and Informatics, IMCS Akademijos 4, 2600 Vilnius, Lithuania
alcapl@ktl.mii.lt, j.gasperovic@algoritmusistemos.lt

Abstract. This paper proposes an approach to evaluate quality in use of IS specification language. It surveys some related works and discusses in detail the components of the proposed quality evaluation framework: quality model, evaluation procedure, and taxonomy of quality characteristics. The main contributions of the paper are evaluation procedure and taxonomy of quality characteristics.

Introduction

The quality of IS significantly depends on specification language, more exactly on the quality in use of specification language. Quality in use is a project related concept. It is defined as the capability of the specification language to enable developers to produce IS specification with effectiveness, productivity, safety and satisfaction in specified contexts of use or, in other words, in a particular project. Interdependencies between the quality of specification language and the quality of specification have been investigated by a number of researchers [8], [9], [10], [11], [12], [15], [16], [21], [23], [24], [25]. However, well-grounded criteria for evaluation of the quality of a specification language are still absent. As a rule, developers estimate specification language subjectively and express different opinions about its advantages and shortcomings. So, the selection of the most appropriate language for the particular project still is a problem.

In this paper specification language is understood as a language of a high abstract level with syntax and semantics appropriate enough to define both the functionality of a system under consideration and its non-functional properties. In general case, the system under consideration can be some real-world business system, information system, software system, hardware system or any other system. We consider only languages that are intended to specify systems, which are relevant to the field of information systems engineering, namely, real-world systems, information systems and software systems. The specifications may be external (requirements specification) or internal (design specification). A specification language that supports the specification process throughout many phases of a life-cycle model is called a wide-spectrum language and a language that supports only one or two phases of a life-cycle model is known as a narrow-spectrum language. We use the term "specification language" in the broad sense that covers both wide-spectrum and narrow-spectrum languages and suppose that wide-spectrum languages should support business modelling as well as requirement specification and design.

The main objective of this paper is to propose a framework to evaluate IS specification language taking into account high-level quality requirements of a particular project. The paper demonstrates reasonability to describe interdependencies between requirements using special graphs proposed in [3]. It discusses the main ideas beyond the proposed framework and describes in detail its components: quality model, evaluation procedure, and taxonomy of quality characteristics. The paper elaborates ideas proposed in [3]. Its main contributions are quality evaluation procedure and the taxonomy of quality characteristics.

The rest of the paper is organised as follows. Section 1 surveys related work. Section 2 describes the main features of a quality model. Section 3 contains description of quality evaluation procedure. Sections 4, 5 and 6 describe the main components of quality model: quality goals, quality assessment function, and quality characteristics tree. Finally, Section 7 concludes the paper.

1. Related Work

First attempts to evaluate the quality of specification languages, mostly diagrammatical languages, have been made already in early days of computing era [18]. From nowadays perspective these attempts look rather naïve, however, historically they were very important, because initiated entirely new research field. First serious research in this field has been done by Wand and Weber [23], [24], [25]. Roughly speaking, the main idea of this research was to evaluate IS specification languages on the basis of collection of models, known as BWW models, based on an ontology defined by the philosopher Mario Bunge. Quality of specification language is evaluated comparing constructs of this language to some collection of "standard" constructs. This methodology allows evaluating ontological completeness and ontological clarity of a language where ontological completeness is understood as the ability of this language to represent all phenomena of interest in the domain of discourse and ontological clarity is understood as the correspondence between the constructs of a language and the constructs defined by the BWW models. The methodology has been improved by Opdahl [16], who proposed how to evaluate what semantic categories language is able to express and how convenient is it to do.

To the same research direction belong the works of Milton, Kazmierczak and Keen [13] and of Mylopoulos [15]. The former proposed instead of Bunge ontology to use Chisholm ontology and argued that the quality of a language should be evaluated not from the viewpoint of its constructs, but from the point of its ability to specify a variety of situations. The latter argued to evaluate quality in terms of three orthogonal dimensions – ontologies, abstraction mechanisms, and tools – and proposed some qualitative metrics and quality evaluation criteria.

Two other research directions are represented by works of Jackson [14], and Sølvberg and his co-authors [8], [9], [10], [11], [12], [20], [21]. The main idea of Jackson is to use attestation techniques or, in other words, he suggests using a library of characteristic situations that should be specified to evaluate specification language quality. Sølvberg and co-authors proposed to use evaluation framework that addresses the quality of specification as well as the quality of specifying process. This framework includes a language quality model represented in a form of quality characteristics tree. The framework provides a systematic structure for evaluating specification languages. Most important ideas beyond this methodology are separation of conceptual and representation issues, quality evaluation from domain, audience and technology points of view, and the use of set-theoretical approach to explain the meaning of quality characteristics. It is very advanced approach, but it focuses rather on the quality of specification itself than on the quality of specification languages and investigates a language quality model only occasionally. The proposed language quality model is only sketched. It is not exhaustive and not homogeneous.

There are also a number of works that do not address the quality of specification languages directly, but has a strong impact on the research in this field. First of all, ISO/IEC 9126 standard [6] should be mentioned among them. Although this standard addresses the quality of software its conceptual basis is significantly wider and can be applied to many other fields. For example, the ISO/IEC 9126 standard has been taken as a baseline for QStudio®[1] [19], which specifies quality concepts for Java®[2] language. This approach defines so-called *Quality Attribute Tree*, which indeed is ISO/IEC 9126 quality model extended by additional sub-characteristics. The *Quality Attribute Tree* is not a proper tree, because many of quality sub-characteristics refine at once several characteristics. The "many to many" relationship introduces no difficulties in using the "tree" for quality assessment, because quality is assessed top down by evaluating the characteristics first and then the sub-characteristics and metrics. So, using stepwise refinement techniques, the notion of code quality is expressed in terms of quality sub-characteristics and mapped further onto programming constructs. In this way quality metrics are attached to sub-characteristics. The advantage of such approach is the possibility to assess by measurement the quality on multiple levels of detail. Another example is EAGLES/ISO methodology [7] that aims to evaluate the quality of natural language processing systems. It supposes that evaluation expresses what some object is worth to somebody. Quality model is constructed according to two different perspectives. The first perspective (object-based perspective) is who likes it. The second perspective (user-based perspective) is what they like. EAGLES augments the ISO/IEC 9126 approach in the sense that it deals with the formulation of stated or implied needs, which are the primary input to the quality requirements definition. The project aims at producing an evaluation package, from which different elements can be taken and combined in different ways to reflect the needs of any particular user. In EAGLES quality requirements definition is based on the union of the implied needs of classes of users; appropriate metrics are selected and measurements are carried out, but any user is left to construct his own rating level definition and assessment criteria definition [7]. ISO/IEC 9126 also has been used as a basis to develop quality characteristics trees for software components [22] and for ERP systems [2].

One more important approach, called the fuzzy model for software quality evaluation (FMSQE), has been proposed by Belchior and developed further by his colleagues [1]. They proposed a hierarchical quality model based on four main concepts: goals, factors, criteria and evaluation processes. Goals represent the general properties that a product should possess. Goals are decomposed in factors; factors can be further decomposed in sub-factors. Factors and sub-factors define different users' perspectives about the quality. Factors (sub-factors) should be decomposed in measurable quality characteristics called criteria. For each criterion one or more alternative evaluation processes, describing a measurement methodology, should be established. In order to obtain the values of factors, both numerical and qualitative measurement results must be aggregated. Measures and aggregate measures are related by quantitative relations. Obtained measures are interpreted using a set of fuzzy functions. Fuzzy functions support the aggregation of measures expressed in different units and are used as a suitable interpretation mechanism able to deal, at the same time, with qualitative measures and numerical data. The proposed approach provides: a membership function mapping the desired quality criterion; a method to calculate the membership function for the aggregate quality; and a final membership function for the whole product.

2. Quality Model

There exists no comprehensive definition of quality. Quality ever depends on context. Quality of IS specification language is also relative. It depends on the requirements of a particular project. Following the ISO 8402 [5] definition of quality, the quality of an IS

1 QStudio is a registered trademark of QA Systems BV, The Netherlands.

2 Java is a registered trademark of Sun Microsystems.

specification language can be defined as the totality of features and characteristics of this language that bear on its ability to satisfy stated or implied project's needs. The language that is excellent for one project may be only acceptable or even unacceptable for some other projects, because each project has its specific priorities. Usually the definition of the term "quality" is formalised by a quality model. Quality models are used in many areas, however, often these models are defined imprecisely, only in the form of a quality characteristics tree and, may be, associated metrics. Although some approaches (e.g., ISO/IEC 9126 [6]) address quality in use, quality requirements usually are not seen as a part of aquality model. They should be defined separately in terms of a quality characteristics tree. It is embarrassing for users, because such requirements are low-level requirements even in the case when they are formulated using abstraction levels provided by quality characteristics tree. In [3], we proposed the main idea how to define context-oriented quality model that includes quality requirements and allows formulating those requirements in the form of high-level quality goals. In this paper we elaborate this idea further. We define context-oriented quality model as the following nine-tuple:

$Q = < \Gamma, \Psi, \Theta, \Delta, \Lambda, F, M, \mu, \xi >$ (1),

where

$\Gamma = <V, A, G, f, l>$ is a weighted AND/OR digraph describing quality goal interdependencies, termed as "goal graph":

- $V = \{v_i | 1 \le i \le N\}$ is a nonempty set of graph vertices (quality goals), $V_1 \subset V$ is a set of vertices that have no incoming arrows (terminal goals);
- $A \subset VxV$ is a nonempty set of ordered pairs of vertices (graph's arrows), $A_1 \subset V_1 x\Psi$ is a nonempty set of ordered pairs that relates terminal goals to the characteristics of internal quality;
- $G = \{g_k | 1 \le k \le N_1\}$ is a nonempty set of weights;
- $f: A \cup A_1 \Rightarrow G$ is a weighting function that maps graph's arrows to weights;
- $l: V \Rightarrow \{AND, OR\}$ is a labelling function that marks vertices with the labels "AND" or "OR".

Ψ is a quality characteristics tree[3] with the set of leaf nodes Ψ_1;

$\Delta = \{\delta_i | 0 \le i \le N_2\}$ is a nonempty set of rating levels;

$\Lambda = \{\lambda_i | 0 \le i \le N_2\}$ is a nonempty set of linguistic terms, used to interpret rating levels;

Θ is a quality assessment function used to evaluate quality goals and defined in the following way:

$$\Theta(v_i) = \begin{cases} \sum_{j=1}^{N_3} \delta_j * g_j, & \text{if } v_i \text{ has the label "AND"} \\ \max_{1 \le j \le N_3} (\delta_j * g_j), & \text{if } v_i \text{ has the label "OR"} \end{cases} \quad (2),$$

where

- $v_i \in V$ is a vertex;
- δ_j is an evaluation value that signs vertices adjacent to the vertex v_i;
- $g_j \in G$ is a weight that signs incoming arrow $a_j \in A \cup A_1$ of the vertex v_i;
- N_3 is the number of incoming arrows for the vertex v_i (goals of the lowest levels are evaluated on the basis of rating levels of appropriate quality characteristics).

$F = \{F_\lambda | \lambda \in \Lambda\}$ is a nonempty set of interpretation functions used to interpret obtained measure of the top-level quality goal v_0. Functions $F_\lambda : \{\Theta(v) | v \in V\} \Rightarrow [0,1]$ are fuzzy functions that are defined following [1]:

3 Following tradition, we use the term "quality characteristics tree", although, it is a hierarchical classification of quality characteristics, represented by tree.

$$F_{\lambda_i}(\Theta(v)) = \begin{cases} \dfrac{\Theta(v) - \delta_{i-1}}{\delta_i - \delta_{i-1}}, \text{ when } \delta_{i-1} \le \Theta(v) \le \delta_i \\ \dfrac{\delta_{i+1} - \Theta(v)}{\delta_{i+1} - \delta_i}, \text{ when } \delta_i < \Theta(v) \le \delta_{i+1} \\ \text{undefined, when } \Theta(v) < \delta_{i-1} \text{ or } \delta_{i+1} < \Theta(v) \end{cases} \quad (3),$$

where

$\{\delta_{i-1}, \delta_i, \delta_{i+1}\} \in \Delta$ are rating levels;

v is a vertex (the function is calculated for the vertex v_0).

$M = \{m_j \mid 1 \le j \le N_4\}$ - a nonempty set of quality metrics;

μ: $\Psi_1 \Rightarrow M$ is one-to-many mapping that relates quality metrics to quality characteristics;

$\xi = \{\xi_\lambda \mid \lambda \in \Lambda\}$ is a nonempty set of rating functions for the set of measured values (scores), where functions ξ_λ: $\Psi_2 \Rightarrow [0,1]$ are defined in analogical way as in (3):

$$\xi_{\lambda_i}(m_j(x)) = \begin{cases} \dfrac{N_2 m_j(x) - \delta_{i-1}}{\delta_i - \delta_{i-1}}, \text{ when } \dfrac{\delta_{i-1}}{N_2} \le m_j(x) \le \dfrac{\delta_i}{N_2} \\ \dfrac{\delta_{i+1} - N_2 m_j(x)}{\delta_{i+1} - \delta_i}, \text{ when } \dfrac{\delta_i}{N_2} < m_j(x) \le \dfrac{\delta_{i+1}}{N_2} \\ \text{undefined, when } m_j(x) < \delta_{i-1} \text{ or } \delta_{i+1} < m_j(x) \end{cases} \quad (4),$$

where

$\{\delta_{i-1}, \delta_i, \delta_{i+1}\} \in \Delta$ are rating levels;

N_2 is the number of rating levels;

$m_j(x)$ – the value of quality characteristic $x \in \Psi_1$ measured using metric m_j;

$\Psi_2 = \{m(x) \mid m \in M, x \in \Psi_1, \mu(x) = m\}$.

A quality model for the particular project is designed in seven steps (Fig. 1).

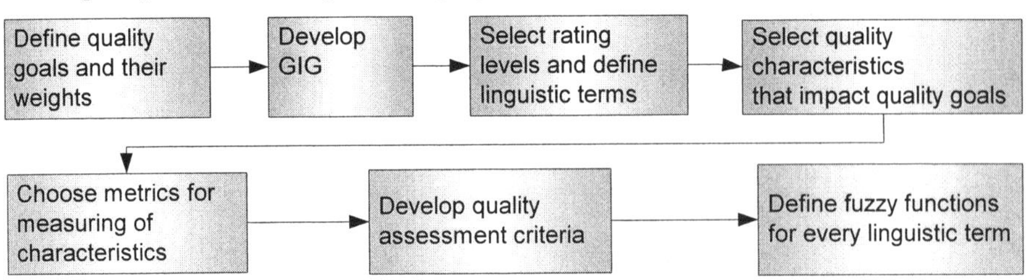

Figure 1. Quality model construction

Quality characteristics tree Ψ is the same for any context-oriented quality model, because it defines internal quality of the language and does not depend on the particular project. In general case, characteristics of internal quality can be measured using several different metrics $m_j \in M$. In the case of different measurements, their values should be aggregated. We suppose that, evaluating quality in use, internal quality has been evaluated already and in this paper we do not discuss measurement and aggregation techniques.

Because for the evaluation of quality in use not all quality characteristics may be important, in the particular quality model only some sub-tree of Ψ may be used. Values of quality characteristics of this sub-tree are mapped to appropriate rating values using (fuzzy) rating functions $\xi_\lambda \in \xi$.

Let us consider the example of a simple quality model Q_1 designed for project P_1 in order to choose the most appropriate IS specification language.

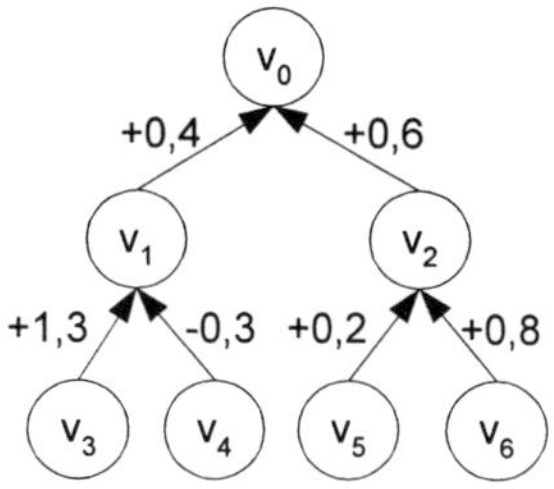

Figure 2. The weighted digraph describing interdependencies between quality goals

Example 1.

1. Goal interdependencies graph Γ is presented in Fig.2.
The set of weights is defined as G={-0.3, +0.2, +0.4, +0.6, +0.8, +1.3}.
Weighting function f is defined as follows:
$f(v_3,v_1)=+1.3$; $f(v_4,v_1)=-0.3$; $f(v_1,v_0)=+0.4$;
$f(v_5,v_2)=+0.2$; $f(v_6,v_2)=+0.8$; $f(v_2,v_0)=+0.6$.

2. Sub- tree of quality characteristics Ψ_1 (Fig.3) has 3 leaf nodes Ch_1, Ch_2, Ch_3.

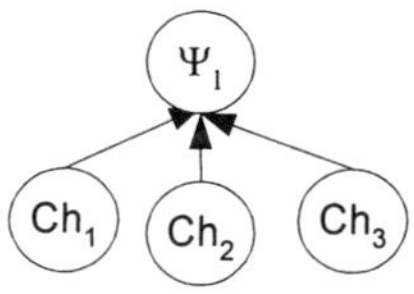

Figure 3. The sub- tree of quality characteristics

3. The set of rating levels is defined as $\Delta = \{0, 1, 2\}$.

4. The set of linguistic terms is defined as $\Lambda = \{U, A, G\}$, where U is *unacceptable*, A is *acceptable*, and G is *good*.

5. According to (2) quality assessment function Θ is calculated in the following way:
$\Theta(v_1)= 1.3\ \Theta(v_3) - 0.3\ \Theta(v_4)$;
$\Theta(v_2)= 0.2\ \Theta(v_5) + 0.8\ \Theta(v_6)$;
$\Theta(v_0)= 0.4\ \Theta(v_1) + 0.6\ \Theta(v_2)$;
$\Theta(v_3)= 1.5\ \xi(Ch_1) - 0.5\ \xi(Ch_2)$;
$\Theta(v_4)= 0.2\ \xi(Ch_2) + 0.8\ \xi(Ch_3)$;
$\Theta(v_5)= 0.3\ \xi(Ch_1) + 0.8\ \xi(Ch_2) - 0.1\ \xi(Ch_3)$;
$\Theta(v_6)= 0.8\ \xi(Ch_1) + 0.2\ \xi(Ch_3)$.

6. The set of quality metrics is defined as $\mu = \{m_1, m_2\}$. Metrics are related to quality characteristics in the following way:
$\mu=\{(m_1,Ch_1), (m_2,Ch_2), (m_2,Ch_3)\}$

7. According to (3) rating functions ξ_λ are calculated in the following way:

$$\xi_U(m_j(x)) = -3m_j(x)+1, \quad \text{when} \quad 0 < m_j(x) \le 1/3$$

$$\xi_A(m_j(x)) = \begin{cases} 3m_j(x), & \text{when } 0 \le m_j(x) \le 1/3 \\ -3m_j(x)+2, & \text{when } 1/3 < m_j(x) \le 2/3 \end{cases}$$

$$\xi_G(m_j(x)) = \begin{cases} m_j(x)-1, & \text{when } 1/3 \le m_j(x) \le 2/3 \\ -3m_j(x)+3, & \text{when } 2/3 < m_j(x) \le 1 \end{cases}$$

8. According to (4), for $\Lambda = \{U, A, G\}$ interpretation functions F_λ in the point v_0 are calculated in the following way:

$$F_U(\Theta(v_0)) = -\Theta(v_0)+1, \quad \text{when} \quad 0 < \Theta(v_0) \le 1$$

$$F_A(\Theta(v_0)) = \begin{cases} \Theta(v_0), & \text{when } 0 \le \Theta(v_0) \le 1 \\ -\Theta(v_0) + 2, & \text{when } 1 < \Theta(v_0) \le 2 \end{cases}$$

$$F_G(\Theta(v_0)) = \begin{cases} \Theta(v_0) - 1, & \text{when } 1 \le \Theta(v_0) \le 2 \\ -\Theta(v_0) + 3, & \text{when } 2 < \Theta(v_0) \le 3 \end{cases}$$

3. Evaluation Procedure

The proposed user-oriented quality model suggests appropriate quality evaluation procedure. This procedure includes five steps (Fig. 4).

1. Characteristics of internal quality, which are relevant to the quality goals V, are identified and the sub-tree Ψ_1 of the tree Ψ is constructed.

2. Rating values for characteristics of sub-tree Ψ_1 are calculated using rating functions from ξ.

3. The rating values for terminal quality goals are calculated using quality assessment function Θ.

4. The rating values are propagated through the goal interdependencies graph Γ using quality assessment function Θ.

5. The rating value of the top-level quality goal $\Theta(v_0)$ is interpreted using interpretation functions $F_\lambda(\Theta(v_0))$.

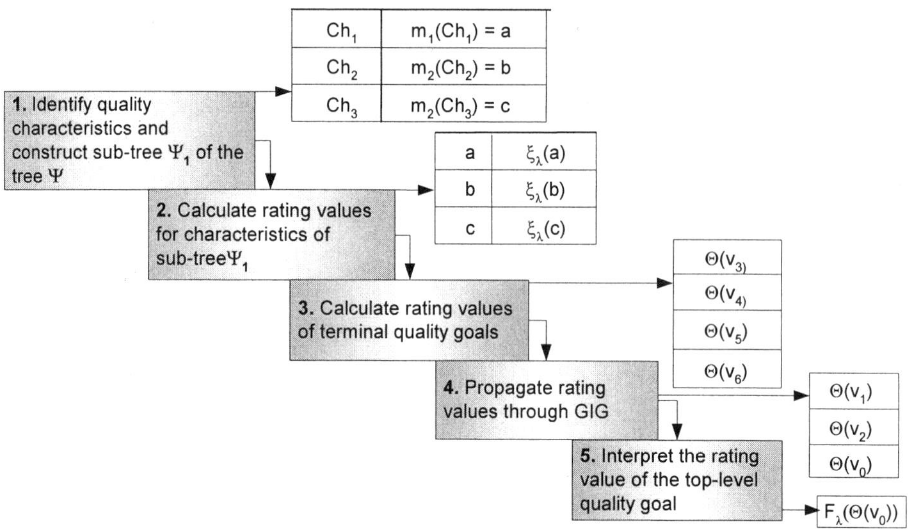

Figure 4. Quality evaluation procedure

Example 2.

The evaluation of IS specification language L_1 according to quality evaluation procedure provided by the quality model Q_1 presented in *Example 1* gives the following results:

1. $m_1(Ch_1) = 0{,}17$; $m_2(Ch_2) = 0{,}47$; $m_2(Ch_3) = 0{,}27$.

2. $m_1(Ch_1)$: $0<m_1(Ch_1)\le 1/3$, $0\le m_1(Ch_1)\le 1/3$

$\xi_U(m_1(Ch_1)) = -3*0{,}17+1=0{,}49$; $\xi_A(m_1(Ch_1)) = 3*0{,}17 = 0{,}51$

$1*0{,}49 + 2*0{,}51 = 0{,}5+1 = 1{,}51$

$m_2(Ch_2)$: $1/3< m_2(Ch_2)\le 2/3$, $1/3\le m_2(Ch_2)\le 2/3$

$\xi_A(m_2(Ch_2)) = -3*0,47+2 = 0,59$; $\xi_G(m_2(Ch_2)) = 3*0,47-1 = 0,41$
$2*0,59+3*0,41 = 2,41$

$m_2(Ch_3)$: $0< m_2(Ch_3)\leq 1/3$, $0\leq m_2(Ch_3)\leq 1/3$
$\xi_U(m_2(Ch_3)) = -3*0,27+1 = 0,19$; $\xi_A(m_2(Ch_3)) = 3*0,27= 0,81$
$1*0,19+2*0,81 = 1,81$

3. $\Theta(v_3)= 1,06$; $\Theta(v_4)= 1,93$; $\Theta(v_5)= 2,2$; $\Theta(v_6)= 1,57$;
4. $\Theta(v_1)= 0,80$; $\Theta(v_2)= 1,70$; $\Theta(v_0)= 1,34$.
5. $F_A(\Theta(v_0)) = -1,34+2=0,66$; $F_G(\Theta(v_0)) = 1,34-1=0,34$.

Thus, the quality of evaluated specification language L_1 for the project P_1 is 66% Acceptable and 34% Good.

4. Quality Goals

In the proposed approach, high-level quality requirements are formulated in the form of quality goals. The user's treatment of quality goals and interdependencies between goals is described using goal interdependency graph (GIG) Γ. The idea of GIG is borrowed from [4], where similar graphs, namely, softgoal interdependency graphs (SIG), are used to define non-functional software requirements. The advantage of GIG, comparing to quality characteristics tree, is that the user can start from business goals, formulate high-level quality requirements and derive further detailed quality requirements, formulated in the terms of low-level quality characteristics. For example, top management may aim to spend money on staff training, minimise the amount of efforts needed to produce specifications, and produce specifications readable by domain experts. In other words, management is looking for an IS specification language that would be simple enough to learn in short terms by not skilled staff, would be efficient and would have high degree of audience appropriateness. GIG allows defining goal interdependences and priorities, helps to expose implicit interdependencies and to refine requirements up to low-level quality characteristics. By refinement, for each goal a set of subgoals to satisfy this goal is introduced. Parent goal may be satisfied by all of its subgoals or by any of them. In addition, some subgoals can contribute positively towards a particular goal and negatively towards another goals.

Although GIGs are very similar to SIGs, they are used for different purposes and in different way. The main problem investigated in [4] is software design problem. SIGs are used in top-down manner with the aim to refine non-functional requirements, including quality requirements, to choose design decisions, which accomplish those requirements in the possibly best way, and to evaluate the impact of chosen decisions in bottom-up way. Our task is to choose such IS specification language, which satisfies quality requirements in the best way. So we need to choose not the design decisions but the most appropriate metrics and measurement procedures and, further, to propagate evaluation results through the GIG in the bottom-up manner.

5. Quality Assessment Function

Each quality goal is evaluated using quality assessment function Θ. This function is used to propagate rating values of quality characteristics through the GIG from bottom towards the top of the graph. Measured or aggregated values of quality characteristics should be mapped to rating values, because, in general case, they are incomparable, expressed in different dimensions. Rating values of the lowest GIG level are used as intermediate to calculate quality assessment function of the highest GIG level. Quality assessment function depends on goals priorities and, consequently, is designed in such way that to evaluate the impact of quality characteristics of the chosen language on the quality goals. Because the arrows are signed by negative and positive priorities, the calculated value may be not exactly integer. It means that we propagate towards top of the graph not the rating levels itself but some values that belong to intervals [v1, v2], where v1 and v2 are adjacent rating levels.

For "AND" vertices the value of quality assessment function is calculated taking into account the weights of all incoming arcs. In this case, positive as well as negative weights are allowed, and it is required that the sum of weights should be equal to 1. For "OR" vertices the value of the Θ is the maximal one from the propagated values. In this case, only positive weights are allowed, and it is required that any weight do not exceed 1.

6. Quality Characteristics

A number of publications on IS specification problems [8], [9], [10], [11], [12], [16], [20], [21], [23], [24], [25] mention different quality characteristics that specification language should have. Some quality characteristics discussed in publications on the quality of programming languages and cognitive dimensions of information artefacts (see, for example, [17]) also are closely related to the quality of specification languages. However, exhaustive set of quality characteristics up to date still has not been proposed. Extensive library research [3] has brought to light that some important quality characteristics even have no names and some important groups of quality characteristics are not thought as a coherent groups. Also in the field of programming languages, where research has already a long history, different researchers evaluate quality of programming languages using slightly different quality characteristics and classify these characteristics in different ways. For example, some researchers argue that reliability of a language is influenced by both readability and writability, others – that it is influenced only by readability. In addition, characteristics of internal quality and characteristics of quality in use often are not entirely separated. On the basis of the results of conceptual analysis of wide spectrum of specification languages, including UML, Z, VDL, Troll, and Alloy, we decided to organise quality characteristics of specification language in a way proposed by ISO/IEC 9126 standard [6]. From six groups of software quality characteristics (functionality, reliability, usability, efficiency, maintainability, and portability) provided by this standard we have left only four, because maintainability and portability are relevant rather to software tools supporting production of specifications than to specification language itself. The current version of the quality characteristics tree is presented in Appendix 1. Let us comment this tree shortly.

Following the idea, proposed by Krogstie and Sølvberg [11] and elaborated in [3], we describe quality characteristics in terms of the linguistic system, defining a formal structure beyond the language, and in terms of the representation system, defining forms of representation of its constructs. Further, we treat the concepts of functionality, reliability, usability, and efficiency in a different way than ISO/IEC 9126 standard and, consequently, refine these characteristics using other lower level sub-characteristics. Functionality of a specification language is understood as the existence of means, required to specify functional and non-functional properties of the system under consideration. Functionality has two dimensions: suitability and flexibility. The first one characterises the ability to express statements about potential systems in a particular realm (real-world systems, IS, software systems, etc.) at a certain level of granularity, the second one – the extent to which a language can be adjusted to specify preliminary not intended properties. We suppose that the specification of the system can be seen as a collection of statements about the properties of this system. Statements can be explicit, or implicit, derivable from other statements. The possibility to produce such specification first of all depends on the completeness of the language or, in other words, on the ability to express all necessary kinds of knowledge at all representation levels: epistemological, ontological, and conceptual [3].

It should be noted that we define two descriptive sub-characteristics of usability: ontology and paradigm. Although these characteristics describe important properties of a language, they are not related directly to the internal quality of this language. However, evaluation of the quality in use often requires considering formalism beyond specification language and assessing of ontological aspects of this language. Besides, these characteristics sometimes may be operationalised, for example, a metric should be developed to measure the degree of procedureness of the language.

7. Conclusions and Further Research

Research on evaluation of specification languages quality, especially on evaluation of quality in use, is still at the beginning. Most of proposed quality models do not separate quality in use from internal quality at all and are only sketched. No commonly accepted agreement exists about the collection of specification language quality characteristics, their names and their taxonomy. It is true for programming languages, too. At the moment, no exhaustive procedure to evaluate quality in use is proposed. The procedure proposed in this paper fills this gap. The paper also attempts to develop exhaustive taxonomy of quality characteristics. Elaboration and operationalisation of the proposed taxonomy, and the procedure for evaluation of internal quality are most important challenges for further research. This research also provides development of metrics to measure characteristics of internal quality, measurement procedures, and measurements aggregation techniques.

References

[1] Xexéo G, Belchior A. e Rocha ARC. Evaluating software quality requirements using fuzzy theory. In: Proceedings of International Conference on Information systems Analysis and Synthesis (ISAS) 96; 1996 July 22-26; Orlando, USA; 1996. Available: http://www.cos.ufrj.br/~xexeo/artigos/isas96/isas96.htm via the INTERNET.

[2] Botella P, Burgués X, Carvallo JP, Franch X, Pastor JA, Quer C. Towards a Quality Model for the Selection of ERP Systems. In: Cechich A, Piattini M, Vallecillo A, editors. Component-Based Software Quality: Methods and Techniques. Lecture Notes in Computer Science 2693; Springer-Verlag; 1998. p 225-245.

[3] Caplinskas A, Lupeikiene A, Vasilecas O. A framework to analyse and evaluate Information Systems specification languages. In: Manolopoulos Y, Navrat P, editors. Lecture Notes in Computer Science 2435. Proceedings of the 6th East European Conference 'Advances in Databases and Information Systems'; 2002 Sept 8-11; Bratislava, Slovakia. Springer; 2002. p. 248-262.

[4] Chung L, Nixon BA, Yu E, Mylopoulos J. Non-functional requirements in software engineering. Kluwer Academic Publishers; 1999.

[5] ISO 8402. Quality management and quality assurance vocabulary. Second edition, 1994-04-01.

[6] ISO/IEC 9126. Information Technology – Software Product Evaluation – Quality Characteristics and Guidelines for their Use. First edition, 1991-12-15, reference number ISO/IEC 9126: 1991(E).

[7] King M, Maegaard B. Issues in natural language systems evaluation. In: Proceedings of the First International Conference on Language Resources and Evaluation; 1998 May 28-30; Granada, Spain; 1998. p. 225-230. Available: http://citeseer.nj.nec.com/514907.html via the INTERNET.

[8] Krogstie J. Evaluating UML using a generic quality framework. In: Favre L, editor. UML and the Unified Process. Hershey, PA, USA: Idea Group Publishing; 2003. p. 1-22.

[9] Krogstie J. A semiotic approach to quality in requirements specifications. In: Liu K, Clarke RJ, Andersen PB, Stamper RK, editors. Proceedings of the IFIP TC8/WG8.1 Working Conference on Organizational Semiotics: Evolving a Science of Information Systems; 2001 July 23-25; Montreal, Quebec, Canada. Kluwer Academic Publishers; 2001. p.231-249.

[10] Krogstie J. Using a semiotic framework to evaluate UML for the development of models of high quality. In: Siau K, Halpin TA, editors. Unified Modelling Language: Systems Analysis, Design and Development Issues. Hershey, PA, USA: Idea Group Publishing; 2001. p. 89-106.
[11] Krogstie J, Sølvberg A. Information Systems Engineering: Conceptual Modeling in a Quality Perspective. Information Systems Groups. NTNU. Trondheim. Norway. 2000. Forthcoming.
[12] Lindland OI, Sindre G, Sølvberg A. Understanding Quality in Conceptual modelling. IEEE Software 1994 March; 11(2): 42-49.
[13] Milton S, Kazmierczak E, Keen C. Comparing Data Modelling Frameworks Using Chisholm's Ontology. In: Bartoli JA, editor. Proceedings of the 4th European Conference on Information Systems, ECIS'98; 1998 June; Aix-en-Provence. France: Euro-Arab Management School; 1998. p. 260-272.
[14] MIT Lab for Computer Science. 1999. Jackson DA. Comparison of Object Modelling Notations: Alloy, UML and Z [manuscript online]. Available from: http://geyer.lcs.mit.edu/~dnj/pubs/alloy~comparison.pdf. Accessed 2004 January 10.
[15] Mylopoulos J. Characterizing Information Modeling Techniques. In: Bernus P, Mertins K, Schmidt G, editors. Handbook on Architectures of Information Systems. Berlin: Springer; 1998. p. 17-57.
[16] Opdahl AL. Applying Semantic Quality Criteria to Multi-Perspective Problem Analysis Methods. In: Dubois E, Opdahl AL, Pohl K, editors. Proceedings of the Third International Workshop on Requirements Engineering: Foundations of Software Quality — REFSQ'97; 1997 June; Barcelona, Catalonia, Spain; 1997. p.49-66.
[17] Pane JF, Myers BA. Usability Issues in the Design of Novice Programming Systems. Pittsburgh, Carnegie Mellon University, School of Computer Science; 1996. Report nr. CMU-CS-96-132. Available: http://www-2.cs.cmu.edu/~pane/cmu-cs-96-132.html via the INTERNET. Accessed 2004 Jan 10.
[18] Peters LJ, Trip LL. Comparing Software Design Methodologies. Datamation 1977; 23(11): 89-94.
[19] QStudio® for Java. The Software Health Tool for Java. QA Systems BV. The Netherlands; 2003. Available: http://www.qa-systems.com/welcome.html via the INTERNET. Accessed 2004 Jan 10.
[20] Seltveit AH. Complexity Reduction in Information Systems Modelling [dissertation]. Trondheim, Norway: IDT, NTH; 1994.
[21] Sindre G. HICONS: A General Diagrammatic Framework for Hierarchical Modelling [dissertation]. Trondheim, Norway: IDT, NTH; 1990.
[22] Simão RPS, Belchior AD. Quality Characteristics for Software Components: Hierarchy and Quality Guides. In: Cechich A, Piattini M, Vallecillo A and . Component-Based Software Quality: Methods and Techniques. Lecture Notes in Computer Science 2693; Springer; 1998. p 184-206.
[23] Wand Y, Weber R. An ontological evaluation of systems analysis and design methods. In: Falkenberg ED, Lindgreen P, editors. Information Systems Concepts: An In-depth Analysis. North-Holland, Amsterdam: IOS Press; 1989. p. 79-107.
[24] Wand Y, Weber R. On the Ontological Expressiveness of Information Systems Analysis and Design Grammars. Journal of Information Systems 1993; 3(4): 217-237.
[25] Wand Y, Weber R. On the Deep Structure of Information Systems. Journal of Information Systems 1995; 5: 203-223.

Appendix 1

Table 1. Specification language quality characteristics definitions

1	**Functionality**	Characteristics of a linguistic system that bears on the existence of the set of features required to specifying properties of a system under consideration.
1.1	Suitability	Characteristics of a linguistic system that bears on ability to express statements about potential systems in a particular realm at a certain level of granularity.
1.1.1	Completeness	Characteristic of a linguistic system that bears on its ability to describe a system under consideration sufficiently and exhaustively.
1.1.1.1	Semantic sufficiency	Characteristics of a linguistic system that bears on its ability to specify all "things" that may be necessary for analysis or design of a system under consideration.
1.1.1.1.1	Ontological sufficiency	Characteristics of a linguistic system that bears on its ability to conceptualise (using ontological primitives provided by linguistic system) any system under consideration.
1.1.1.1.2	Epistemological sufficiency	Characteristics of a linguistic system that bears on its ability to express epistemological primitives, (i.e. to define conceptual primitives on the basis of a given set of ontological ones and to combine defined concepts further in order to form more complex concepts).
1.1.1.2	Expressibility	Characteristic of a linguistic system that bears on its ability to express any statement about any system under consideration.
1.1.1.3	Reasoning power	Characteristic of a linguistic system that bears on its ability to derive new statements about properties of system under consideration.
1.1.1.4	Composability	Characteristic of a linguistic system that bears on its ability to compose language constructs and features together.
1.1.2	Expressive adequacy	Characteristics of a linguistic system that bears on its ability to formulate statements about any system under consideration in adequate terms and express them with required degree of precision.
1.1.2.1	Ontological adequacy	Characteristic of a linguistic system that bears on its ability to express its ontological commitments within this system itself (i.e. use adequate terms).
1.1.2.2	Epistemological adequacy	Characteristic of a linguistic system that bears on its ability to express epistemological primitives directly (i.e. use such epistemological schemes as generalisation, aggregation, etc.)
1.1.2.3	Selective power	Characteristic of a linguistic system that bears on its ability to distinguish details with the needed degree of precision. (i.e. describe two different instances of a concept and two properties of an instance of a concept in a distinguishable way).
1.1.2.4	Generalitive power	Characteristic of a linguistic system that bears on its ability to describe system at different levels of granularity (i.e. suppress irrelevant

		details while preserving essential properties of the system).
1.2	Flexibility	Characteristics of a linguistic system that bears on the extent to which a language can be adjusted to specify preliminary not intended properties.
1.2.1	Universality	Characteristic of a linguistic system that bears on the ability to apply its constructs in different realms (i.e. real-world systems, IS, software systems, etc.).
1.2.2	Adaptability	Characteristic of a (general-purpose) language (both linguistic system and representation system) that bears on its ability to configure syntax and semantics to in order to adapt this language for arbitrary realm.
1.2.3	Extensibility	Characteristic of a (general-purpose) language (both linguistic system and representation system) that bears on its ability to extend this language with new features.
2	**Reliability**	Characteristics of a linguistic system that bear on its ability to ensure correctness of specifications.
2.1	Preventability of errors	Characteristic of a language (both linguistic system and representation system) that bears on the number of errors in the specification.
2.1.1	Semantic unambiguity	Characteristic of a linguistic system that bears on unambiguity of the concepts of language.
2.1.2	Notational unambiguity	Characteristic of a representation system that bears on unambiguity of the structure of language constructs.
2.1.3	Distinguishibility	Characteristic of a representation system that bears on absence of subtle distinctions in syntax, which may be overlooked or confused.
2.1.4	Simplicity	Characteristic of a language (both linguistic system and representation system) that bears on the comprehensibility of language.
2.1.4.1	Conceptual simplicity	Characteristic of a linguistic system that bears on the comprehensibility (including conceptual cleanliness) of the concepts defined by language.
2.1.4.2	Functional simplicity	Characteristic of a linguistic system that bears on a number of features of language.
2.1.4.3	Representational simplicity	Characteristic of a representation system that bears on the comprehensibility (including syntactical transparency) of the constructions of language.
2.1.5	Semantic power	Characteristic of a linguistic system (level of language) that bears on the semantic power of the concepts defined by this system.
2.1.6	Orthogonality	Characteristic of a linguistic system that bears on the degree in which the concepts of the language interfere with each other.
2.1.7	Uniformity	Characteristic of a language (both linguistic system and representation system) that bears on the degree of internal standardisation of the syntax (syntactical uniformity) and the semantic (semantic uniformity) of the construction of the language.
2.2	Verifiability	Characteristic of a linguistic system that bears on the ability (including executability) to check correctness of produced specifications.

3	**Efficiency**	Characteristics of a linguistic system and a representation system that bear on the amount of efforts needed to produce specification.
3.1	Expressive efficiency	Characteristic of a linguistic system that bears on the efforts necessary to express the statements about the phenomena.
3.2	Representational efficiency	Characteristic of a representation system that bears on efforts necessary to represent the statements about the phenomena.
3.3	Semantic power	Characteristic of a linguistic system (level of language) that bears on the semantic power of the concepts defined by this system.
3.4	Orthogonality	Characteristic of a linguistic system that bears on the degree in which the concepts of the language interfere with each other.
3.5	Permissiveness	Characteristic of a language (both linguistic system and representation system) that bears on its ability to express and represent things in several different ways.
3.6	Viscosity	Characteristic of a language (both linguistic system and representation system) that bears on the possibility to narrow the change impact on the produced specification.
3.7	Technological efficiency	Characteristics of a language (both linguistic system and representation system) that bear on the technological appropriateness of a language.
3.7.1	Manageability	Characteristics of a language (both linguistic system and representation system) that bear on appropriateness of the language to deal with large and complex specifications.
3.7.1.1	Decomposability	Characteristic of a language (both linguistic system and representation system) that bears on the possibility to divide the produced specification into relatively autonomous and independent subparts.
3.7.1.2	Genericity	Characteristic of a linguistic system that bears on the possibility to use generic types.
3.7.2	Processability	Characteristic of a language (both linguistic system and representation system) that bear on appropriateness of the language to manipulate specifications by software.
4	**Usability**	Characteristics of a linguistic system and a representation system that bear on the audiences' effort to understand and to learn a language.
4.1	Understandability	Characteristics of a linguistic system that bear on the audiences' effort to understand its conceptualisation.
4.1.1	Ontology	Characteristic of a linguistic system that describes language ontology.
4.1.2	Paradigm	Characteristic of a linguistic system that describes mathematical theory beyond the language.
4.1.3	Naturalness	Characteristic of a linguistic system that bears on the correspondence between the language ontology and the common sense.
4.1.4	Semantic unambiguity	See 2.1.1
4.1.5	Notational unambiguity	See 2.1.2
4.1.6	Distinguishibility	See 2.1.3
4.1.7	Uniformity	See 2.1.7

4.2	Learnability	Characteristics of a linguistic system and a representation system that bear on the audiences' effort for learning of language application.
4.2.1	Simplicity	See 2.1.4
4.2.2	Semantic power	See 3.3
4.2.3	Naturalness	See 4.1.3
4.2.4	Uniformity	See 2.1.7
4.2.5	Commonality	Characteristic of a language (both linguistic system and representation system) that make the language adhere to wide-accepted standards or conventions.

Modeling and Model Transformations

Databases and Information Systems
J. Barzdins and A. Caplinskas (Eds.)
IOS Press, 2005

Model Transformation Language MOLA: Extended Patterns

Audris KALNINS, Janis BARZDINS, Edgars CELMS
Institute of Mathematics and Computer Science, University of Latvia,
29 Raina Boulevard, Riga, LV-1459, Latvia

Abstract. The paper describes a new graphical transformation language MOLA for MDA-related model transformations. Transformations in MOLA are described by combining traditional control structures, especially loops, with pattern-based transformation rules. Besides an overview of the basic MOLA, the paper describes an extension of MOLA – powerful patterns, which may include transitive closure. The paper shows how the usage of these patterns simplifies control structures for typical MDA tasks.

1. Introduction

The increased use of modeling techniques nowadays requires effective support for model transformations. Perhaps, one of the most actual areas in software engineering today is the Model Driven Architecture (MDA) [1]. MDA is a software development approach in which models are the primary artifacts. According to the MDA, the different types of models are defined (most usually in UML [2,3] notation), mapping between models and towards different targets is formalized, and guidelines are automated, in order to improve efficiency and guarantee that the process of the transformation between models is properly followed. Model-to-model transformation is therefore a key technology for MDA. While the current Object Management Group (OMG) standards such as the Meta Object Facility (MOF) [4] and the UML provide a well-established foundation for defining different types of models, no such well-established foundation exists for transformations between them. The need for standardization in this area led to the MOF 2.0 Query/Views/Transformations (QVT) request for Proposals (RFP)[5] from OMG.

To a great degree the success of the MDA initiative and of QVT in particular will depend on the availability of a concrete syntax for model-to-model transformations that is able to express non-trivial transformations in a clear and compact format that would be useful for industrial production of business software [6].

The submissions by several consortiums have been already made, e.g. [7, 8, 9], and it is somewhat surprisingly, that only a few of them use a natural graphical representation of their language. Currently none of them has reached the status of a complete model transformation language. Several proposals for transformation languages have been provided outside the OMG activities. The most interesting and complete of them seem to be UMLX [10] and GReAT [11].

According to our view, and many others [6], model transformations should be defined graphically, but combining the graphical form with text where appropriate. Graphical forms of transformations have the advantage of being able to represent mappings between patterns in source and target models in a direct way. This is the motivation behind visual languages such as UMLX, GReAT and the others proposed in the QVT submissions. Unfortunately, the currently proposed visual notations do not provide easy readable descriptions of model transformations.

The common setting for all transformation languages is such that the model to be transformed – the source model is supplied as set of class and association instances conforming to the source metamodel. The result of transformation is the set of instances conforming to the target metamodel – the target model. Therefore the transformation has to operate with instance sets specified by a class diagram (actually, the subset of class notation, which is supported by MOF).

Approaches that use graphical notation of model transformations draw on the theoretical work on graph transformations. Hence it follows that most of these transformation languages define transformations as sets of related rules. Each rule consists of a pattern and action part, where the pattern has to been found (matched) in the existing instance set and the actions specify the modifications related to the matched subset of instances. This schema is used in all of abovementioned graphical transformation languages. Languages really differ in the strength of pattern definition mechanisms and control structures governing the execution order of rules.

The most detailed pattern definition is in the GReAT language. There it is possible to match a set of instances to one element of the pattern (variable cardinality patterns). However, the patterns are still limited in depth but this is compensated by a very elaborated rule control structure specified graphically by dataflow-like diagrams. UMLX has a similar but slightly weaker pattern mechanism. The control structure is completely based on recursive invocations of rules. In the proposal by QvT-Partners [7] graphical patterns are combined with extensive use of textual constraints. The control structure is based on recursive invocation of rules. In the DSTC/IBM/CBOP proposal [12] (now merged with [7]) patterns are specified in a textual (Prolog-like) form, the most interesting feature of this language is the possibility to include a transitive closure in patterns.

This paper proposes a new graphical transformation language MOLA (Model Transformation Language). The main design goal for MOLA has been to make the transformation definitions natural and easy readable, by relying on simple iterative (non-recursive) control structures, based on traditional structured programming. In addition, as far as it improves readability, the intention was to make each rule more powerful. In particular, this requires the strengthening of pattern mechanism. The MOLA project actually consists of two parts – the basic and extended MOLA. The basic MOLA uses simple patterns and more relies on control structures – it has a more procedural style. The main element there is a graphical loop concept, which can easily be combined with a transformation rule. The main new feature of the extended MOLA is the possibility to define looping patterns of "unlimited depth" (in addition to variable cardinality), thus incorporating the mechanism of transitive closure in patterns. In the result, very simple control structure – a sequence of simple loops then is sufficient for many transformation jobs. Certainly, such a pattern definition requires an adequate definition of the matching procedure, which is also described in the paper. High execution efficiency for the procedure is guaranteed in the typical case when the pattern cardinalities are adapted to the metamodel multiplicities, to which the instance set conforms (the "uniqueness principle" is observed). Patterns in MOLA are defined as directed graphs, to enable the required control over the matching procedure – also a new feature for pattern definition. As a consequence of larger and more powerful patterns, a typical step of a transformation frequently can be described by one rule. This natural non-recursive style of transformation definitions in MOLA has been tested on several real MDA jobs, at the same time using the features of MOLA to keep each rule not too complicated (namely the right balance there ensures the human readability of transformations). As far as we know, the extended MOLA is the only graphical transformation language, which supports transitive closure in patterns. The ideas for pattern definition in MOLA have been partially inspired by the authors' previous experience in defining mappings for generic modeling tools based on metamodels [13].

This paper describes the basic elements of MOLA. The main emphasis is on the extended pattern concept. Language description is based on a typical MDA example (used also in [7, 8, 10, 14]) – transformation of a simplified class diagram into a database definition. Section 2 describes the general structure of MOLA, section 3 - the example. Section 4 describes the basic MOLA – rules with simple patterns and the new concept of loop. The complete pattern mechanism in extended MOLA, including cardinality constraints, looping patterns and the corresponding matching procedure is described in section 5.

2. Overview of Language Elements in MOLA

The MOLA language is a natural combination of pattern-based model transformation rules with control structures from traditional structured programming, both specified in a graphical form.

MOLA is meant for transforming models built according to one metamodel – the **source metamodel** (SMM) to models conforming to another metamodel – the **target metamodel** (TMM). In a special case, SMM and TMM may coincide. Both the source and target models actually are treated as instance sets of the corresponding metamodel classes and associations.

A **transformation definition** in MOLA consists of the both metamodels and the transformation program. A **transformation program** in MOLA is a sequence of **statements**. A statement is a graphical area, delimited by a rectangle – in most cases, a gray rounded rectangle. The statement sequence is shown by dashed arrows. The program starts with the UML start symbol and ends with an end symbol.

The simplest kind of statement is a **rule**, which performs an elementary transformation of instances. A rule contains a **pattern** – a set of elements representing class and association instances, built in accordance with the source metamodel. Pattern elements can have **attribute constraints** (OCL expressions). The pattern specifies what kind of instance group must be found in the source model, to which the rule must be applied. A rule has also an **action** specification – new class instances to be built, instances to be deleted, association instances (links) to be built or deleted and the modified attribute values (as assignments). Both for the pattern and action part the UML object (instance specification) notation is used.

The most important statement type in MOLA is the **loop**. Graphically a loop is a rectangular frame, containing a sequence of statements. This sequence starts with a special **loop head** statement. The loop head is also a pattern, but with one element – the **loop variable** highlighted (by a **bold** frame). Informally a loop variable represents an arbitrary instance of the given class, which satisfies the conditions specified by the pattern. Actually there are two types of loops in MOLA, differing in semantics details. The first type (denoted by a **simple** frame) is executed once for each valid loop variable instance, therefore it is called **FOREACH** loop. Mainly this type of loop will be used in the paper. The second type (denoted by a **3-d** frame) is executed while there is at least one loop variable instance satisfying the pattern conditions – it is called the **WHILE** loop. Loops can be nested to any depth. A loop head can contain actions – it is also a rule. Such a combined statement will be widely used in the examples of this paper.

Other control structures in MOLA are the **branch** construct – several frames started by pattern statements as conditions and the **subprogram** concept together with the **subprogram call**, where the parameters can contain references to instances used in the calling program. However, these control structures actually will not be used in the paper – the aim of this paper is to demonstrate how extended patterns in MOLA allow to use a very simple control structure – a sequence of simple loops.

Section 4 discusses in detail the syntax and semantics of basic MOLA on an example. Then the extended patterns are introduced – the section 5 describes the extended MOLA.

The general execution schema in MOLA is simple – when a source model is supplied, statements are applied to it in the specified order. A statement is always applied to the whole instance set which is being gradually transformed by the rule actions. The potential execution efficiency is ensured by the corresponding features of the pattern language, where it is easy to specify that only "useful" pattern matches occur.

During the transformation process one more optional metamodel – the **intermediate metamodel** (IMM) may be used. IMM contains both SMM and TMM as a subsets, and additional elements – classes, attributes and associations necessary for performing the transformation. There is a special kind of additional associations – **mapping associations** which in fact are present in every transformation. These associations physically implement the mapping from the elements of source model to target elements, therefore they link classes from SMM to the corresponding classes in TMM. See more on the role of mapping associations in 4.2. Namely due to a large set of mapping associations it is recommended to use IMM for non-trivial transformations (it is also permitted in MOLA to define mapping associations "on the fly" – directly in rules, if IMM is not used). Another important elements of IMM are **computed attributes** – "temporary" attributes added to classes of IMM for storing intermediate values. There may be several kinds of computed attributes, the most used here are the rule-local ones. Names of rule-local attributes start with "?" in the IMM, see more on them in 5.2. Non-local temporary attributes (with the scope of several statements) can be also defined/created and destroyed by special statements.

3. Example - the Class Model to Relational Model Transformation

The paper will be based on a typical MDA example, considered also in [7, 8, 10, 14] – the transformation of a simplified class diagram into a semantically equivalent relational database definition. For all the different versions of the example the one in [7] is used here. This version permits to demonstrate the easy definition of transitive closure in extended MOLA. The SMM for this task is shown in Fig.1.

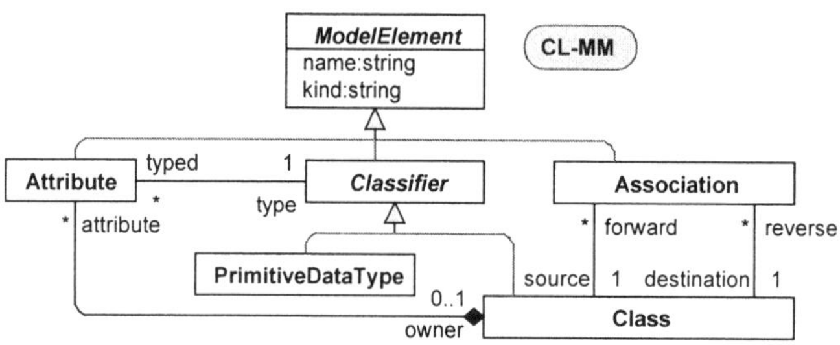

Figure 1. Source metamodel for simplified class diagram

All elements can have a *name*. The metaattribute *kind* is applicable to metaclasses *Class* and *Attribute*, only a *Class* where its value is "*persistent*" must be transformed into a database *Table*, the value of *kind* equal to "*primary*" determines that an *Attribute* actually is a part of a primary key (all other values of *kind* are irrelevant). The *type* of an *Attribute* can be either a *PrimitiveDataType*, or another *Class*. Fig.2 shows the TMM - a simplified relational database definition metamodel.

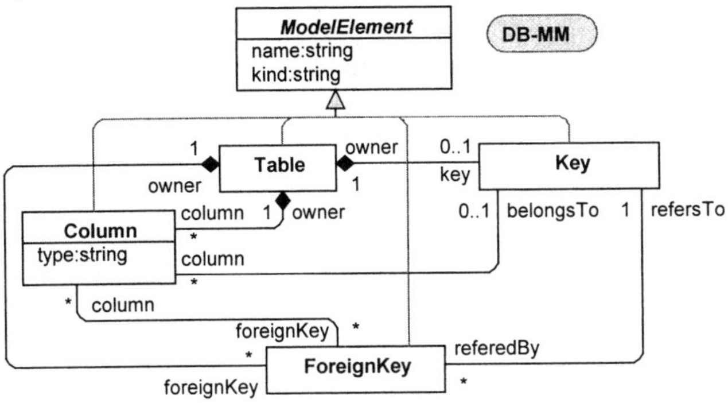

Figure 2. Target metamodel for database definition

A *Table* consists of *Columns*, and it can have a (primary) *Key*, which contains some of the *Columns*. The *ForeignKey* for a *Table* always refers to a *Key* of another *Table*.

The informal transformation algorithm is quite straightforward. For each *persistent* class a table and its key must be built. The primitive-typed attributes of this class become columns of the table (with the same name and type name). The columns which correspond to primitive-typed attributes of *primary* kind become parts of the key. In addition, for class-typed attributes, the primitive attributes of the target class are also transformed to columns of the table for the original class (as "indirect attributes"), and this process of finding indirect attributes is continued down until no more indirect attributes can be found (so-called class flattening – a transitive closure-like process). A column for an indirect attribute has a compound name – the concatenation of all attribute names down to the primitive one. An association is converted into a foreign key for the source class (table), and this foreign key refers to the key for the destination class. In addition, new columns are added to the source table (and to the foreign key) – one (equally named and typed) for each column of the corresponding key. The problem of ordering the columns in keys is ignored in the example. Association multiplicities also are not used in this simplified example – only the association direction matters. It should be noted that the processing of indirect attributes and associations is independent – it may be done in any order.

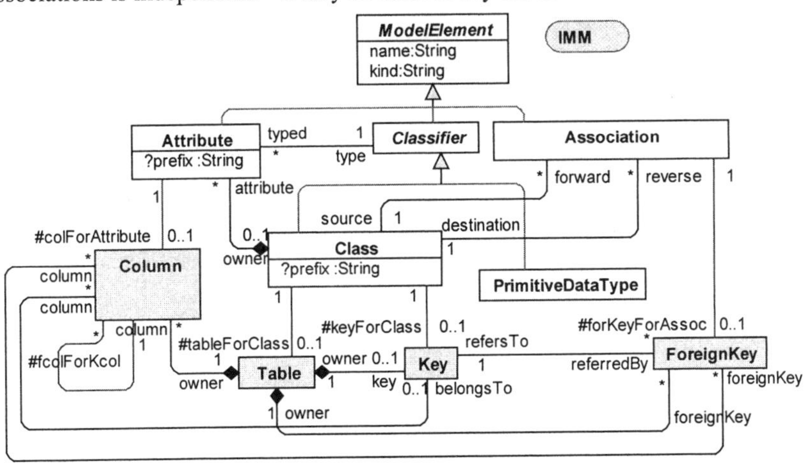

Figure 3. Intermediate metamodel for the transformation

The one remaining element to be described is the IMM – see Fig.3. In addition to a copy of SMM and a copy of TMM (the classes of TMM are in a darker color), it contains mapping associations from source to target elements, e.g., from *Class* to *Table*. These mapping associations will be used in rules in sections 4 and 5, they have the # character prefixed to role names and are in red color in Fig.3. IMM contains also a computed attribute *prefix* of rule-local kind (names of rule-local attributes start with "?" in MOLA), it will be used in section 5.2.

4. Structure of Simple Loops and Rules

As it was stated, both loops and rules rely on patterns in MOLA. In this section the structure of simplest patterns will be described in detail. These patterns, for which only a fixed-size match is possible, have a very simple matching algorithm. The patterns in this section actually are weaker than those described in [10, 11], the goal of this section is just to demonstrate the general principles of MOLA in a very simple case. Non-trivial patterns of MOLA will be described in section 5.

4.1 Basic Patterns

A pattern in MOLA specifies the instance set which can be matched to it. From a syntax point of view, it is similar to UML 2.0 collaborations or structured classifiers. The main **element** of a pattern is a source **metamodel class,** specified in UML **instance notation**. Each element has an optional **instance name** and the **class name,** the same class may be used several times in a pattern. In totality, they must be unique within a pattern. Each element matches to an appropriate instance of that class. Since a typical use of pattern in MOLA is in a loop head, we start with this case. There one pattern element – the loop head (a bold one) has a special meaning. All other elements of the pattern are used to specify, namely which instances of this class in the source model can be used as loop variable instances. The other pattern elements (which may correspond also to target metamodel classes) also must match to an appropriate instance – they specify the context of a loop variable.

In addition to elements, a pattern contains **pattern associations** – selected metamodel associations between the used classes and **attribute constraints** – OCL constraints specified within elements (in braces). The specified association instances must exist between the matched model instances and the attribute constraints must evaluate to true. Pattern associations can have also a {NOT} constraint – this means that no specified instance can be linked to the "main match" by the given link.

Thus a pattern in a loop head specifies which instances of the given class in the source model qualify as valid instances for the loop variable. The other pattern elements have the "**exists**" semantics – there must be (or must not be) an appropriate instance in the selected match, but there is no need to find all possible matches for them.

The loop variable in a pattern in fact plays the role of its **root** – the match is started from it.

Fig.4 shows the simplest pattern consisting just of the loop variable. This pattern says that an instance of *Class* in the source model (i.e., the instance set corresponding to the class diagram to be transformed into a database definition) matches to this pattern, if its *kind* has the value *"persistent"* (*kind* – a string-typed attribute of *Class*).

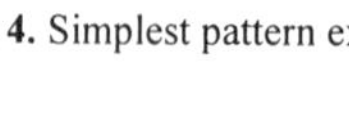

Figure 4. Simplest pattern example

Fig. 5 shows a more non-trivial pattern, involving several elements and associations. In this pattern only these instances of *Attribute* qualify as a loop variable instances, which have an *owner* link to a Class instance, which in turn has a *#tableForClass* link to a *Table* instance, and also have a *type* link to a *PrimitiveDataType* instance. The *Table* class is from the target metamodel, and the *#tableForClass* is a mapping association – this means that these instances have to be already built by previous statements. Pattern associations typically specify only one of the role names – that leading away from the root.

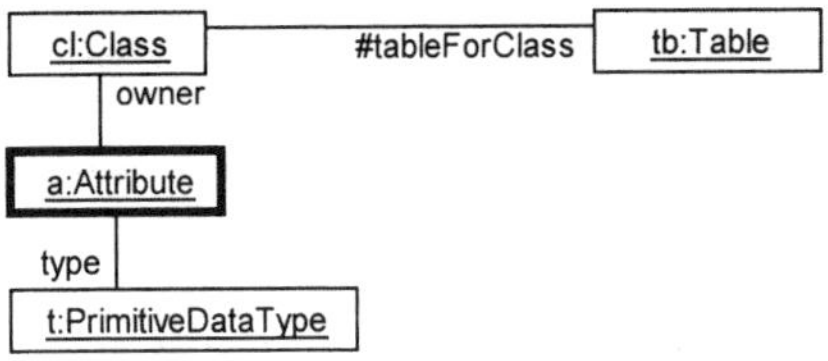

Figure 5. Associations in a pattern

The simple patterns described so far do not require a more formal match definition. However, for extended patterns in section 5 such a definition will be used.

Here one principle of a good programming style in MOLA should be given. To achieve a high execution efficiency, pattern associations leading away from the loop variable (the pattern root) should have the **0..1 multiplicity** at this end in the corresponding metamodel. This means that we test the **existence of one possible instance**. In addition, in the case of existence, the match is unique then and we can reference this instance for various purposes, e.g., to use its attribute values in a deterministic way. Actually, all examples in the paper use this principle. For other multiplicities the extended patterns in section 5 serve well.

Patterns can use also the **reference** notation – an element whose name is prefixed by the @ character – this means that an already selected instance (by a previous statement, typically a loop head) must be used. This way patterns can be structured – similarly as, e.g., in [11]. See an example in Fig. 8.

4.2 Actions of the Rule, Complete Rule Examples

Pattern matching is only one part of the rule application. Another one is to perform the actions specified in the rule (on the basis of the current match). These actions modify the current instance set – typically, the target model. The following actions can be specified in a rule:

- building new class instances
- building new association instances (connecting new as well as existing instances)
- changing the attribute values – both for new and existing class instances
- deleting instances

The **action specification** (the "RHS part") of a statement has a structure similar to the pattern. It also consists of elements to build (in the instance notation) and associations linking the new elements between themselves or to the pattern elements. Syntactically the action part is distinguishable by dotted lines and the line color – it is in **red**. Actions can also specify the deletion of an existing element (matched by the pattern) – this is shown by dashed lines.

The most typical action is the **building** of a **new class instance**. Building of class instance in MOLA is always accompanied by building of one special association – the **mapping association**, which in the rule must be linked to a pattern element (e.g., an association with the role name *#tableForClass* is linked to *cl:Class* in Fig.6). The role name of this association is specified in the intermediate metamodel (here – Fig.3) if that exists, but anyway its name must be prefixed by the # character. At the instance level it means that one instance of the new class is built and linked by the mapping association to the existing instance of the corresponding pattern element.

One goal of the mapping association is to serve for matching in the patterns of next rules. It is very typical in MDA model transformations that the transformation of a "higher level" element – package, class etc. determines how its subordinates – classes, attributes etc. must be transformed. The mapping association is namely the element linking such subordinate rules and ensuring their consistency. In addition, the mapping association reifies physically the mapping between the source and target model (and serves for tracing), hence such a name is used for this concept in MOLA. In our simple patterns, the cardinality of the mapping association is 1 – 1, but in more advanced patterns of MOLA it may have cardinality 1 – 1..* (and serve for determining the instance set of the new class which must be built).

The remaining action element is the **assignment** of **attribute values** (done by Pascal-like assignment statements). For an attribute to be set the new value is defined by an OCL style expression, which can contain one extension – attributes from pattern elements may be referenced, just by prefixing them with the instance name. The semantics is straightforward – take the attribute value from the existing instance matched to the element. The attribute assignments can be done for the "new" instances, but attributes of existing instances (in the pattern elements) can also be modified this way.

Fig.6 shows a complete example of a statement in MOLA. This is the first statement in the program for building the database definition from a class diagram.

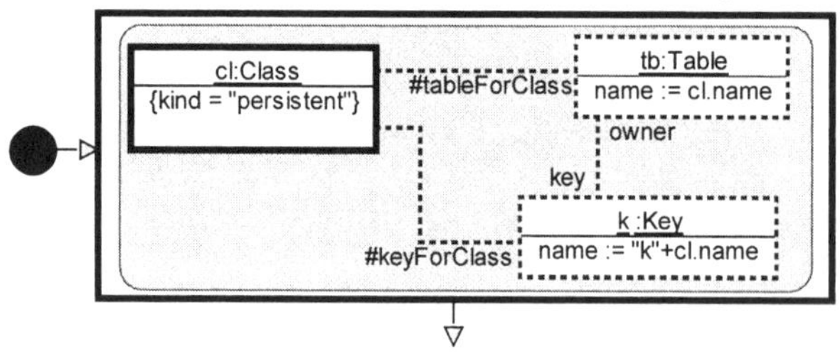

Figure 6. Simple statement in MOLA

This statement is a FOREACH loop consisting of its loop head only, this loop head is also a rule which builds new instances. It does the first job in the transformation process – builds instances of both *Table* and *Key* for any *Class* instance whose *kind* has the value *persistent*. The *name* attribute in each of the new instances is set to the specified value – to the *name* value in the matched *Class* instance. In addition, the two mapping associations are built, as well as an association instance with the roles *key* – *owner* between the new instances.

Fig. 7 shows the next two statements of the transformation program – both FOREACH loops too. The first one builds columns corresponding to primitive-typed attributes of persistent classes from the source model. Its pattern (discussed in section 4.1) selects the appropriate table in the target model (built by the previous statement) and the rule associates the new column to it.

The other statement in Fig.7 has the intention to attach the *belongsTo* association to each *Column* which corresponds to an *Attribute* with *kind* = *primary* (the other end of the association must be the *Key* for the relevant *Table*).

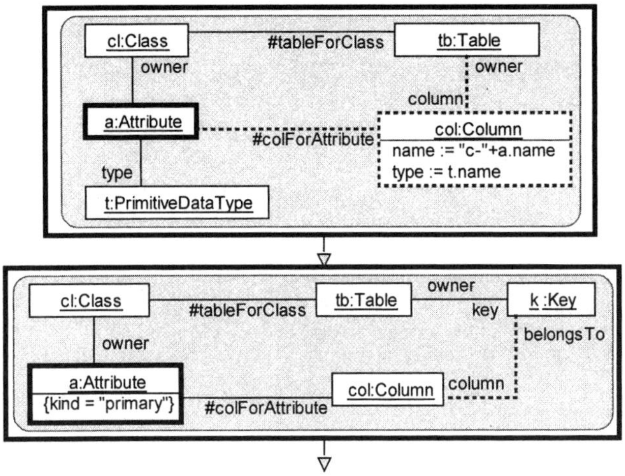

Figure 7. Statements transforming primitive attributes to columns and associating key columns

One more task to be done is to process associations in the source model. Fig.8 shows the corresponding program statement – a nested FOREACH loop. The top level loop builds the foreign key for each association in source model and associates it to the relevant primary key (built by the first statement). The nested loop builds a new column (in the table corresponding to the source class) corresponding to each column in the referred primary key. The pattern of this loop uses three references to instances selected in the top loop – all prefixed by the @ character. We remind that this means that namely these referenced instances must be used in any match for the subordinate pattern. Thus there is no need to repeat the corresponding selection conditions in the nested loop – its pattern becomes simple.

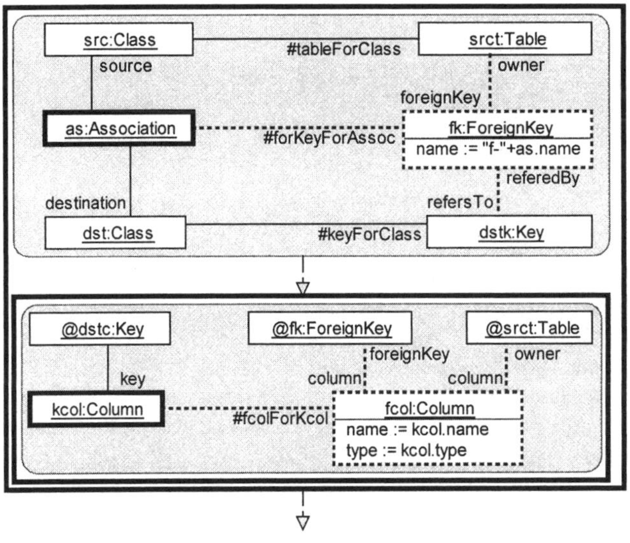

Figure 8. Statement processing source model associations

Certainly, the building of columns for a foreign key could be done by a separate independent loop, but then its pattern would be more complicated. In general, a certain "**breadth first programming style**" – each action a separate top level loop – is in most cases usable for typical MDA jobs. But sometimes nested loops help to structure complicated patterns.

The remaining task – flattening class-typed attributes can also be implemented in the basic MOLA described so far. However, since the flattening is a true transitive closure task, and in its most complicated form – find all possible paths in the source model from a class to its indirect primitively-typed attributes and compute a name along each of the paths, it requires creating copies of attribute instances, building temporary associations and attributes and using depth three loops. In other words, the standard algorithm for building all paths in a graph has to be implemented. An alternative more readable solution is to use extended patterns to be considered in the next section.

5. Extended Patterns

The patterns considered so far have one limiting property – only one instance can be matched to an element. Since this is too restrictive for some tasks in real transformations, especially those related to transitive closure, various ways to extend the pattern notation will be considered in this section. First, patterns with cardinality constraints will be considered – they may have unlimited number of instances associated with one pattern element, but the matching depth is still limited. An efficient match building procedure for this case is defined in a completely different way – as a stepwise algorithm on a graph. The efficiency of this procedure is guaranteed if uniqueness principle is observed – pattern cardinalities match to the metamodel multiplicities. Finally, the looping patterns with unlimited matching depth are introduced – namely these patterns are more powerful than those in [10, 11] and permit to perform nontrivial actions, including transitive closure, in one rule.

5.1 Cardinality Constraints for Navigation Associations

In section 4 the simplest case was considered where each pattern element was matched to a single instance and each association to a single association instance linking the matched class instances. In order to have larger fragments of the instance set mapped to the pattern, with several instances associated to one pattern element (what is really required by transformation rules), the extended MOLA uses **cardinality constraints** attached to pattern associations (actually something similar is used also in [10, 11]). In addition, the associations with cardinality constraints are treated as **directed** graph edges – using the UML navigability notation.

One of the constraints - **ALL**. It is used when the association has * or 1..* multiplicity at the appropriate end in the metamodel. It corresponds to * in UML – take all what you can, but nothing bad, if none. Another constraint is **OPT**. It is used with associations having 0..1 multiplicity at the appropriate end and means – take the instance if it exists, but the pattern does not fail if none exists. Actually OPT is a decorative version of ALL for the 0..1 case – to improve the readability. And we remind that empty cardinality constraint actually means **just one**. **NOT** also can be used as a cardinality constraint – there is none. In fact, there are constraints in MOLA which correspond to all possible UML multiplicities, but currently we don't need the other.

Now, more precisely, what is a valid extended MOLA pattern. Here we consider only the case when the pattern is used as a loop head. Then there is a loop variable, which serves as a pattern **root**. The pattern consists of **two** parts, having the **loop variable** as the **sole common node**. The first part, which uses undirected associations (and no additional cardinality constraints), is the same as before. It expresses (as before) the conditions for selecting valid instances for the loop variable. The other – the **extension** part uses **directed** associations and cardinality constraints and is used for matching to a set of instances. It must be a **directed acyclic graph (DAG)** starting from the loop variable as a root (a more complicated case with loops is considered in the next section). This part can be built by taking classes from the metamodel (in fact, IMM), converting them to pattern elements (adding instance names) and adding metamodel associations as directed edges. The extension part must be **distinguishable by associations** – if several edges with the same role name leave a node, they must lead to different classes (this requirement is essential for having an efficient match procedure, a weaker version of this restriction will be given in 5.2). The next step is to add appropriate cardinality constraints to the **navigable ends** of associations – ALL if the multiplicity in the metamodel is * or 1..* and OPT or nothing (one) if multiplicity is 0..1 or 1. When cardinality constraints are set this way, we say that the "**uniqueness principle**" is observed (the pattern fits to the metamodel). The pattern in Fig. 9 obviously satisfies the uniqueness principle – ALL is at the *attribute* end of the association from *Class* (where the multiplicity in IMM is *), all other multiplicities are 0..1. To emphasize the fact that we expect many instances of *Attribute* to be matched, a decorative element in the pattern – the **multiobject** notation (from UML collaborations) can be used for the *Attribute* element.

We make here also one assumption – the extension part edges leaving the root node have the constraint ALL or OPT (the extension part should express an unlimited, but optional at the same time part of the match).

Let us consider an example for the usage of ALL constraint in an extended pattern – the first statement from Fig.7 but defined in an alternative way – in Fig.9.

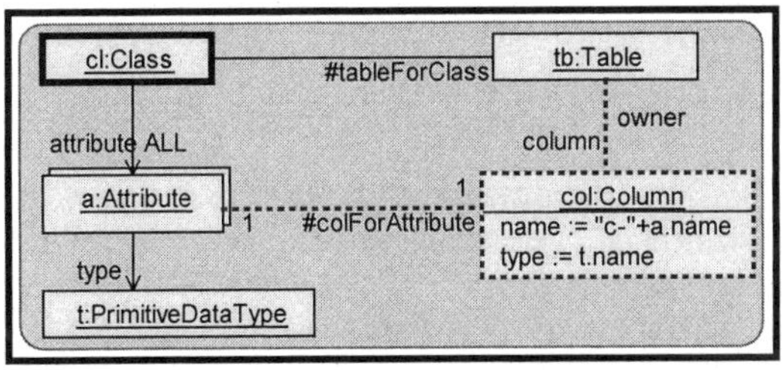

Figure 9. Rule with cardinality constraint

The loop variable accepts as valid those instances of *Class*, for which a *Table* has been built. Now, when the loop is executed for a valid instance of the loop variable, the following new action is performed. For the extension part of the pattern (containing the elements *a:Attribute* and *t:PrimitiveDataType*) a temporary instance **DAG** is built containing all matches for the given root instance. For the given example this DAG is very simple – **all** those *Attribute* instances for the given *Class* instance (root), which are linked to a *PrimitiveDataType* instance, together with the corresponding association instances (*attribute* and *type*). The result is indeed a DAG, but not a tree, because a *PrimitiveDataType* instance can be used for several *Attribute* instances. Here the original instances from the source model are used as nodes for the DAG (we can assume that the nodes and edges of the DAG are highlighted, e.g., by "coloring" them green), but in some cases node copies are built – see section 5.2.

When the instance DAG is built, the rule actions are performed. Here a *Column* instance is built for each *Attribute* instance in the DAG (i.e., an instance associated to the element *a:Attribute*). This is specified by the mapping association (*#colForAttribute*) which now has an explicit multiplicity 1-1. Then the DAG is discarded (the highlighting is removed).

Now we will define the match building for an arbitrary pattern. The most natural way is to use a **procedural match definition.** We treat the pattern (its extension part) as a DAG from the root and build the instance DAG starting from its root – the current loop variable instance. Valid instance nodes are added to it layer by layer, in strict accordance with the pattern – so that each instance node can be assigned to a pattern node at that layer and the edges also match. Here the number of layers is fixed (determined by the pattern). Cardinality constraints must be taken into account – if the constraint is ALL, it doesn't matter how many edges exit a node in the current layer, but for the default constraint (just one) there must be the corresponding edge to a node in the next layer. If that edge is not found, the node must be removed from the current layer as invalid. The pattern edges with ALL constraint actually generate fan-out cases in the instance DAG.

The formal definition of the **matching procedure** (generating a complete valid match – the instance DAG) is the following:

1. mark the current instance of the loop variable as the only instance in the layer one of the instance DAG and associate it to the pattern root.
2. take a node in the current layer of the instance DAG. For each directed association leaving the pattern node, to which the instance node is associated, find **all** association instances from the current node and select those where the target instance satisfies the corresponding attribute constraint; add this "filtered neighborhood" to the next layer. Repeat this for all nodes in the layer. If a node in the next layer has been reached twice, mark it only once (the path history is not important in this mode).
3. assign instances in the next layer to the corresponding elements in the pattern (it can be done uniquely due to the required distinguishability by associations).
4. check cardinalities – for each node in the current layer and for each navigation association (which has the default cardinality constraint - i.e., "just 1") from the associated pattern node check whether there is an instance of this association. If there is none, remove the instance node from the current layer, and recheck the previous layers (for layer one it cannot occur due to our assumption). ALL and OPT constraints require no check.
5. repeat steps 2, 3, 4 for each layer of the pattern

The semantics of ALL guarantees that always the maximal match is selected – no subset of a match can be a valid match. Even more, for a given root the procedure result is **deterministic** – it is due to the "uniqueness principle" for the pattern, that from several possible instances all are selected, and one instance must be selected from possible one. Namely this would permit also an efficient match implementation in MOLA.

5.2 Looping Patterns

In this section we introduce the final elements of extended patterns in MOLA. First, we permit patterns to have **directed loops** in the extension part when a pattern is built on the basis of a metamodel fragment. This extension is essential for defining a **transitive closure in a pattern**. Features will also be provided for defining a closure involving all possible paths.

The requirement introduced in 5.1 that the extension part must be **distinguishable by associations** is still in place. This requirement is sufficient for the example in Fig. 10 and many similar ones. A weaker restriction – the K-distinguishability – sufficient for any reasonable MDA task will be considered at the end of section (however, it makes the matching procedure more complicated).

Certainly, the **uniqueness principle** from 5.1 must be observed when assigning cardinality constraints to pattern edges – violating this principle would lead to a much more complicated and inefficient matching procedure.

Though a pattern now may have loops, the instance graph for it will be required to be a DAG anyway. From the theoretical point of view, we may be interested in finding instance-level loops via patterns, but no MDA related job was found where it makes sense. Therefore loops at the instance level will be simply forbidden by the matching procedure.

One more remark refers to nodes of the instance DAG. In 5.1 a simple case was considered where the original model nodes were used (just highlighted). But this implies that the path history cannot be stored in the instance DAG – if a node is reachable via two or more paths, data from the path cannot be stored in the node. The only way for storing this data is to make copies of the original instances and store them in the DAG. In an extended MOLA statement it can be specified which **pattern nodes must be copied** (such a node is marked by a **square icon**) during the building of the DAG (it makes sense only for the looping nodes). The temporary copies in the DAG are related to their originals and "inherit" all attributes and association instances from them. When the statement completes, the temporary copies are discarded. Copying selected pattern nodes is the easiest way to implement transitive closures where all paths from a node must be traversed – as the one in Fig. 10.

The same **matching procedure** from the previous section is usable, but with the following extensions:

- step 2. New instance which is already present in the DAG on the path (from the root to the current instance) is not added to the next layer – a safeguard against infinite loops. If the target instance corresponds to a pattern element in the "copy list", make a copy of the instance and of the relevant associations, relate the copy to the original.
- step 5. Repeat steps 2, 3, 4 until no instances are placed in the next layer (the repetition is no longer limited by the number of layers in the pattern due to possible loops)

A typical use of a looping pattern is for performing a **transitive closure** along a metamodel association. Transitive closure is directly supported in the textual languages [12, 14], but no other graphical pattern languages [7, 10, 11] support it.

Fig.10 shows an example of a looping pattern. It is the last statement in the class to database transformation program and performs a recursive "flattening" of the class diagram – adding indirect attributes (columns) to a class (table), whose direct attributes have another class as a type. In this example actually a transitive closure on the *attribute* association is performed. The only loop in the pattern is formed by the *type* association from *a2*. The *type* edge can lead either to *t* or to *c2*, the situation is distinguishable because they are of different classes. Both of the edges have the OPT cardinality constraint. During the pattern match, just one of these possible continuations will occur (because an *Attribute* always must have a *type*). If no *type* leads back to a *c2*, then looping along this path is finished. If all the looping is finished, a large instance tree (it is indeed a tree due to instance copying, except the "terminal" instances of *t*) with the root *Class* as a root and certain number of *Attribute* instances as leaves is built as the result of the match. The tree represents all possible paths via indirect attributes from the given class instance - due to the copying of *c2* and *a2* two paths never join. The leaves have a primitive type – they will be used for columns. However, it should be noted that both leave and non-leave *Attribute* instances are assigned to *a2* – they must be sorted out to find leaves only. This tree represents graphically the transitive closure of the *attribute* association.

Looping patterns typically involve complicated **assignments** to **computed attributes** of metamodel classes.

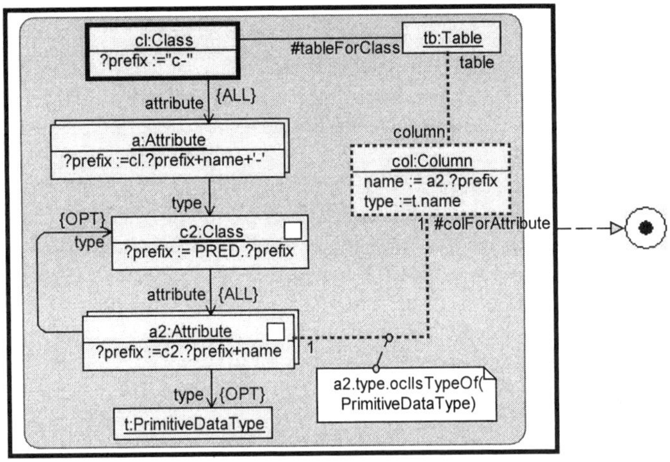

Figure 10. Rule with looping pattern

There are several kinds of computed attributes in MOLA, differing in their scope. Here the most used attributes are **statement-local attributes**, their scope being actions within one statement execution (here – one iteration of the loop, during which the extended match is built). Their values are discarded after the loop iteration is complete. Their main use is for finding various qualified or compound names, typically appearing in MDA tasks.

If the intermediate metamodel is used (here – Fig. 3), rule-local attributes are defined in it for all relevant classes, their **names start with "?"**. Their computation is performed immediately after the pattern match – when the instance DAG is complete. Rule-local attributes are computed in the same order as the match itself was built – starting from the root and moving away from it. If the instance DAG contains copies, these attributes are located in copies – not in the original instances. The assignment statement for a computable attribute in its expression part can contain the value of this attribute in the predecessor node and any values of normal (source) attributes in the current node (these are unqualified). The attribute value from the predecessor node can be qualified either by its instance name or by a **PRED** keyword. The use of PRED is required in cases when a node may have several nodes as predecessors – due to loops in patterns. Each node in the path must have the corresponding assignment statement, otherwise the attribute computation is terminated there. Several attributes may be computed simultaneously in a rule.

The example in Fig.10 contains assignments for the single computed attribute – *?prefix*, which is contained in *Attribute* and *Class*. The computation starts in the root, where the constant value – the string "*c-*" is assigned. When the value is propagated through an *Attribute*, the value of its *name* and the constant "-" is concatenated to it. The propagation through the *Class* node does not change the value. It is easy to see, that in the result the value of *?prefix* for each leaf of the match tree is the concatenation of all *Attribute* names along that path, separated by "-" and prefixed by "*c-*" – namely the value specified in this task as a *Column* name for indirect attributes. The obtained values are used namely for this purpose – they are used in the assignment for the *name* of the new *Column* instance. A *Column* is built only for those instances assigned to *a2* which are of primitive type (are leaves in the tree) – this is specified by the OCL constraint at the mapping association.

Fig. 10 completes the example transformation program in extended MOLA – the complete transformation consists of Fig. 6, 7, 8 and 10.

We conclude the section with a weaker restriction for patterns, than the distinguishability by associations. Namely, if in a pattern an association with the same role name can lead to several nodes, these nodes must be **K-distinguishable** – their neighborhood of order <=K (in the sense of directed graphs) must contain **mutually exclusive** elements – different local constraints, different mandatory (just one) associations, mandatory associations leading to different classes etc. The distinguishability by associations can be considered to be 0-distinguishability.

For most MDA examples – e.g., more complicated versions of the example in this paper, and many similar ones, typically K is 1 or 2. In order to use patterns with K-distinguishable elements, a **look-ahead** (in the instance set, but along the pattern edges) not longer than K has to be included in the matching step 3.

6. Conclusions

The paper describes the basic principles of the graphical model transformation language MOLA. There are two innovative elements in MOLA. One is natural combination of simple control structures with pattern based rules. The other one is the powerful pattern mechanism supporting variable cardinality and looping patterns, thus enabling transitive closure in patterns and simplifying even more the control structure of the language. The complete language MOLA is tested on several real world MDA examples, such as converting statecharts to FSM and realistic class-to-database transformation including class inheritance, transformation of business process models to workflows etc. The results show that in most cases more compact and readable rule definitions have been obtained, when compared to e.g., pure recursive style in [10]. The extended patterns permit to strike a right balance between complexity of rules and control structures governing them, thus providing the required transformation readability. Simple recursions, typical e.g., to statechart flattening job, can be specified in a readable way using the WHILE loop in basic MOLA.

The extended patterns, though more complicated than patterns of basic MOLA, are defined in a way to permit also an efficient implementation.

Acknowledgements

Authors of the paper are grateful for valuable discussions and comments provided by their colleagues at the IMCS Modeling and Model Transformations seminar. This research was partially funded by Science Council of Latvia under the project Nr.02 0002.

References

[1] OMG: MDA Guide Version 1.0.1, http://www.omg.org/docs/omg/03-06-01.pdf
[2] OMG: Unified Modeling Language: Superstructure. Version 2.0 (Final Adopted Specification). http://www.omg.org/cgi-bin/doc?ptc/2003-08-02 (2003)
[3] Booch G., Jackobson I., Rumbaugh J. The Unified Modeling Language. Reference Manual, Addison-Wesley, 1999.
[4] OMG: Meta Object Facility (MOF) Specification. Version 1.4. http://www.omg.org/cgi-bin/doc?formal/2002-04-03
[5] OMG: Request For Proposal: MOF 2.0/QVT. OMG Document ad/2002-04-10, http://www.omg.org/cgi-bin/doc?ad/2002-04-10
[6] Bettin J. Ideas for a Concrete Visual Syntax for Model-to-Model Transformations. Proceedings of the 18th International Conference, OOPSLA'2003, Workshop on Generative Techniques in the context of Model Driven Architecture, Anaheim, California, USA, October 2003.
[7] QVT-Merge Group. MOF 2.0 Query/Views/Transformations RFP, Revised submission, version 1.0. OMG Document ad/2004-04-01, http://www.omg.org/cgi-bin/doc?ad/2004-04-01
[8] Compuware, SUN. MOF 2.0 Query/Views/Transformations RFP, Revised Submission. OMG Document ad/2003-08-07, http://www.omg.org/cgi-bin/doc?ad/2003-08-07

[9] Interactive Objects Software GmbH, Project Technology, Inc. MOF 2.0 Query/Views/Transformations RFP, Revised Submission. OMG Document ad/2003-08-11, http://www.omg.org/cgi-bin/doc?ad/2003-08-11
[10] Willink E., A concrete UML-based graphical transformation syntax - The UML to RDBMS example in UMLX. Workshop on Metamodelling for MDA, University of York, England, 24-25 November 2003.
[11] Agrawal A., Karsai G, Shi F. Graph Transformations on Domain-Specific Models. Technical report, Institute for Software Integrated Systems, Vanderbilt University, ISIS-03-403, November 2003.
[12] DSTC/IBM/CBOP. MOF Query/Views/Transformations RFP, Second Revised Submission. OMG Document ad/2004-01-06, http://www.omg.org/cgi-bin/doc?ad/2004-01-06
[13] Celms E., Kalnins A., Lace L. Diagram definition facilities based on metamodel mappings. Proceedings of the 18th International Conference, OOPSLA'2003, Workshop on Domain-Specific Modeling, Anaheim, California, USA, October 2003, pp. 23-32.
[14] Kleppe A., Warmer J., Bast W. MDA Explained. The model driven architecture: practice and promise. Addison-Wesley, 2003.

Databases and Information Systems
J. Barzdins and A. Caplinskas (Eds.)
IOS Press, 2005

MDA: Correctness of Model Transformations. Which Models Are Schemas?

Karlis PODNIEKS
Institute of Mathematics and Computer Science, University of Latvia, 29 Raina Boulevard, Riga, LV-1459, Latvia

Abstract. How to determine, is a proposed model transformation correct, or not? In general, the answer may depend on the model semantics. Of course, a model transformation is "correct", if we can extend it to a "correct" instance data transformation. Where should model semantics be defined? Assume, model syntax and semantics are defined in the same meta-model. Then, how to separate syntax from semantics? The answer could be the definition of model schemas proposed in the paper.

Outline

The paper is structured as follows. Section 1 discusses the correctness of model transformations that are solving the UML to RDBMS transformation problem. Section 2 introduces an extension of this problem, in which model syntax and semantics are defined in the same meta-model. Section 3 is the core of the paper – it proposes a general definition of model schemas. Section 4 applies this definition to XML-schemas. Section 5 considers the relationship between schemas and model constraints. Finally, Section 6 discusses the related work.

1. Which UML to RDBMS Transformations Are Correct?

The Object Management Group (OMG) has issued a Request for Proposal for a Query/Views/Transformations (QVT) language that would allow defining of mappings between different information models [1]. "In defining mappings from model to model, the question of correctness of the mapping arises. ... The more complex form of correctness is that of semantic correctness; does the result of transformation *mean* the same thing as the input?" [2].

Indeed, let us consider a fragment of the example problem used by MOF QVT submitters, the so-called UML to RDBMS transformation problem ([3], Section 5.1.6). Figure 1 represents fragments of the input and output meta-models. Figure 2 represents an example input model – an interpretation of the UML meta-model.

The transformation problem is expressed as follows. The input model is an interpretation of the input meta-model. It consists of persistent and transient classes owning attributes. Attributes may be primitive (having a primitive data type), or complex (having a transient class as a type). The output model is an interpretation of the output meta-model. It consists of tables owning columns. The transformation in question must: a) Transform each persistent class into a single table. b) Transform each class attribute of a primitive type into a column of the corresponding table. c) "Drill down" class attributes of complex types to leaf-level primitive attributes; transform these primitive attributes into columns of the corresponding table.

How to determine, is a proposed transformation of this kind "correct", or not?

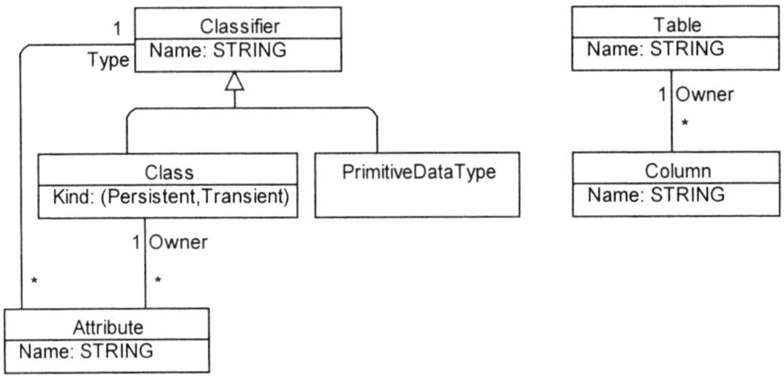

Figure 1. Fragments of UML and RDBMS meta-models

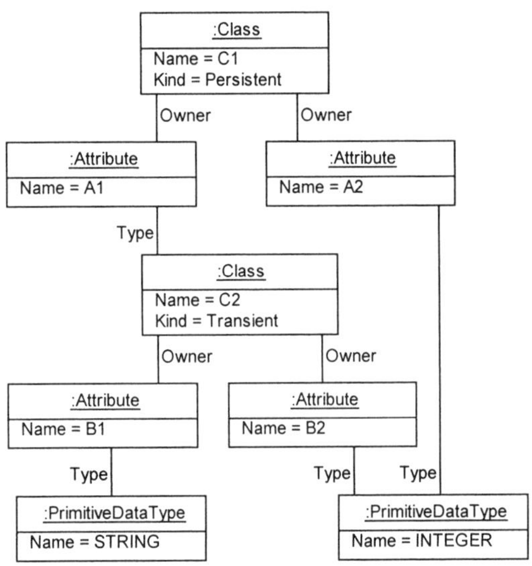

Figure 2. Example input model – an interpretation of the UML meta-model

Let us consider two different transformations that are transforming (uniformly) any UML model into a RDBMS model. These transformations will be demonstrated for the example input model represented in Figure 2. The generalization is obvious.

"Absolutely lossless" transformation. The following trivial transformation *T1* should be regarded as "absolutely lossless" – it transforms Figure 1 into Figure 3:

a) *T1* transforms a persistent class named *C1* into a table named *t_C1_Persistent*.

b) If the class *C1* owns an attribute *A1*, and the type of *A1* is a transient class *C2*, and *C2* owns an attribute *B1* of a primitive type *STRING*, then *T1* creates a column named *c_A1_C2_Transient_B1_STRING*.

c) Similarly, in all the other situations.

T1 is an "absolutely lossless" transformation, because it is reversible – **no information gets lost** during the transformation. Indeed, having an output model, generated by *T1*, we can restore all elements of the input model.

Note. Of course - with the exception of the transient classes that are not used as attribute (or sub-attribute) types in persistent classes.

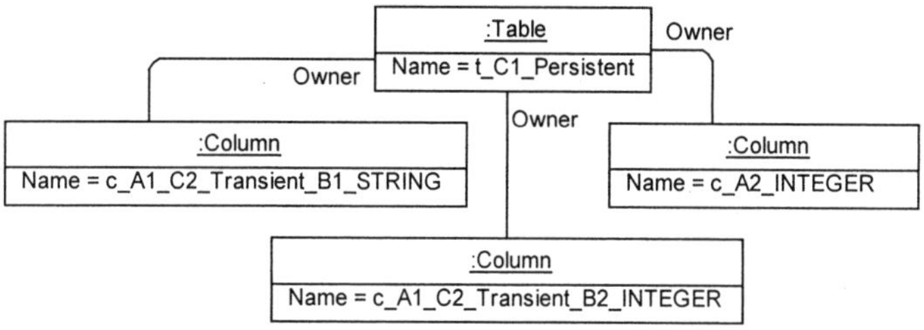

Figure 3. Example output model created by the transformation *T1*

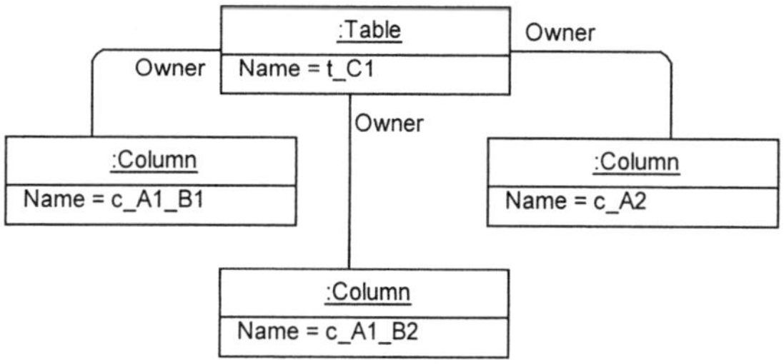

Figure 4. Example output model created by the transformation *T2*

Practical transformations do not need to be "absolutely lossless". Of course, none of the actual MOF QVT proposals is using this "absolutely lossless" transformation *T1* (see, for example, [3]). Instead of *T1*, they are using another transformation *T2*, which transforms Figure 2 into Figure 4, and thus, differs from *T1* as follows:

b) If the class *C1* owns an attribute *A1*, and the type of *A1* is the class *C2*, and *C2* owns an attribute *B1*of a primitive type *STRING*, then *T2* creates a column named *c_A1_B1*.

When compared to the *T1*'s version of the column name *c_A1_C2_Transient_B1_STRING*, the transformation *T2*, in its version *c_A1_B1*, omits the intermediate class name and attribute (*C2_Transient*), and the primitive type name (*STRING*). Thus, *T2* is not completely reversible - the information about names of transient classes and primitive types **gets lost during the transformation**. Why should we regard this widely used *T2* as a "correct" transformation? In which sense, the result of *T2* "*means* the same thing as the input" [2]?

The intended semantics of the UML to RDBMS transformation is as follows. We do not need this transformation by itself. We need it as a **basis for instance data (database contents) transformations**. The input UML model can be regarded as a schema of an object-oriented database, and the output RDBMS model – as a schema of a relational database. Thus, in fact, to solve the UML to RDBMS transformation problem completely, we must provide not only the model (i.e. database schema) transformation. To make it useful, we must provide also the instance data (i.e. database contents) transformation that would allow converting (without loss of information) the contents of any object-oriented database into the contents of a relational database. The solution of this problem is a well-known topic described in database textbooks for students.

And, of course, for the above small fragment of the problem (Figure 1), the database contents transformation *D2* extending the schema transformation *T2* is trivial. In the resulting database created by *D2*, we do not represent the instances of intermediate complex attributes (like as *A1* in Figure 2), and links connected to them. But, nevertheless, *D2* **is completely reversible**. Indeed, we can restore easily the contents of the input (object-oriented) database from the contents of the output (relational) database, by using, additionally, the information contained in the database schemas:

a) For each row of the table *t_C1*, create an instance of the class *C1*.

b) For a cell corresponding to the column *c_A1_B1*, and containing the value *"123"*, create (if not created before) an instance of the attribute *A1*, and link it to the corresponding *Owner* instance of *C1*, create (if not created before) the corresponding *Type* instance of the class *C2*, and create an instance of the attribute *B1* containing the value *"123"*, and link it to the corresponding *Owner* instance of *C2*.

c) Similarly, in all the other situations.

Thus, by referring to database schemas, we can restore all the input database information, missing in the output database. And thus, the pair *T2+D2* can be regarded as a lossless transformation.

Note. Of course, in the MDA context, many transformations do not need to be lossless. In MDA, transformations may lose information; they may merge parts of several models, add new information via user interfaces etc. In MDA, a model transformation is acceptable, if it performs its task.

2. Where Should Model Semantics Be Defined?

As we now see, it may happen that specifying the correctness of model transformations may be impossible, if we restrict the problem to the model syntax, and ignore **model semantics**.

In [4], after considering several model management operators, the author concludes (see his Section 3.10): "The model management operators defined in Section 3 are purely syntactic. That is, they treat models and mappings as graph structures, not as schemas that are templates for instances… Still, in most applications, to be useful, models and mappings must ultimately be regarded as templates for instances. That is, they must have semantics. Thus, there is a semantic gap between model management and applications that needs to be filled."

Mathematical theories are formalized by using the first order predicate logic, i.e. by using some first order language, the axioms of predicate logic and by assuming the necessary specific axioms of a particular theory. The generally acknowledged technique of exploring the "semantics" of formal mathematical theories is the notion of **interpretation** (see any logic textbook: sorts, their interpretation domains, interpretations of constant letters, function letters and predicate letters, standard interpretations of logical connectives and quantifiers). In computer science, finite interpretations (i.e. interpretations with finite domains) are, in general, more important than the infinite ones.

In modeling, usually, the formal aspects of models are specified by using meta-models. Meta-models are serving here as "theories of models". Hence, by applying the widely approved technique of mathematical logic, we could propose to think of **models as interpretations of their meta-models**.

Frequently, meta-models are represented as UML class diagrams (together with sets of constraints written in OCL). Of course, these diagrams (together with their constraints) can be represented as first order formal theories (if necessary, small subsets of set theory may be involved). Thus, a correct semantics of meta-model diagrams can be obtained automatically by applying the above-mentioned standard notion of interpretation.

In [5], the authors argue (at the end of their Section 5.3): "there are many reasons to avoid stating that "a model is an instance of a meta-model because its elements are instances of meta-model-elements". Indeed, the relationship between a theory and its interpretations is much more complicated than the relationship between a class and its instance objects.

Note. Sadly enough, in mathematical logic and in computer science, the term "model" has opposite meanings. In mathematical logic, "a model of a theory" is an interpretation under which all axioms of the theory become true (i.e. model is a kind of "reality" modeled in the theory). In computer science, "model" means some formal structure that can replace a fragment of reality in our reasoning about this fragment (i.e. model is a kind of "theory" modeling a fragment of reality). In such a situation, speaking of models is misleading, and to avoid this, the term "interpretation" could be used instead: an interpretation of a theory (of a meta-model) is an interpretation under which all axioms (constraints) of it become true.

In many cases, meta-models define mainly the allowed syntax of the corresponding models, and not their full semantics. Where should model semantics be defined?

As an example, let us consider the models (database schemas) corresponding to the meta-models represented in Figure 1, and represented in Figures 2, 3, 4. In Figure 4 we see the "table" *t_C1* (what's a table?), which owns three "columns" – *c_A1_B1*, *c_A1_B2*, and *c_A2* (what's a column?). Of course, we know that, in fact, each table is a collection of **rows** (not mentioned in schemas); each row consists of **cells** (not mentioned in schemas); each cell carries a value and corresponds to one of the columns (assigned to the table); and, in each row, each column (assigned to the table) is represented by at most one cell. But, of course, this knowledge cannot be derived from Figure 4, which represents (according to its meta-model of Figure 1) only the data specific to the database consisting of a single three-column table *t_C1*, and not the general semantics of relational databases.

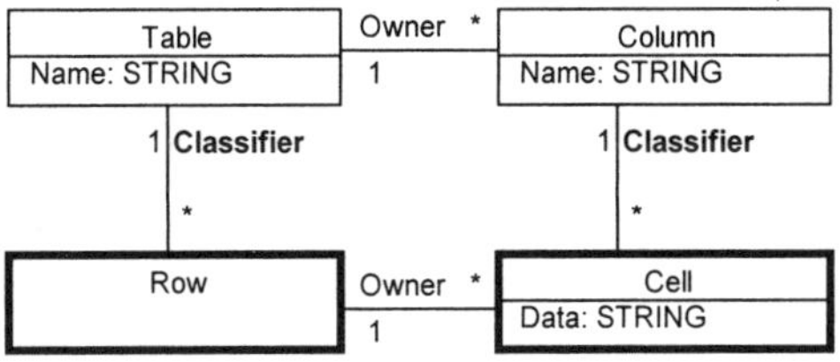

Figure 5. Fragment of RDBMS semantics meta-model (compare with Figure 1)

However, this knowledge **can be derived** from the RDBMS semantics meta-model represented in Figure 5 (with the following non-graphical constraint added: in each row of a particular table, each column – of this table - is represented by at most one cell).

Note. The above-mentioned non-graphical constraint can be stated in an OCL-like language as the following two statements:

Row is *Owner* of *Cell* --> *Row.Classifier* is *Owner* of *Cell.Classifier*

CellA.Owner = *CellB.Owner* --> *CellA.Classifier* <> *CellB.Classifier*

The first statement may be put alternatively as a diagram commutativity condition:

Cell.Owner.Classifier = *Cell.Classifier.Owner*.

By applying the standard notion of interpretation to Figure 5 we obtain, in fact, "two in one" – the database schema (i.e. table names with column names assigned to them), together with the database contents (i.e. cell values arranged in rows, columns and tables). This corresponds very well to the situation in the popular RDBMSs, where database schema is regarded as "internals" of each database ("database definition"). For example, in

SQL we do not use CREATE SCHEMA statements; instead, we are using CREATE TABLE statements (this allows defining of tables and their relationships "on the fly").

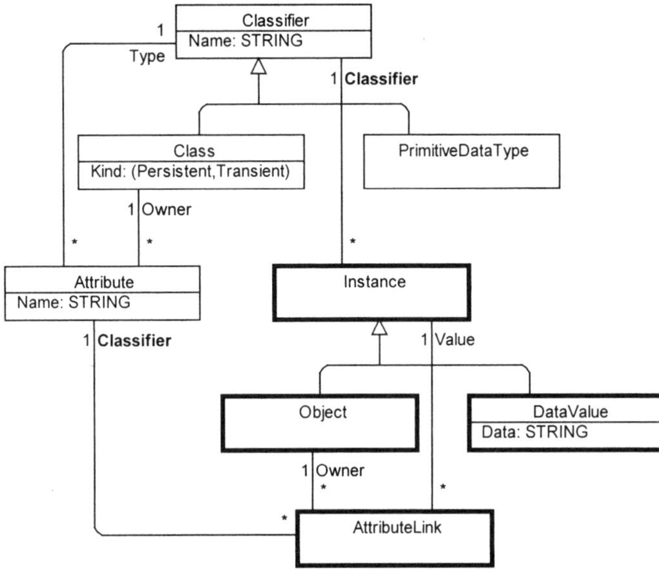

Figure 6. Fragment of UML semantics meta-model (compare with Figure 1)

Similarly, we may follow the official UML semantics definition [6], and, to capture the intended class diagram semantics missing in the UML fragment meta-model of Figure 1, extend it as represented in Figure 6 – with the following two constraints added: a) in each object of a particular class, each attribute – of this class - is represented by at most one attribute link, b) each attribute link is linked to its value – an object, or a data value, depending on the type of the corresponding attribute. After this, by applying the standard notion of interpretation, we are forced to have, again, "two in one" – classes together with their instance objects.

Note. The above-mentioned non-graphical constraints can be stated in an OCL-like language as follows:

Object is *Owner* of *AttributeLink* --> *Object.Classifier* is *Owner* of *AttributeLink.Classifier*

AttributeLinkA. Owner = *AttributeLinkB.Owner* -->
AttributeLinkA.Classifier <> *AttributeLinkB.Classifier*

Instance is *Value* of *AttributeLink* --> *Instance.Classifier* is *Type* of *AttributeLink.Classifier*

Object.Classifier is *Class*

DataValue.Classifier is *PrimitiveDataType*

The first and the third of these constraints may be put, alternatively, as diagram commutativity conditions:

AttributeLink.Owner.Classifier = *AttributeLink.Classifier.Owner*

AttributeLink.Value.Classifier = *AttributeLink.Classifier.Type*

If OMG, in its Request for Proposal for QVT language [1], would have used the Figure 5, 6 style meta-models (instead of the Figure 1 style ones), then the QVT partners would be forced to demonstrate that their proposed languages are good enough for simultaneous transformations of models and instance data.

Thus, once again, where should model semantics be defined? Now, we see a possible solution: we may **define model semantics directly in the meta-model**.

3. Which Models Are Schemas?

Is there a systematic way allowing to separate, in a meta-model, the "data" elements (boxes in bold) from the "schema" elements (regular boxes)?

Instances of "schema" elements (in Figures 5, 6 - *Table, Column, Class, Attribute, PrimitiveDataType*) are present in databases "by schema". For example, any relational database contains exactly those tables that are listed in its schema, and each table contains exactly the columns listed for it in the schema. Collections of "data" element instances, on the contrary, may differ in different databases having a common schema. For example, two common-schema relational databases containing a table named *Customers*, may contain different collections of rows of this table.

Under which conditions, a part of a meta-model could be regarded - meaningfully - as a "schema" for the rest of it?

From the above two examples the following two theses can be derived.

Thesis 1. Schema represents only the information specific to a particular model, and not the general semantics of the model-type to which it belongs.

Thesis 2. Schema defines a classification of data elements that conforms to the associations existing between these elements.

Let us try justifying Thesis 2. Simultaneously, its precise meaning will be elaborated (see Definition 1 below). Let us assume that our meta-models are defined by means of UML class diagrams with non-multiple generalizations and binary associations only.

First, the following restriction seems to be reasonable: each data element should be "named in the schema" or, more precisely, each data element should have a mandatory many-to-one or one-to-one association with some unique schema element. For data element instances, such an association defines a kind of **classification**. Thus, let us always call this distinguished association *Classifier*. For example, in Figure 5, *rows* (data element instances) are classified by *tables* (schema element instances), and *cells* (data element instances) are classified by *columns* (schema element instances). If some data element would not be linked in such a way to a schema element, then, in which sense could we speak about a "schema"?

In Figure 6, a somewhat more complicated situation appears: the data element class *Instance* consists of two subclasses – *Object* and *DataValue*. At the schema level, this triple is represented by the *Classifier* class consisting of two subclasses – *Class* and *PrimitiveDataType*. Both of these "triades" conform to the classification defined by the *Classifier* association in the following sense: *Object.Classifier* is always *Class*, and *DataValue.Classifier* is always *PrimitiveDataType*.

Thus, we have arrived at the following condition for schemas:

Condition 1 (the simplest case, Figure 5). Each data element has a mandatory many-to-one or one-to-one association (called *Classifier*) with some unique schema element (i.e. different data elements are classified by different schema elements).

Condition 1 (the full version, Figure 6). Subclasses of schema elements are schema elements. Subclasses of data elements are data elements. In the generalization hierarchy, each top-level data element has a mandatory many-to-one or one-to-one association (called *Classifier*) with some unique schema element (i.e. different data elements are classified by different schema elements). The data element generalization hierarchy is mapped into the schema element generalization hierarchy in the following sense: a) *Data_Element.Classifier* is always *Data.Element.Map*; b) If *Data_Element_1* is a subclass of *Data_Element_2*, then *Data_Element_1.Map* is a subclass of *Data_Element_2.Map*.

The next step: **associations** connecting data elements also should be represented in the schema. For example, in Figure 5, the association *Owner* connects data elements *Cell* and *Row*. The association *Owner* between *Column* and *Table,* in fact, represents this *Cell-Row*-association in the schema part of Figure 5 - because the following constraint holds (see above):

Row is *Owner* of *Cell* --> *Row.Classifier* is *Owner* of *Cell.Classifier*

Indeed, on the one hand, each cell is owned by some row, which is classified by some table. On the other hand, each cell is classified by some column, which is owned **by the same table** that classifies the row.

Thus, we have arrived at the following

Condition 2 (Figure 7). For each *Data_Association* there is a unique *Schema_Association*, to which it conforms in the following sense:

Data_Element_1 is in *Data_Association* with *Data_Element_2* -->
Data_Element_1.Classifier is in *Schema_Association* with *Data_Element_2.Classifier*

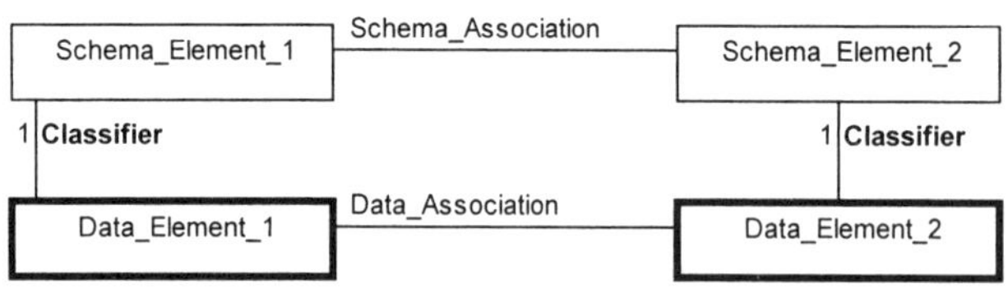

Figure 7. Representing a data association in a schema

Here, *Data_Association* conforms to the classifications of *Data_Element_1*-s and *Data_Element_2*-s: if x is associated with y via *Data_Association*, then the classifiers of x and y must be associated via *Schema_Association* (but not necessarily conversely!).

Some more specific situations are possible:

a) x and y are associated via *Data_Association*, **iff** the classifiers of x and y are associated via *Schema_Association*. Then, in fact, *Data_Association* is completely derivable from *Schema_Association*. Hence, such a *Data_Association* can be removed from the meta-model without loss of information.

b) *Data_Association* and *Schema_Association* both are many-to-one. Then Condition 2 may be put equivalently as a diagram commutativity condition:
Data_element_1.Data_Association.Classifier =
Data_element_1.Classifier. Schema_Association

c) **"Table"**. Both associations are many-to-one, and, additionally, *x1.Data_Association* = *x2.Data_Association* --> x1.*Classifier* <> *x2.Classifier*. Then, Figure 7 is isomorphic to Figure 5, i.e. we can regard instances of *Schema_Element2* as tables ("table views"), consisting of columns named by instances of *Schema_Element1*. Instances of *Data element_2* represent rows, and instances of *Data_Element_1* – cells.

Now, what about the associations, connecting data elements and schema elements, other than the *Classifier* of Condition 1? First of all, we must reject - as "non-schematic" - the associations that are connecting one instance of a data element with several instances of the same schema element. Thus, it remains to consider only many-to-one and one-to-one associations of data elements with schema elements (as *DS_Association* in Figure 8). In fact, each such *DS_Association* defines a different classification of instances of *Data_Element_1* (than the one defined by the *Classifier* association).

If *DS_Association* conforms to the "canonical" classification of *Data_Element_1*, i.e. if there is a schema association *S_Association* such that

Data_Element_1.DS_Association = *Data_Element_1.Classifier.S_Association*,

then, in fact, *DS_Association* is completely derivable from *S_Association*. Such a *DS_Association* can be removed from the meta-model without loss of information.

If, on the contrary, *DS_Association* does not conform to the "canonical" classification, then we may "refine" this classification by introducing a new *Schema_Element_12* consisting of pairs (*Schema_Element_1*, *Schema_Element_2*) and by re-defining the *Classifier* association as follows (Figure 8):

(*Schema_Element_1*, *Schema_Element_2*) classifies *Data_Element_1*, iff
Schema_Element_1 classifies *Data_Element_1* (in the old sense) &
Data_Element_1 is in *DS_Association* with *Schema_Element_2*.

After this, *DS_Association* becomes completely derivable from the new *Classifier* association and *S_Association* defined as follows (Figure 8):

(*Schema_Element_1*, *Schema_Element_2*)
is in *S_Association* with *Schema_Element_2*.

And, as such, we can remove *DS_Association* from the meta-model.

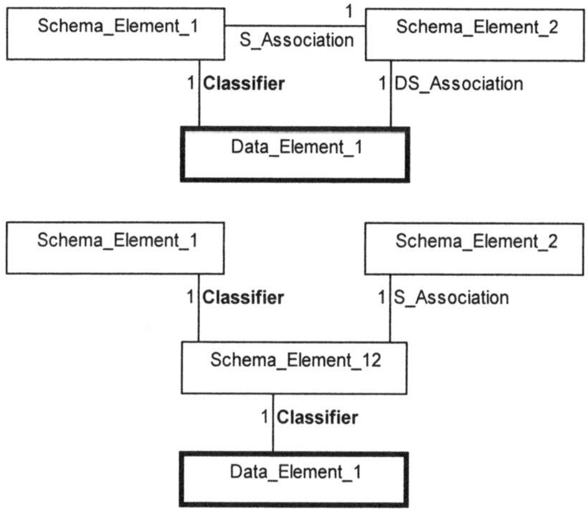

Figure 8. Representing a data-schema association in a schema

Thus, we have arrived at the following

Condition 3. *Classifier* is the only kind of associations between data elements and schema elements.

Now, we can formalize the notion of schemas:

Definition 1. Assume a meta-model that is defined by means of a UML class diagram with non-multiple generalizations and binary associations only. Assume, its classes are divided in two disjoint subsets - schema elements and data elements in such a way that the above Conditions 1, 2, 3 hold. Then, let us say that this **meta-model defines schemas**.

Note. For some of the data associations, the above-mentioned stronger "Table" version of Condition 2 may be appropriate.

Definition 1 conforms to Thesis 2. Indeed, by this definition, schemas are classifications of data instances that conform to associations existing between these elements. Thesis 2 can be used as a guideline when trying to define schemas in new situations (for an example, see Section 4 below).

Problem A. Does Definition 1 conform to Thesis 1?

Definition 2a. Assume a meta-model that defines schemas. **Model** (**database**) is an interpretation (in the sense of predicate logic) of the meta-model (i.e. an interpretation that includes both the schema part and the data part).

Definition 2b. Assume a meta-model that defines schemas. **Model schema** (**database schema**) is an interpretation of the schema part of the meta-model (i.e. each model includes its schema).

Now, in this context, let us consider model transformations. If two meta-models define schemas, then we may consider two kinds of transformations.

Definition 3a. Assume two meta-models MMA and MMB that define schemas. **Model transformation** is an algorithm transforming each MMA-model into an MMB-model in such a way that the schema part is transformed into the schema part, and the data part – into the data part.

Definition 3b. Assume two meta-models MMA and MMB that define schemas. **Schema transformation** is an algorithm transforming the schema part of each MMA-model into the schema part of an MMB-model.

To be considered as lossless, model transformations do not need to be reversible with respect to the schema part of input models. This is not necessary, because schema element instances belong to the input model "by schema", i.e. they can be restored from the meta-data - from the input schema. Thus, we can propose the following uniform definitions of lossless transformations.

Definition 4a. Assume two meta-models MMA and MMB that define schemas. Let us consider a **model transformation** D that converts any MMA-model into an MMB-model. Then D is called **lossless**, iff, there is a reverse transformation (algorithm) that restores, from any of the results of D (i.e., MMB-models), the entire **data part** of the corresponding input MMA-model.

Definition 4b. A **schema transformation** is called **lossless**, iff it can be extended to a lossless model transformation.

Both of the UML to RDBMS schema transformations considered above (the "theoretical" transformation *T1* and the "practical" transformation *T2*) can be extended to lossless model transformations (see Section 1). Thus, according to the Definition 4b, both schema transformations are lossless.

Problem B. Which reversible schema transformations (like as the above "theoretical" transformation *T1*) can be extended to lossless model transformations?

Problem C. How complicated is the task of detecting, does a lossless model transformation exist for two model schemas, or not?

4. XML-Schemas

In order to verify the above concept of schemas, let us consider XML and XML-schemas [7]. The bottom part of Figure 9 represents a simplified meta-model of "unconstrained XML", where tagged elements may be mixed up freely without any typing.

If we wish to obtain here a kind of schemas, then, according to Thesis 2, we must introduce some **classification** of XML elements. Of course, the solution is obvious: **XML-tags** define the natural XML-element classification. In this way we obtain a simplified XML-schema meta-model represented in Figure 9 - with the following non-graphical constraints added:

"For types: **no loops** are allowed via *Parent* association."

XML_Record_Type is *Parent* of *XML_Element_Type* <-->
    XML_Element_Type is *Field* of *XML_Record_Type*

XML_List_Type is *Parent* of *XML_Element_Type* <-->
    XML_Element_Type is *Member* of *XML_List_Type*

XML_ElementA is *Parent* of *XML_ElementB* -->
    XML_ElementA.Classifier is *Parent* of *XML_ElementB.Classifier*
        XML_Atom.Classifier is *XML_Atom_Type*
        XML_Sequence.Classifier is *XML_ Sequence_Type*

This schema constrains XML-documents: now, an element cannot contain arbitrary sub-elements, it may contain only sub-elements of specific types pre-scribed by the schema.

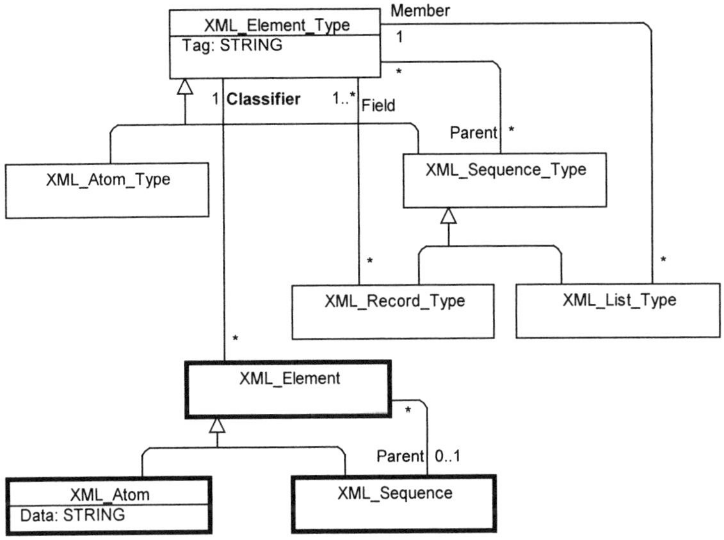

Figure 9. Simplified meta-model of XML-schemas

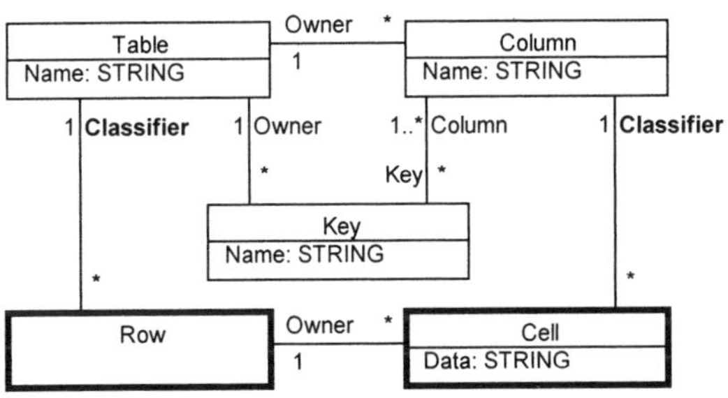

Figure 10. Extended fragment of RDBMS semantics meta-model

5. Constraints

As an example, let us consider key-constraints in relational databases. A table may possess zero or more keys, each consisting of one or more columns of this table. The corresponding meta-model is represented in Figure 10 - with the following commutativity constraint added: *Key.Column.Owner = Key.Owner.*

The meaning of the key-constraint defined by *Key* can be expressed as follows: in a table, if the cell data of two rows coincide for all columns that belong to *Key*, then these rows are equal. Or, expressed in an OCL-like language:

For All *Table, RowA, RowB*:
If *RowA.Owner = Table & RowB.Owner = Table & KeysAreEqual*(*RowA*, *RowB*)
Then *RowA = RowB*,

where *KeysAreEqual*(*RowA*, *RowB*) is the following expression:

For All *Column, CellA, CellB*:
If *Column.Key = Table & CellA.Owner = RowA & CellB.Owner = RowB &*
CellA.Classifier = Column & CellB.Classifier = Column
Then *CellA.Data = CellB.Data*

Thus, as a schema element, *Key* can serve only as an "enumerator" of constraints of a specific kind. The definition of the **meaning** of these constraints involves more than schema – it involves data elements and their *Data* attributes. This is not surprising (see Thesis 1): relational schemas do not define the general semantics of relational databases. Each schema represents only the information specific to a particular database.

Note. In a similar way, arbitrary functional dependencies can be specified.

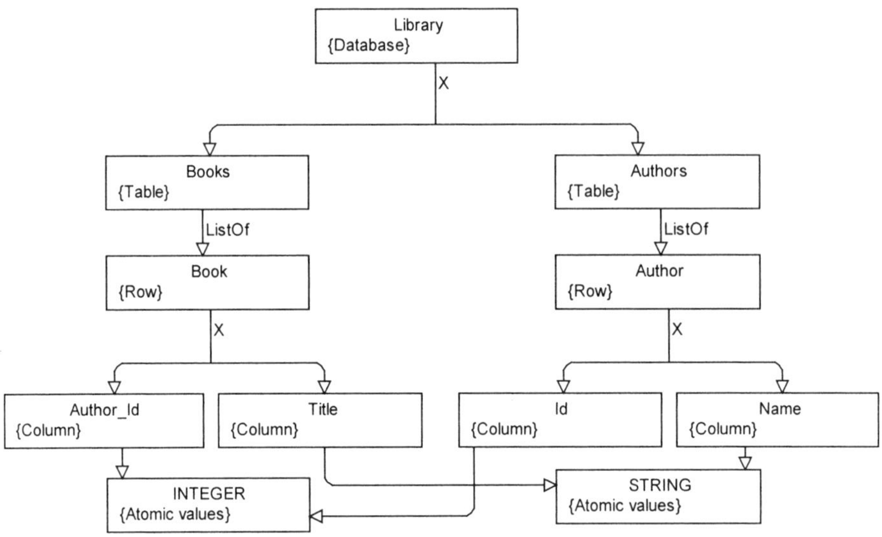

Figure 11. Example database schema according to [12]

6. Related Work

About the significance of diagram commutativity in modeling semantics – see [8].

In [9] an elegant theory of "graph schemas" for unstructured data is developed. An unstructured set of data may conform to several schemas, each of which, in its way, constrains data, thus allowing for query optimization. Despite the different setting, in their Section 5 the authors arrive at a version of the above Thesis 2: "Nodes in a schema have the potential to classify nodes in a database".

After the first versions of [10, 11], in [12] an extremely general algebraic definition of database schemas (called "abstract schemas") is proposed. To explain the basic idea, let us define a schema for a relational database *Library* consisting of two tables *Books* and *Authors* (S means "sort", see also Figure 11):

S(*Library*) = *Books* X *Authors*;
S(*Books*) = ListOf(*Book*); S(*Authors*) = ListOf(*Author*);
S(*Book*) = *Author_Id* X *Title*; S(*Author*) = *Id* X *Name*;
S(*Author_Id*) = INTEGER; S(*Title*) = STRING;
S(*Id*) = INTEGER; S(*Name*) = STRING;

Each named database element is defined here as a sort (i.e. domain) of allowed values. The last four elements are atomic, the other ones are complex, and their domains are defined by using a fixed set of type constructors (ListOf constructs lists of values, X – records of values). A database is defined then as any value of the sort *Library*.

In general, a database schema of this kind can be defined as an acyclic oriented graph of the kind represented in Figure 11 (single root, atomic sorts as leafs, accessible nodes only).

This notion of schema conforms to the above Thesis 1: schema represents only the data specific to the *Library* databases, and not the general semantics of "algebraic" databases (i.e. schemas do not define the meaning of INTEGER, STRING, ListOf and X). However, if we would restrict our databases to the relational databases only, then mentioning **rows** in the schema would become obsolete: **all** relational tables consist of rows and cells.

How about Thesis 2? Of course, sorts define a classification of data instances. Since the data model of "abstract schemas" does not include associations (they are implemented by using primary and foreign keys and the corresponding constraints), this is enough to conclude that the notion of "abstract schemas" conforms to Thesis 2.

Acknowledgements

I'm grateful for valuable discussions provided by my colleagues, especially Janis Barzdins, Audris Kalnins and Martins Opmanis. Science Council of Latvia has funded the work reported in this paper under project Nr.02-0002. Pictures were prepared by using the graphical business-modeling tool GRADE (www.infologistik.com, www.gradetools.com).

References

[1] OMG Document ad/02-04-10 (MOF 2.0 Query / Views / Transformations RFP). Available at www.omg.org.

[2] Gerber A., Lawley M., Raymond K., Steel J., Wood A. Transformation: The Missing Link of MDA. In: Proceedings of Graph Transformation: First International Conference (ICGT 2002), October 7-12, 2002, Barcelona, Spain, Lecture Notes in Computer Science, vol. 2505, Springer-Verlag, 2002, pp. 90-105.

[3] QVT Partners. Initial Submission for MOF 2.0 Query / Views / Transformations RFP. Version 1.0 (2003.03.03). Available at qvtp.org.

[4] Bernstein Ph. A. Applying Model Management to Classical Meta-Data Problems. In: Proceedings of the Conference on Innovative Database Research (CIDR), 2003, pp. 209-220.

[5] Bezivin J., Gerbe O. Towards a Precise Definition of the OMG/MDA Framework. In: ASE'01, Automated Software Engineering, San Diego, USA, November 26-29, 2001. Available online.

[6] OMG Document - formal/03-03-09 (UML 1.5 chapter 2 - UML Semantics). Available at www.omg.org.

[7] W3C Recommendation - XML Schema Part 1: Structures, May 2, 2001. Available at www.w3c.org.

[8] Johnson M., Dampney C. N. G. On the Value of Commutative Diagrams in Information Modelling. In: Proceedings of 3rd International Conference on Algebraic Methodology and Software Technology (AMAST'93), Springer Workshops in Computing, 1993, pp.45-58.

[9] Buneman P., Davidson S. B., Fernandez M. F., Suciu D. Adding Structure to Unstructured Data. In: Proceedings of 6th International Conference on Database Theory (ICDT'97), Lecture Notes in Computer Science, vol. 1186, Springer-Verlag, 1997, pp. 336-350.

[10] Diskin Z. Abstract Metamodeling, I: How to Reason about Meta- and Metamodeling in a Formal Way. In: Proceedings of the 8th OOPSLA Workshop on Behavioral Semantics, Denver, USA, November, 1999.

[11] Alagic S., Bernstein Ph. A. A Model Theory for Generic Schema Management. In: Proceedings of International Workshop on Database Programming Languages (DBPL '01), Lecture Notes in Computer Science, vol. 2397, 2002, pp.228-246.

[12] Goguen J. A. Data, Schema and Ontology Integration. In: Workshop on Combination of Logics: Theory and Applications (CombLog'04), Lisbon, Portugal, July 28-30, 2004. Available online.

Databases and Information Systems
J. Barzdins and A. Caplinskas (Eds.)
IOS Press, 2005

A Model Driven Approach to Business Rules Model Transformations

Sergejus SOSUNOVAS, Olegas VASILECAS
Information Systems Laboratory, Vilnius Gediminas Technical University
Vilnius, Lithuania

Abstract. Business rules based development has became very popular during the last several years. Two major groups in the business rules community can be identified clearly: the first one concentrates on taxonomies of business rules and the second one analyses implementation issues of business rules in inference systems. This paper provides a novel approach to implement business rules. Object Management Group's model driven architecture is used to implement business rules using meta-model based transformations. The issues of transformation of business rules, which are formally specified in UML/OCL to the event-condition-action rules of active database management system are analysed. The paper presents results of transforming of one kind of business rules expressed in OCL to event-condition-action rules. The formal transformation specification in TRL language is presented.

1. Introduction

An unavoidable technological change challenges nowadays enterprises. In order to stay profitable and competitive enterprises have to accept this challenge and to adopt new technologies in their business environment as fast as possible. However utilization of new technology does not always mean change of the way business goes. Technological changes may result in increasing the speed of business or the quality of information what affects the quality of business decision making. The main business practices and experience possibly will stay the same.

Enterprise transition to the modern information processing and management requires to establish a new project, and to execute appropriate phases like business analysis and requirement elicitation. Larger part of the enterprises already has executed the number of business analysis and requirement specification phases in the previous projects hence the new one project could use already created business models and requirements specifications. Nevertheless existing models were created with the old technology in mind and could not be very useful. Hence existence of technological independent business oriented models will simplify enterprise adaptation for the future technologies.

The problem of technological change and reuse of models is addressed by the object management's group (OMG) model driven architecture (MDA). MDA is based on the existence of the several layers of models and transformation of models of platform independent layer to the models of platform specific layer. At the last layer applications are generated. Model transformation is defined by transformation rules of the transformation specification. These transformation rules refer to the meta-models of model being used. Meta-models and common meta-model creation requirements defined by OMG meta object facility (MOF) and formal specification of business rules using UML/OCL [1], [2] enables model transformation in process of enterprise systems development using business rules (BR) approach.

A modern enterprise system consists of business, information and software systems [3]. It is usual to create models of business systems, elicit business needs, specify information system (IS) requirements, and then using IS needs to specify software systems requirements. At each of before mentioned steps some kind of models should be created. Transformation between these models usually is executed manually by appropriate IT staff. It is feasible to map the models of these systems to the corresponding MDA levels and automate transformation.

BR are the general part of all these systems. BR in the business systems context are defined as a directive, intended to influence or guide business behaviour, in support of business policy that has been formulated in response to an opportunity, threat, strength, or weakness [4]. Whereas in the IS context BR is statement that defines or constrains some aspect of the business [4]. Usually BR are implemented in the software code. It is possible to claim that automatic transformation of BR from one system to another will facilitate system development.

The rest of the paper proceeds as follows. Section 2 maps business, information and software system models to the corresponding MDA levels. Section 3 presents theoretical foundations of BR in IS specified in OCL transformation to the event-condition-action (ECA) rules of software systems using MDA. Transformation example of one of BR type and supporting meta-models are described in Section 4. Section 5 provides formally specified transformation rules. Section 6 summarises the proposed approach.

2. Related Work

System development using MDA framework implies creation of models of following types: platform independent models (PIM) and platform specific model (PSM). Models of PIM enable specification of the several, usually selected by the user, platform independent and technological free perspectives of the modelled system. To simplify the complexity of PIM for different user groups (e.g. business staff, IT staff) several abstraction levels can be outlined: computation independent models (CIM) and PIM per se [5]. Though CIM and PIM overlap to some extent, the boundaries of these levels are defined in terms of problems and solutions they represent. CIM assumes existence of problem statement and its description, as far as PIM includes a number of platform independent solutions [6].

Informational needs and related problems are identified during analyses of business systems. Business system - the place where the problems reside, is a set of interrelated business processes performed to achieve certain long duration goals [3]. The usage of CIM to model business systems simplifies understanding of these systems, elimination of the problems and its sources. Supporting processes of IS fulfilling the needs of business systems. IS concentrate on information and information processing methods therefore these systems are technology free and can be modelled by PIM. IS are supported by business software systems, the latter ones employs the power of different technologies to provide information processing facilities for IS. The technological solutions of business software systems are modelled by PSM.

Technical environment enabling an execution of the code of business software systems itself is a system. This environment – computer system – includes necessary hardware, networks, system and middleware software, and a realisation of business software system. Computer system can be modelled using platform dependent model introduced in [6].

BR are one of the integral part of the business system. BR exist and are used in each of before mentioned systems however the form of BR may be different in every particular system. In the business system BR exists in the form of mission statement and business goals usually expressed in natural language. Besides the number of rules exist in unexpressed yet form.

BR modelling in business systems is challenging issue because of a business variety and the lack of general business vocabulary. Following an industry demand for general business vocabulary and BR meta-model OMG issued request for proposals for business semantics and BR [7]. Submitted responses [8], [9], [10] present three different approaches for BR modelling at CIM level. Meta-models for a limited number of BR are presented in [11], [12]. Additionally, BR can be specified using RuleML [13] language. The degree BR are involved in business, information and business software systems development may be different. Business rules approach [14], [15] assumes that BR are the main component of the development of these systems and provides some BR modelling methods.

BR in IS appear to be more information-centric and usually refer to some data items. BR from business system have to be transformed to corresponding constructs of IS, hence CIM have to be transformed to PIM. During this transformation the form of BR changes and additional semantic value is gained. Usually during transformation of BR specified in business system in natural language are formalised. In the context of PIM BR are most often formally expressed in OCL. This language was created to specify business constraints at IBM insurance and is powerfully enough to express different types of BR [16], [17], and [18].

OCL is declarative language however in order to implement BR on some platform it is needed to transform them to imperative form. The notion of event, when violation of BR occurs and an action is important to model BR using PSM. ECA rules are widely used in implementations of different platforms and it is feasible to transform BR specified in OCL to the ECA rules. ECA model as the result of transformation can be used to generate applications or triggers of active database management system that is more effective than implementation by means of declarative assertions [16].

Model transformation is essential part of the MDA framework. Transformation specification, a set of transformation rules, is used to define transformation of one model to another model, assuming that models are based on the meta-models complying with the MOF. Meta-model transformation approaches [19], [20] are implemented in a number of transformation specification languages: UMLX [21], ATL [21], BOTL [23], TRL [24]. Despite of numerous publications on this topic, the transformation of BR between different enterprise systems components, for example business systems and IS, is not addressed suitably.

3. Transformation Decisions

OCL became part of the UML specification from version 1.1. It was developed in the IBM and is based on the first order logic. OCL is designed to specify constraints on object-oriented models (e.g. UML). The semantics of UML itself is defined using OCL.

Every OCL expression begins specifying its applying context. The context may be a UML model classifier (e.g. class name, class attribute) or a method. Depending on the context it is possible to use an invariant, a keyword: inv, when context is classifier or pre- and post condition, the keywords pre and post, when the context is a method. There is a condition at the end of each OCL expression. The invariant condition must be satisfied by each instance of the model. When the context is method pre condition must be satisfied before executing the method, post condition must be true just before the end of the execution. Despite of the word “constraint” in the title of the OCL not only constraint BR can be specified. The UML 2.0 specification will include the notion of actions which will expand the OCL application domain for other BR types. One of the possible appliances of the action clause is to specify events being sent during the execution of the method, allowing to specify actions of BR, or to specify what would happen when the invariant fails.

In this paper it is assumed that constraint BR are specified using OCL invariants. OCL invariants are designated to constraint UML class diagrams. Events, initiating part of ECA rules, are generated after some action executed on data item. In order to transform OCL clauses to the ECA rules it is necessary to transform UML class diagram to the database schema. In this paper "one-class-one-table" approach is used, each UML class is transformed to the table of relational database (RDB). Corresponding transformation of interclass relations is executed. The formal transformation specification is presented in [24]. However MDA allows easily change UML-RDB transformation principles do not affecting OCL-ECA transformation.

OCL is a purely declarative language and can be expressed in database in the following way: using assertion statements [25], using views [26]. In order to transform OCL clauses to the ECA rules it is necessary to answer the following questions [27]:

- What event initiates a trigger?
- What is trigger granularity?
- What is trigger condition?
- What is trigger initiation time?

OCL specification defines that every instance of the model must satisfy OCL constraints. Hence there are three strategies to implement OCL constraints: toleration, avoidance and combined. Constraint violation toleration strategy implies checking of all implemented OCL constraints from time to time. Usually user initiates checking and takes appropriate action if the violation is detected. In violation avoidance strategy user is alerted if violation occurs, the last transaction triggered violation is cancelled. Often in the case of complicated BR user from the business environment may be not competent enough to solve the problem which initiates violation of BR. Then combined strategy should be used. According to this strategy two types of BR are distinguished. Violation of BR of first type could be solved by the user and transaction must be cancelled. Respectively violation of the BR of the second type must be only detected and transaction must be stopped for a while.

This description of OCL clauses implementation uses the violation avoidance strategy. It is necessary to watch a value of every attribute used in the OCL clause. Hence each invariant instance must be evaluated every time those attributes are created changed, deleted. Therefore triggers implementing constraint must be initiated on database column creation, modification and deletion actions.

Trigger initiation time implies the time when the trigger condition must be evaluated, before execution of trigger initiated action or after. But if the trigger condition is executed before the action it does not have access to the created/modified data. It necessary to check condition after trigger initiated action.

Inserting n records in a time trigger can be executed one or several times, depending on the trigger granularity. There are two types of trigger granularity statement level and row level. Using violation tolerance strategy, when all constraints are evaluated on the snapshot of the database it would be feasible to use statement level triggers, however avoidance strategy must be implemented using row level triggers.

OCL clauses are effect free, which means that specification doesn't describe what action should be executed if the violation of constraint occurs. Implementing OCL clauses in ECA rules of active database systems the system can just alert user about constraint violation or cancel transaction. During database transaction execution it is possible temporary violations of constraints, therefore it is needed to check constraints just before the end of transaction, so called deferred triggers must be used.

4. Example of Transformation

In the following example BR belonging to the comparative evaluators' family will be transformed to the ECA rules. The comparative evaluators' BR family contains business rules of several types: equal-to, not-equal-to, greater-than, greater-than-or-equal-to, less-than. An example of comparative business rule: "Cigarettes can not be sold to the customers younger than 18 years old". This business rule expressed in OCL (context is presented in Fig. 2):

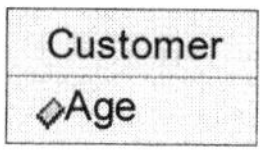

Figure 1. UML model for the OCL example

```
Context Customer
Inv AgeConstraint: Age >=18
```

In order to transform such business rule to the ECA rule according to MDA framework two meta-models are needed, one describing the semantics of OCL and the second describing semantics of ECA rules. OCL specification already contains general meta-model, for our purposes that meta-model is too big and the view of it must be created. Simplified OCL meta-model used for transformation of comparative evaluators' BR and complete to present such kind of BR is depicted in Fig. 3 and Fig. 4.

ECA meta-model which is partly based on meta-model presented in [11] and extended to include ability to represent conditional statements is depicted in Fig. 6. Event of ECA rule is usually related to some data structure, in simplified ECA meta-model this data structure is "Table" from [24] (Fig. 5).

As it is common for MOF compliant meta-models almost all elements are inherited from one single element. In ECA meta-model case this element is called "ModelElement" (Fig. 7) and its one attribute "name", used in the child elements in a number of ways (e.g. to represent operation sign, and elements names).

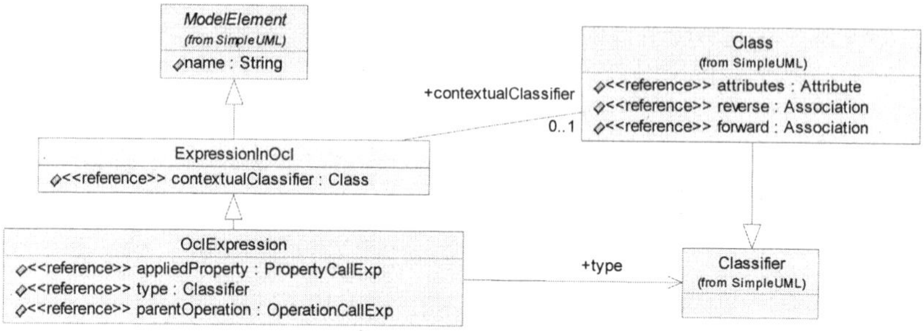

Figure 2. Simplified OCL meta-model (First part)

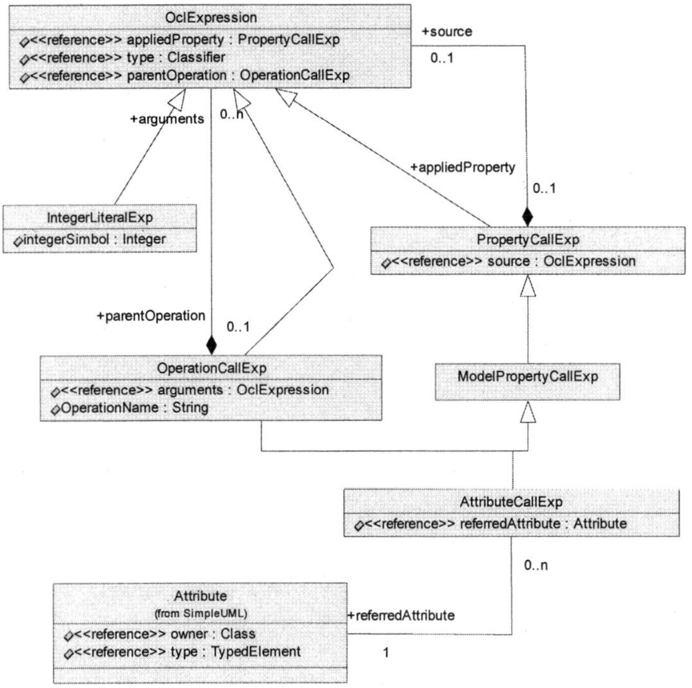

Figure 3. Simplified OCL meta-model (Second part)

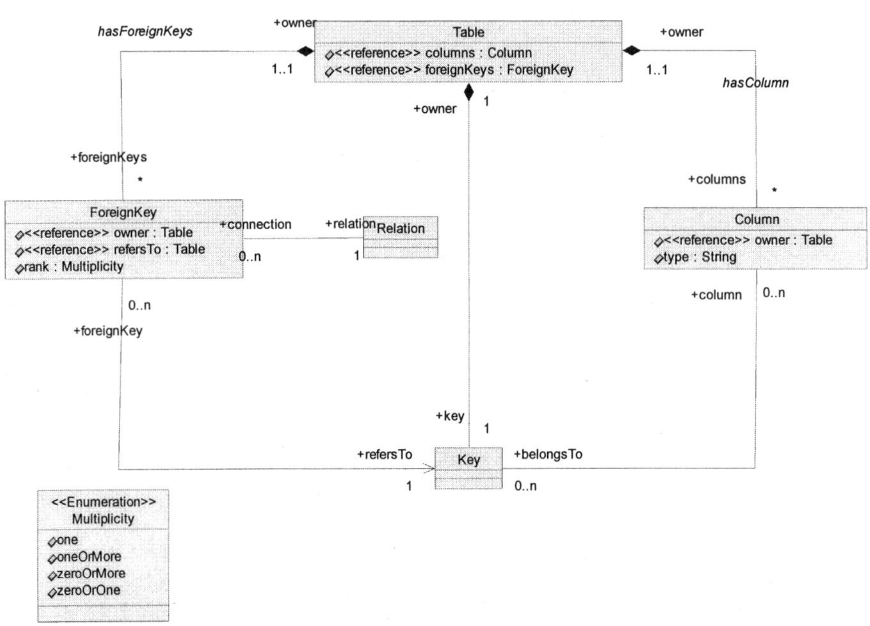

Figure 4. Simplified relational data base meta-model [24]

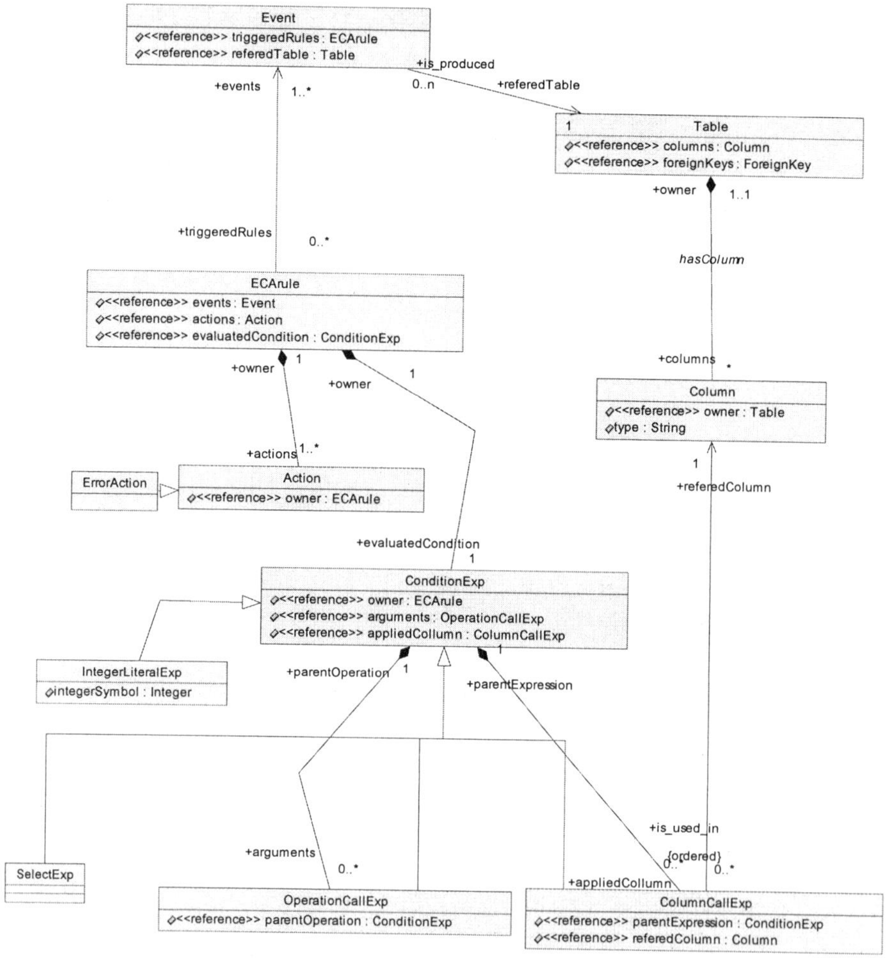

Figure 5. Simplified ECA rule meta-model

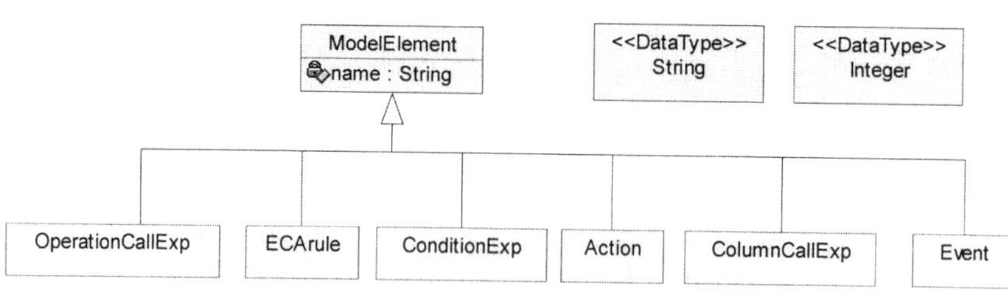

Figure 6. Classes of ECA simplified meta-model

5. Transformation Specification

There is no standard transformation language for the specification of transformation rules of meta-model based transformations. However the general structure of transformation rules proposals is quite similar. Every transformation language first declares the meta-model or the view of meta-models being used in transformation. There is a pattern in the beginning of each transformation rule specifying what source meta-model elements will be used for the transformation and destination meta-model constructs will be produced. The main differences between different transformation languages are: the type of transformation rules (declarative or imperative), the notion of source-destination models, rule usage strategies, rule organisation, transformation directions [19].

Transformation rule language (TRL) [24], is used to define model queries and transformations according to MOF 2.0 meta-model principles. This language is an OCL extension, hence bigger part of methods including model navigation methods are taken from OCL. TRL is organised in modules and packages. Three main packages are: query, transformation and view.

TRL extends OCL in a number of ways aimed to enable and facilitate transformation. Low coupling between different transformation specifications allows easily apply different transformation approaches for the different parts of the same meta-model. Ability to refer to the already transformed elements simplifies the development of specification.

OCL-ECA transformation specification presented in this paper is defined using TRL. This specification extends UML-DBMS transformation specification defined in [24]. UML-DBMS transformation specification is necessary to refer to the "Column" and "Table" already transformed elements.

OCL-ECA transformation specification (formal definition):

```
Unit UML_to_SimpleRDB
source SUML_OCL
target SimpleARDB

EntryPointRule Create_UML_Relation
creates Relation{
  allOfType(AssociationEnd).Create_UML_SurogatKey(#context.name ,#result)
  #result.connection=#context.connection.Create_UML_ForeignKey()
  allOfType(ExpressionInOcl).Create_ECA()}
Rule Create_UML_SurogatKey(){
  #context.type.Create_UML_Key()}
Rule Create_UML_ForeignKey()
creates ForeignKey
return #result{
  #result.refersTo=#context.type.Create_UML_Key()
  #context.association.connection.Surrogate_Foreign(#context.type.name ,#result)
  #result.name=#context.name
  #context.multiplicity.range.Create_Mult(#result)}
Rule Create_Mult(#result_mult){
  if(#context.lower.equals(1)) {
   if(#context.upper.equals(1)) {
     #result_mult.rank=one } else {
     #result_mult.rank=oneOrMore}
  } else {if(#context.upper.equals(1)) {
     #result_mult.rank=zeroOrOne
   } else {#result_mult.rank=zeroOrMore}}}
Rule Create_UML_Key()
creates Key
return #result{
```

```
    #result.name=#context.name
    #result.owner=#context.Create_UML_Table()
    allOfType(Stereotype).Create_Unique(#result ,#context)}
  Rule Create_UML_Table()
  creates Table
  return #result{
    #result.name=#context.name
    #result.columns=#context.feature.Create_UML_Column()}
  Rule Create_UML_Column()
  creates Column
  return #result{
    #result.name=#context.name
    #result.type=#context.type.name}
  Rule Create_Unique(#key_column ,#key){
   if(#context.name.equals(identifier)) {
     #context.extendedElement.Create_Unique_Second(#key_column ,#key)
   } else {}}
  Rule Create_Unique_Second(#key_column ,#key){
   if(#context.owner.name.equals(#key.name)) {
     #key_column.column=#context.Create_UML_Column(#key)
   } else { }}
  Rule Set_UML_Associations(#uml_class_name ,#sardb_reffered_tables){
   if(#context.type.name.equals(#uml_class_name)) {
     #context.association.connection.Surrogate_Foreign(#context.type.name
,#sardb_reffered_tables)
   } else { }}
  Rule Surrogate_Foreign(#uml_class_name ,#sardb_reffered_tables){
   if(#context.type.name.equals(#uml_class_name)) {
   } else {
     #sardb_reffered_tables.owner=#context.type.Create_UML_Table() }}
  Rule Create_ECA()
  creates ECArule
  return #result{
   #result.name=#context.name
   #context.contextualClassifier.Create_Events(#result.events)
   #result.actions=#context.Create_Actions()
   if(#context.bodyExpression.isOfType(OperationCallExp)) {
     #result.evaluatedCondition=#context.bodyExpression.Create_OperationCallExp()
   } else {  }}
  Rule Create_OperationCallExp()
  creates OperationCallExp
  return #result{
   #result.name=#context.referredOperation.specification
   #context.arguments.Create_Arguments(#result.arguments)}
  Rule Create_Arguments(#arguments){
   if(#context.isOfType(IntegerLiteralExp)) {
     #arguments.add(#context.Create_IntegerLiteralExp())
   } else {
     #arguments.add(#context.Create_ColumnCallExp())}}
  Rule Create_IntegerLiteralExp()
  creates IntegerLiteralExp
  return #result{
   #result.integerSymbol=#context.integerSymbol}
  Rule Create_ColumnCallExp()
  creates ColumnCallExp
  return #result{
```

```
  #result.referedColumn=#context.referredAttribute.Create_UML_Column()}
Rule Create_Actions()
creates ErrorAction
return #result{
  #result.name=Violation_of_OCL_constraint}
Rule Create_Events(#events){
  #events.add(#context.CreateInsert_Event())
  #events.add(#context.CreateUpdate_Event())}
Rule CreateUpdate_Event()
creates Event
return #result{
  #result.name=Update
  #result.referedTable=#context.Create_UML_Table()}
Rule CreateInsert_Event()
creates Event
return #result{
  #result.name=Insert
  #result.referedTable=#context.Create_UML_Table()}
```

6. Conclusions

Transformation development process has some similarities with classical system development processes. First the transformation requirements are specified, then transformation is designed in a declarative manner. At the high abstract level the mapping between appropriate components of destination and source models is defined. Transformation is implemented specifying imperative transformation rules. In order to achieve necessary quality the transformation correctness should be evaluated. It is feasible to apply iterative transformation development process.

Developing of the transformation rules is very close to the programming of usual software. So well known approaches to the organisation of business rules can be applied. However because of the specificity of transformation rules (e.g. high granularity) the new programming techniques and tools should be developed.

MDA enables uncomplicated migration of the BR based systems to new platforms. Modelling of BR in platform independent language OCL allows precise and unambiguous specification of BR. Presented transformation of OCL clauses to ECA rules and then to triggers of active DBMS is a common and efficient approach to implement BR. BR implemented as triggers are more easily maintained and changed.

References

[1] OMG UML 2.0 OCL Draft Adopted Specification. OMG document: ptc/03-08-08. http://www.omg.org/cgi-bin/apps/do_doc?ptc/03-08-08.pdf

[2] OMG. OMG/MOF Meta Object Facility (MOF) Specification. OMG document: formal/02-04-03, http://www.omg.org/cgi-bin/apps/do_doc?formal/02-04-03.pdf.

[3] Čaplinskas, A., Lupeikienė, A., Vasilecas, O. Shared Conceptualisation of Business Systems, Information Systems and Supporting Software. Databases and Information Systems II. Kluwer Academic Publishers, 2002, pp. 109-120.

[4] Ross, R., G. Principles of the Business Rule Approach. Addison Wesley, 2003.

[5] Siegel, J. Developing in OMG's Model-Driven Architecture. OMG document: 01-12-01, 2001. http://www.omg.org

[6] Bézivin, J., Gerard. S., Muller, P.-A., Rioux L. MDA components: Challenges and opportunities. Proceedings of Metamodelling for MDA workshop, York, England, November, 2003. http://www.cs.york.ac.uk/metamodel4mda/onlineProceedingsFinal.pdf

[7] OMG. Business Semantics of Business Rules RFP. OMG document: br/03-06-03, 2003. http://www.omg.org/cgi-bin/apps/do_doc?br/03-06-03.pdf.

[8] Adaptive, Business rule Solutions LLC et all. Business Semantics of Business Rules. OMG document: bei/2004-01-04, 2004. http://www.omg.org/cgi-bin/doc?bei/04-01-04

[9] Fujitsu Limited. Business Semantics of Business Rules. OMG document: br/2004-01-01, 2004. http://www.omg.org/cgi-bin/doc?br/04-01-01

[10] International Business Machines. Business Semantics of Business Rules. OMG document: br/01-01-02, 2004. http://www.omg.org/cgi-bin/doc?br/04-01-02

[11] Herbst, H. A Meta-Model for Business Rules in Systems Analysis. CAiSE'95 proceeding. Berlin, Springer, 1995, pp. 186-199.

[12] Moriarty, T. Business Rule Management Facility: System Architect 2001. On-line paper. http://www.intelligententerprise.com/000801/metaprise.jhtml?_requestid=441609

[13] Boley, H. The Rule Markup Language: RDF-XML Data Model, XML Schema Hierarchy, and XSL Transformations. Proceedings of INAP2001 LNCS 2543. Invited Talk, Tokyo, Springer-Verlag, 2003, pp. 5-22.

[14] Ross, R., G. The business Rule Book (2nd ed.). Business Rule Solutions, Houston, 1997.

[15] Von Halle, B. Business rules applied: building better systems using the business rules approach. John Wiley & Sons, New York, 2001.

[16] Wagner, G., Tabet, S., Boley, H. MOF-RuleML: The Abstract Syntax of RuleML as a MOF Model. Integrate 2003 proceeding. http://www.omg.org/docs/br/03-09-01.doc

[17] Sosunovas, S., Vasilecas, O., Using UML/OCL for business rules modelling. Proceedings of Sixth Lithuanian young scientist conference, Vilnius Gediminas Technical University, Technika, Vilnius, 2003, pp: 139-148.

[18] Sosunovas, S., Vasilecas, O. The meta-model based transformation of business rules models. In A. Caplinskas, V. Denisov at al (eds.). Proc. of the Conference "Information Technology'2004", Kaunas University of Technology, 2004, pp. 585-592. (in Lithuanian)

[19] Czarnecki, K., Helsen., S. Classification of Model Transformation Approaches. OOPSLA'03 Workshop on Generative Techniques in the Context of the MDA proceedings, 2003, http://www.softmetaware.com/oopsla2003/mda-wokshop.html

[20] Gardner T., Griffin C., Koehler J., Hauser R. A review of OMG MOF 2.0 Query/View/Transformations toward the final Standard. OMG document: ad/03-08-02. http://www.omg.org/cgi-bin/apps/do_doc?ad/03-08-02.pdf

[21] Willink, E. D. UMLX - A Graphical Transformation Language for MDA, Workshop on Model Driven Architecture Foundations and Applications, University of Twente, The Netherlands, 2003, pp. 13-24.

[22] Bézivin, J., Dupé, G., Jouault, F., Pitette, G., Rougui, J., E., First experiments with the ATL model transformation language: Transforming XSLT into XQuery, 2nd OOPSLA Workshop on Generative Techniques in the context of Model Driven Architecture, October 23-30 2003, California, USA. http://www.softmetaware.com/oopsla2003/bezivin.pdf

[23] Brauna, P., Marschall, F. The Bi-directional Object-oriented Transformation Language. Technical report, Technische Universität München, TUM-I0307, 2003, http://wwwbib.informatik.tu-muenchen.de/infberichte/2003/TUM-I0307.pdf

[24] Alcatel, Softeam, Thales, TNI-Valiosys, Codagen Technologies Corp. Response to the MOF 2.0 Query/View/Transformations RFP. OMG document: ad/2003-08-05, 2003, http://www.omg.org.

[25] Demuth, B., Hussmann. H., Using OCL constraints for relational Database Design. LNCS 1723, Springer-Verlag Berlin Heidelberg, 1999, pp. 598-613.

[26] Demuth, B., Hussmann, H., Loecher. S. OCL as a Specification Language for Business rules in Database Applications Design. LNCS 2185, Springer-Verlag Berlin Heidelberg, 2001, pp. 104-117.

[27] Badawy, M., Richta. K. Deriving triggers from UML/OCL specification. In M. Kirikova at al (Eds.). Proceedings of ISD 2002 conference on Information Systems Development: Advances in Methodologies, Components and Management. Kluwer Academic/Plenum Publishers, 2002, pp. 305-316.

Databases and Information Systems
J. Barzdins and A. Caplinskas (Eds.)
IOS Press, 2005

Reverse Engineering of Relational Databases to Object Databases

Irina ASTROVA
Tallinn University of Technology
Ehitajate tee 5, 19086 Tallinn, Estonia

Abstract. Whereas reverse engineering of relational databases to object databases is a widely studied subject, a majority of the work on schema transformation has been done on analyzing key correlation. Data and attribute correlations are considered rarely and thus, have received little or no analysis. This is partially evident from that the existent approaches can extract only a small subset of semantics embedded within a relational database; for example, they can fail in discovering inheritance and optimization structures. Another problem with the existing approaches is that they can require much user interaction for semantic interpretation, thus giving less opportunity for automation. As an attempt to resolve these problems, we propose a novel approach, which is based on an analysis of key, data and attribute correlations, as well as their combination. Our approach transforms a relational database schema into an object database schema in the context of Unified Modeling Language (UML). It has two important advantages over the existing approaches in that: (1) it allows the user to extract more semantics from a relational database, and (2) it reduces user interaction, which is mainly required to confirm the extracted semantics and to give them names.

1. Introduction

Reverse engineering of relational databases to object databases is defined as a process that obtains semantics about the legacy database, then converts the schema from relational to object, and finally represents the results as an object database schema [1]. There are several motives that require reverse engineering of relational databases to object databases; for example, to gain an understanding of semantics of the legacy database [2], document the legacy database structure (i.e. schema) [3] and migrate the database from relational to object [4].

Regardless of these motives, the main task of reverse engineering of relational databases to object databases is *schema transformation*, that is, the translation of the relational database schema into a semantically equivalent database structure expressed in the new (e.g. object) technology [5]. In particular, the schema transformation consists of deriving an object database schema from a relational database schema, by replacing a construct (possibly empty) in the relational database schema with a new construct (possibly empty) in the object database schema [1]. Mapping a relation into a class, replacing a primary/foreign key with a generalization relationship are examples of the schema transformation.

2. Related Work

Over the past ten years, researchers have proposed different approaches to reverse engineering of relational databases to object databases, as shown in Table 1. The existing approaches differ in their specific assumptions, inputs and methodological characteristics. However, a majority of that research has been done on analyzing key correlation. Data and attribute correlations are considered rarely and thus, have received little or no analysis.

Table 1. Approaches to reverse engineering of relational databases to object databases

Approach	Assumptions	Input	Output	Characteristics
Albert et al.'s [6]	3NF	- Relational database schema - Data instances	Object database schema	- Provides a simple and an automatic schema transformation - But does not cope with inheritance and optimization structures
Premerlani et al's [3]	Non-3NF	- Relational database schema - Data instances	Object database schema (in the context of OMT)	- Emphasizes on an analysis of candidate keys, rather than primary keys - Copes with optimization structures - But requires much user interaction
Ramanathan and Hodges' [7]	3NF	- Relational database schema - Key dependencies	Object database schema (in the context of OMT)	- Requires less input information - But does not cope with optimization structures
Behm et al.'s [4]	3NF	- Relational database schema - Data instances - Key dependencies - Inclusion dependencies - Types of relationships (strong or weak)	Object database schema	- Includes both schema transformation and data mapping - But considers only data equality and data inclusion
Tari and Stokes' [8]	3NF	- Relational database schema - Data instances - Key dependencies	Object database schema	- Emphasizes on an analysis of external keys, rather than primary keys - Analyzes key and data correlations, as well as their combination - But does not distinguish between data equality and data inclusion

This is partially evident from that the existent approaches can **extract only a small subset of semantics embedded within a relational database**, thus producing an object database schema that looks rather "relational". For example, [4, 6, 8, 7] fail in discovering optimization structures. Even though, more recently, new tools have appeared to provide an automatic mapping of the relational model to the object model, they still have limitations. For example, Rational Rose [9] maps an inheritance structure to a composite aggregation. After the schema transformation, the user can replace the composite aggregation with a generalization, also known as an inheritance hierarchy.

Another problem with the existing approaches is that they can **require much user interaction for semantic interpretation**, thus giving less opportunity for automation. For example, even though [3] discovers inheritance and optimization structures, it represents a fairly informal process of schema transformation with weakly ordered steps that entail a lot of involvement from the user, backtracking and reodering.

As an attempt to resolve the problems, we propose a novel approach, which is based on an analysis of key, data and attribute correlations, as well as their combination. Our approach transforms a relational database schema into an object database schema in the context of Unified Modeling Language (UML) [9, 10]. Our choice of UML yields an important advantage. One notation is used to represent both schemata. The existing approaches usually use Peter Chen's Entity-Relationship (ER) notation for relational database schema representation and the OMT (Object Modeling Technique) for object database schema representation.

3. Relational Database Schema vs. Object Database Schema

Before going into detail of schema transformation, we give a formal description of: (1) relational database schema, (2) object database schema, and (3) relationship between the two.

3.1 Relational Database Schema

This is the source for the schema transformation. Formally, a relational database schema (based on the relational model) consists of:

- A set of relations R
- A set of attributes A_R
- A set of relational database types T_R that includes only simple types
- A function *attr*: $R \rightarrow 2^{A_R}$ that returns attributes of relations: $\forall r(r \in R \wedge attr(r) \subseteq A_R)$
- A function *type:* $A_R \rightarrow T_R$ that returns types of attributes: $\forall a(a \in A_R \wedge \exists t(t \in T_R \wedge type(a)=t))$
- A function *key:* $R \rightarrow 2^{A_R}$ that returns primary keys of relations: $\forall r(r \in R \wedge key(r) \subseteq attr(r))$ [4].

3.2 Object Database Schema

This is the target for the schema transformation. Formally, an object database schema (based on the object model) consists of:

- A set of classes C
- A set of attributes A_O
- A set of object database types T_O that includes both simple and composite types
- A function *attr:* $C \rightarrow 2^{A_O}$ that returns attributes of classes: $\forall c(c \in C \wedge attr(c) \subseteq A_O)$
- A function *type:* $A_O \rightarrow T_O$ that returns types of attributes: $\forall a(a \in A_O \wedge \exists t(t \in T_O \wedge type(a)=t))$ [4].

3.3 Relationship Between Schemata

We describe how the relational database schema is related to the object database schema using:

- A function *rel:* $C \rightarrow R$ that returns relations from which classes have been derived. Generally, each class corresponds to a single relation: $\forall c(c \in C \wedge \exists r(r \in R \wedge rel(c)=r))$
- A function *tt:* $T_R \rightarrow T_O$ that translates relational database types into object database types (e.g. `VARCHAR` into `String`), because of "impedance mismatch" between the two type systems [11]: $\forall a(a \in A_R \wedge \exists t(t \in T_O \wedge tt(type(a))=t))$ and $T_R \subset T_O$.

4. Our Approach

The schema transformation uses a relational database in third normal form (3NF)[1] as the main input. It goes through five steps: (1) classification of relations, (2) mapping relations, (3) mapping attributes, (4) mapping relationships, and (5) establishing cardinalities. We illustrate each of these steps for the schema transformation using the UML notation [9, 10].

4.1 Classification of Relations

Relations are classified into one of the three categories: (1) base relations, (2) dependent relations, and (3) composite relations.

4.1.1 Base Relations

If a relation is independent of any other relation in the relational database schema, it is a base relation. Formally, a relation $r \in R$ is a *base relation*, if $\neg\exists r_1 \in R$ such that $K_1 \subset K$, where $K=key(r)$ and $K_1=key(r_1)$.
In Figure 1, `Department` is a base relation, as it has no foreign key. An attribute `Department_ID` in `Employee` is a foreign key to `Department`. However, `Employee` is also a base relation, as `Department_ID` is not part of its primary key. Yet another example of base relation is `Project`.

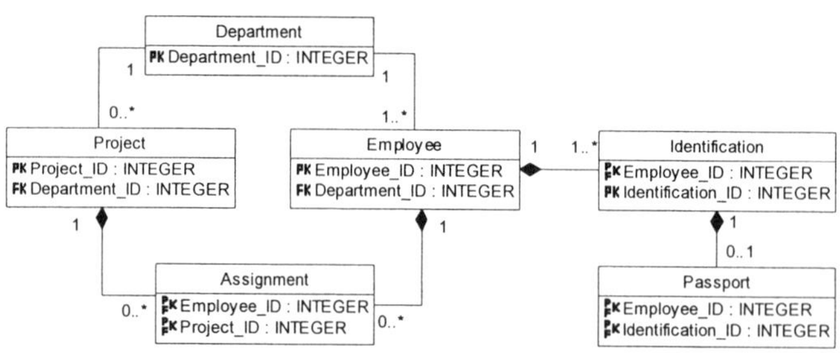

Figure 1. Classification of relations

4.1.2 Dependent Relations

If a primary key of a relation depends on another relation's primary key, it is a dependent relation. Formally, a relation $r \in R$ is a *dependent relation*, if $\exists r_1, r_2 \ldots r_n \in R$ such that $K_1 \subseteq K_2 \ldots \subseteq K_n \subseteq K$, where $K=key(r)$, $K_i=key(r_i)$, $i \in \{1\ldots n\}$ and $n \geq 2$.

In Figure 1, `Identification` is a dependent relation, as it gets `Employee's` `Employee_ID` that is part of its primary key.

---

[1] The reason for starting schema transformation with 3NF relations is that such relations are the "best" structures, which reflect object concepts [8]. "More normalized" relations, such as 4NF and BCNF, may break relations at the level of loosing the original structures of classes. On the other hand, "less normalized" relations, such as 1NF and 2NF, may leave relations containing many tuples, which are difficult to assembly into objects during the data mapping process.

4.1.3 Composite Relations

All other relations fall into this category. Formally, a *composite relation* is a relation that is neither base nor dependent.

In Figure 1, `Assignment` is a composite relation, as its primary key is composed of primary keys of `Employee` and `Project`[2].

4.2 Mapping Relations

Mapping relations is straightforward and usually poses no difficulties in the schema transformation, as each relation becomes a class, with the exception of composite relations. A composite relation is difficult to map, because depending on its structure, it can correspond up to three different constructs in an object database schema:

- A binary or higher degree association, when all its attributes are primary/foreign keys[3]
- An association class, when it contains an additional attribute that is not a primary key or primary/foreign key
- A qualified association, when it contains an additional primary key.

In Figure 2, a base relation `Employee` maps to a class `Employee`. Similarly, a base relation `Project` maps to a class `Project`. A composite relation `Assignment` becomes a binary association, as it consists entirely of primary keys of all its associated relations: `Employee` and `Project`.

Figure 2. Mapping relations: a binary association

In Figure 3, not only does a composite relation `Employment` comprise primary keys of all its associated relations (`Employee` and `Department`), but it also contains the date, when an employee started to work for the department. Therefore, `Employment` becomes an association class, the class that acts as an association, but yet has an attribute `startDate`.

---

[2] Because both dependent and composite relations require a composite key – the key that has more than one attribute in it – sometimes it may be difficult to distinguish between them. In Figure 1, `Passport` "looks" like a composite relation, as its primary key is composed of primary keys of `Employee` and `Identification`: key(`Passport`)=key(`Employee`)∪key(`Identification`). However, it is a nesting of keys that classifies `Passport` as a dependent relation, as opposed to it being a composite relation: key(`Employee`)⊆key(`Identification`)⊆key(`Passport`).

[3] A composite relation can also map to a generalization (namely, a multiple inheritance), when all its attributes are primary/foreign keys. But as there is no really good way to represent a multiple inheritance in a relational database schema, a composite relation will map to an association. The user can then replace that association with a generalization.

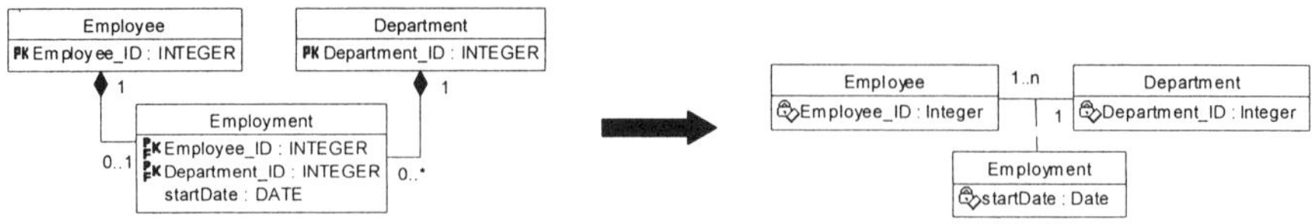

Figure 3. Mapping relations: an association class

In Figure 4, a primary key attribute `deptName` in a composite relation `Belongs_to` is a qualifier, meaning that a company has many different departments, identified by a (unique) department name. Therefore, `Belongs_to` becomes a qualified association.

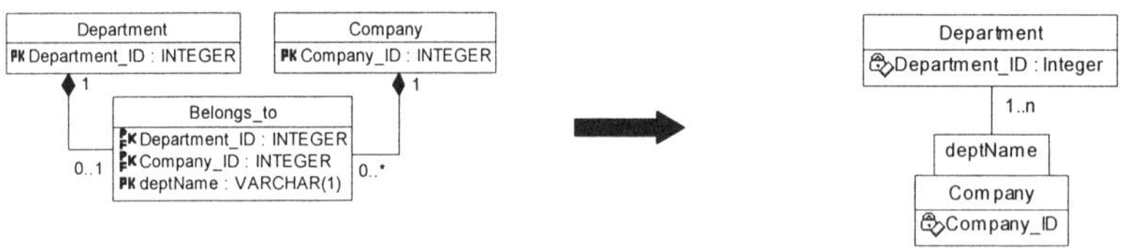

Figure 4. Mapping relations: a qualified association

4.3 Mapping Attributes

Each attribute in a relation becomes an attribute in a class, with the exception of foreign keys and primary/foreign keys. A foreign key and primary/foreign key are ignored, as they refer to another relation.

In Figure 5, we add an attribute `name` (of type `String`) to a class `Employee`.

Figure 5. Mapping attributes

4.4 Mapping Relationships

In the relational database schema, relationships are represented through foreign keys and primary/foreign keys[4]. For example:

- In Figure 1, an employee *works for* a department. This relationship is represented by an attribute `Department_ID` in `Employee`. `Department_ID` is a foreign key to `Department`

[4] Relationships can also be represented through composite relations, but we have already handled them in the prior step of schema transformation (see Section 4.2). For example, in Figure 2 an employee *is assigned to* a project. This relationship is represented through a composite relation `Assignment`. Here an attribute `Employee_ID` refers to `Employee`, while `Project_ID` to `Project`.

- In Figure 1, a passport *is* an identification document. This relationship is represented by an attribute `Identification_ID`, which appears as a primary/foreign key in `Passport`.

A foreign key and primary/foreign key can correspond to a binary association, generalization or composite aggregation, depending on the types of key, data and attribute correlations.

4.4.1 Key, Data and Attribute Correlations

Formally, given two relations $r_1 \in R$ and $r_2 \in R$:

- The types of *key correlation*: key equality ($K_1=K_2$), key inclusion ($K_1 \subset K_2$), key overlap ($K_1 \cap K_2 \neq \varnothing$, $K_1-K_2 \neq \varnothing$, $K_2-K_1 \neq \varnothing$) and key disjointedness ($K_1 \cap K_2=\varnothing$)
- The types of *data correlation*: data equality ($r_1[K_1]=r_2[K_2]$), data inclusion ($r_1[K_1] \subset r_2[K_2]$), data overlap ($r_1[K_1] \cap r_2[K_2] \neq \varnothing$, $r_1[K_1]-r_2[K_2] \neq \varnothing$, $r_2[K_2]-r_1[K_1] \neq \varnothing$) and data disjointedness ($r_1[K_1] \cap r_2[K_2]=\varnothing$)
- The types of (non-key) *attribute correlation*: attribute equality ($A_1=A_2$), attribute inclusion ($A_1 \subset A_2$), attribute overlap ($A_1 \cap A_2 \neq \varnothing$, $A_1-A_2 \neq \varnothing$, $A_2-A_1 \neq \varnothing$) and attribute disjointedness ($A_1 \cap A_2=\varnothing$),

where $K_1=key(r_1)$, $K_2=key(r_2)$, $A_1=attr(r_1)-K_1$, $A_2=attr(r_2)-K_2$, $r_1[K_1]=\pi_{K1}(r_1)$ and $r_2[K_2]=\pi_{K2}(r_2)$.

4.4.2 Analysis of Key Correlation

To map relationships, we start with analyzing key correlation. Then we proceed on to an analysis of data and attribute correlations, because additional ("hidden") semantics, such as inheritance and optimization structures, are not solely contained in keys, but also in tuples (i.e. data) and attributes.

First, consider a relationship between `Project` and `Task` in Figure 6, when **key inclusion** holds on it[5]. That relationship maps to a composite aggregation (also known as a non-shared aggregation, an aggregation by value or just a composition), because the identification of instances in `Task` requires the primary key of `Project` (i.e. `Project_ID`). That is, `Project` has an identifying relationship with `Task`.

Figure 6. Mapping relationships: key inclusion

Second, consider a relationship between `Project` and `Task` in Figure 7, when **key disjointedness** holds on it[6]. That relationship maps to a binary association, as it associates two relations that are independent of each other, or use a non-identifying relationship. (Indeed, the relationship cannot be identifying, because the identification of instances in `Task` does not require any attribute in `Project`'s `Project_ID`.)

---

[5] If tasks were uniquely identified only within a project, then the primary key of `Project` (i.e. `Project_ID`) would be combined with an attribute `Task_ID` to define the primary key of `Task`: `key(Project)`⊂`key(Task)`.

[6] If an attribute `Task_ID` in `Task` were always unique, even between projects, then `Task` would get `Project`'s `Project_ID` that was not part of its primary key: `key(Project)`∩`key(Task)=`∅.

Figure 7. Mapping relationships: key disjointedness

4.4.3 Analysis of Data and Attribute Correlations

In mapping relationships, while the analysis of key correlation is crucial for deriving semantics from the relational database schema, data and attribute correlations are also important to analyze. For example, key equality can determine an inheritance structure. But the type of inheritance (either single or multiple) is determined with the types of data and attribute correlations. The importance of the analysis of data and attribute correlations is perhaps best explained by example.

First, consider a relationship between `SoftwareProject` and `Project` in Figure 8, when **key equality, data equality** and **attribute disjointedness**[7] hold on it. This is an example of vertical partitioning, where attributes of a single (logical) relation have been split into two relations, having the same primary key. Therefore, we combine `SoftwareProject` and `Project` into a single class, say `SoftwareProject`, whose attributes are the union of the attributes of the two relations.

Figure 8. Mapping relationships: key equality, data equality and attribute disjointedness

Second, consider a relationship between `SoftwareProject` and `Project` in Figure 9, when **key equality, data disjointedness** and **attribute equality** hold on it. This is, again, an example of optimization structure; but horizontal partitioning, where data of a single (logical) relation have been split into two relations, having the same attributes. Therefore, we combine `SoftwareProject` and `Project` into a single class, say `SoftwareProject`, whose data are the union of the data of the two relations.

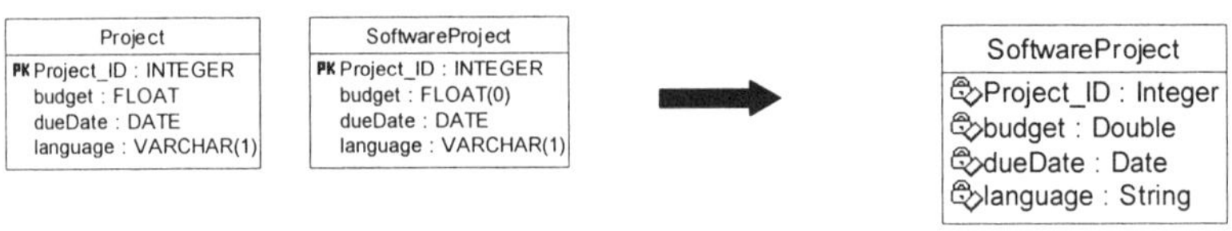

Figure 9. Mapping relationships: key equality, data disjointedness and attribute equality

Third, consider a relationship between `SoftwareProject` and `Project` in Figure 10, when **key equality** and **data inclusion** hold on it. That relationship maps to a single inheritance, as all data of `SoftwareProject` are also included in `Project`; i.e. a

[7] By attribute "disjointedness", we mean that the two relations have no common attributes other than primary keys, foreign keys and primary keys/foreign keys.

software project is a project. But the converse is not true, as some projects can be hardware projects, for example.

Figure 10. Mapping relationships: key equality and data inclusion

Fourth, consider a relationship between `SoftwareProject` and `HardwareProject` in Figure 11, when **key equality, data overlap** and **attribute disjointedness** hold on it. This is an example of a multiple inheritance, as some data are common to both relations. Therefore, we "discover" a new class, say `HardwareSoftwareProject`, which has generalization relationships to both `SoftwareProject` and `HardwareProject`.

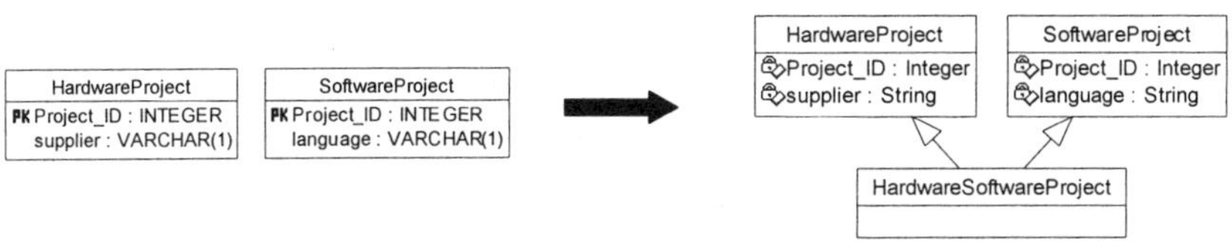

Figure 11. Mapping relationships: key equality, data overlap and attribute disjointedness

Fifth, consider a relationship between `SoftwareProject` and `HardwareProject` in Figure 12, when **key equality, data disjointedness** and **attribute overlap** hold on it. This example also illustrates generalizations, but a single inheritance. Because some attributes (e.g. `budget` and `dueDate`) are common to both relations, we can see that `SoftwareProject` and `HardwareProject` are part of the inheritance hierarchy, but there is no relation corresponding to their superclass. Therefore, we "discover" a new class, say `Project`, which both `SoftwareProject` and `HardwareProject` inherit from.

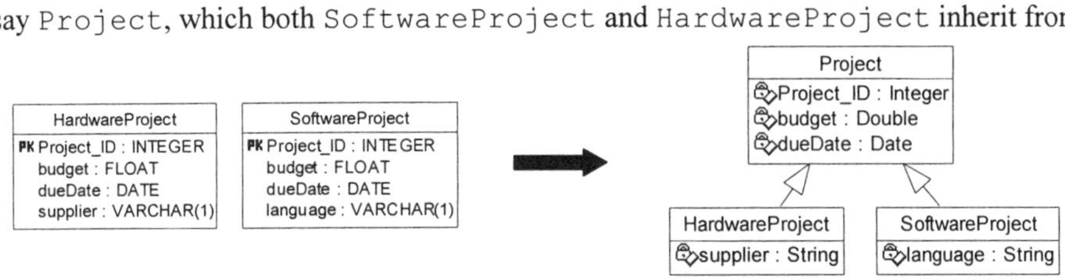

Figure 12. Mapping relationships: key equality, data disjointedness and attribute overlap

Sixth, consider what a relationship between `SoftwareProject` and `HardwareProject` in Figure 13 might look like, if **key equality, data overlap** and **attribute overlap** held on it. Because some attributes and data are common to both relations, this is an example of the "diamond-shaped" inheritance hierarchy, in which a class has generalization relationships to two superclasses and the two superclasses refer in turn to a common superclass.

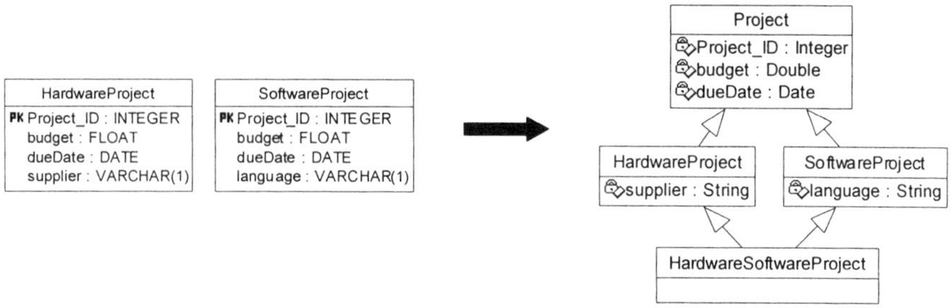

Figure 13. Mapping relationships: key equality, data overlap and attribute overlap

Finally, consider a relationship between `Employee` and `Project` in Figure 14, when **key overlap** and **data equality** (or **data inclusion**) hold on it. Because all the information about departments in `Project` is also included in `Employee`, a closer examination of that relationship reveals that a relation `Department` is missing. Therefore, we "retrieve" `Department`, which has composite aggregations with both `Employee` and `Project`. That is, `Employee` and `Project` are indirectly related to each other through `Department`. For example, an employee works on a project controlled by a department he or she belongs to.

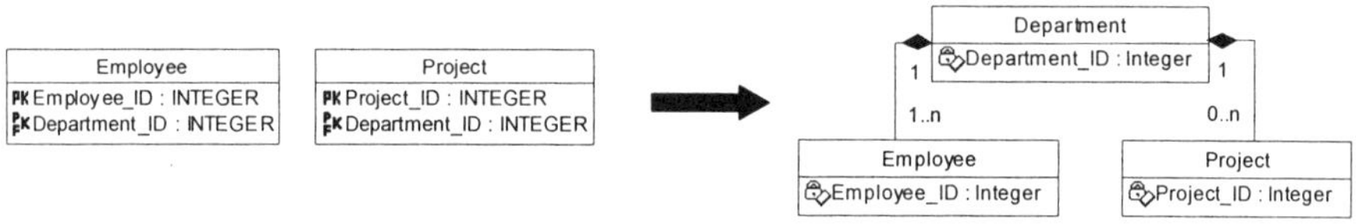

Figure 14. Mapping relationships: key overlap and data equality (or data inclusion)

4.5 Establishing Cardinalities

Cardinality (or multiplicity) is the number of data instances that can participate in a relationship. To establish cardinalities, we need to consider foreign key values, as it is the ability of foreign key to be null and unique that determines if the relationship is zero-or-one-to-one, one-to-one, one-to-many or many-to-many, as shown in Table 2.

Table 2. Cardinalities for a foreign key in r_2, which refers to r_1

Foreign key values	r_1	r_2
Nullable and unique	0..1	0..1 or 1
Nullable and not unique	0..1	0..* or 1..*
Not nullable and not unique	1	0..* or 1..*
Not nullable and unique	1	0..1 or 1

In Figure 15, assuming that in `Employee`, a foreign key attribute `Department_ID` cannot be null, we can see that every instance of `Employee` is associated with exactly one instance of `Department`. In addition, assuming that `Department_ID` is not unique, we can see that more than one instance of `Employee` can be associated with each instance of `Department`. Therefore, we establish a one-to-many relationship between `Department` and `Employee` (or many-to-one relationship between `Employee` and `Department`).

Figure 15. Establishing cardinalities

4.6 Alternatives

The schema transformation is formally defined as a set of rules (see Appendix). Each rule defines how a construct in the relational database schema is mapped to a (semantically equivalent) construct in the object database schema. However, generally, there can be several alternatives to the rules. Some examples below show these alternatives more clearly.

In Figure 16, `Department` has a cardinality of 0..1. Alternatively, we can introduce a subclass of `Employee`, say `Manager`, to represent those employees that manage departments. In some situations, however, the user may not be willing to introduce a new subclass. Therefore, we represent the relationship between `Department` and `Employee` as a zero-or-one-to-one association and express the constraint – that an employee can manage at most one department – with the cardinality alone.

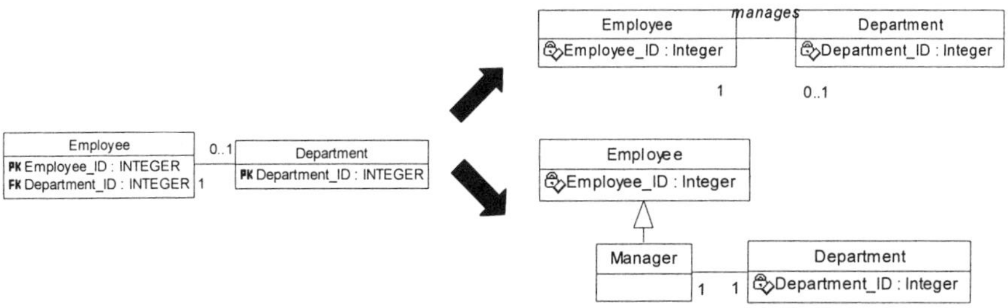

Figure 16. Mapping relationships: key disjoint and data equality. We can recast a zero-or-one-to-one association to a combination of inheritance and one-to-one association

Figure 17 shows an association relationship between `Project` and `Task`. Alternatively, that relationship can be modeled as an aggregation, because logically projects are at a higher level than tasks. In particular, projects consist of tasks.

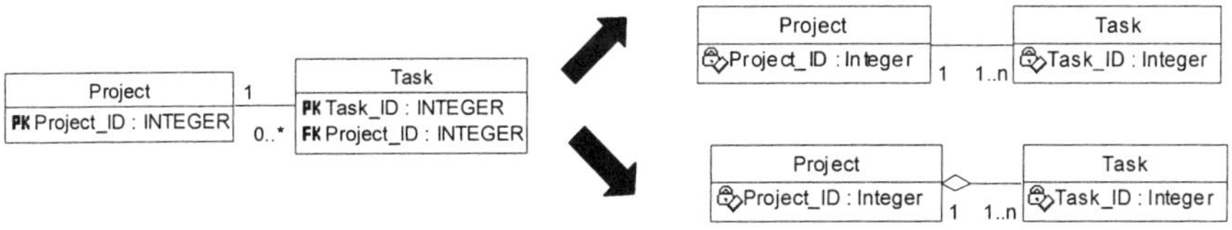

Figure 17. Mapping relationships: key disjoint and data inclusion. We can recast an association to an aggregation

5. Future Work

A relationship between base (or dependent) relations is the most common kind of relationships. However, a relationship between composite relations gives rise to some special construct – a *constrained association* – that we need to consider, when transforming the relational database schema into the object database schema. Constrained associations may arise in many situations, such as:

- *When two composite relations are connected by a time sequence*. For example, consider a relationship between `Works_on` and `Works_for` in Figure 18, when **key overlap** and **data inclusion** hold on it. The former specifies that an employee works on a project, while the latter specifies that an employee works for a department. Of course, an employee cannot work on a project, if he or she has not worked for a department. Indeed, the employee is first employed by the department, and then he or she is assigned to the project. Therefore, the relationship between `Works_on` and `Works_for` maps to a constrained association. In particular, `Works_on` is a subset of `Works_for`

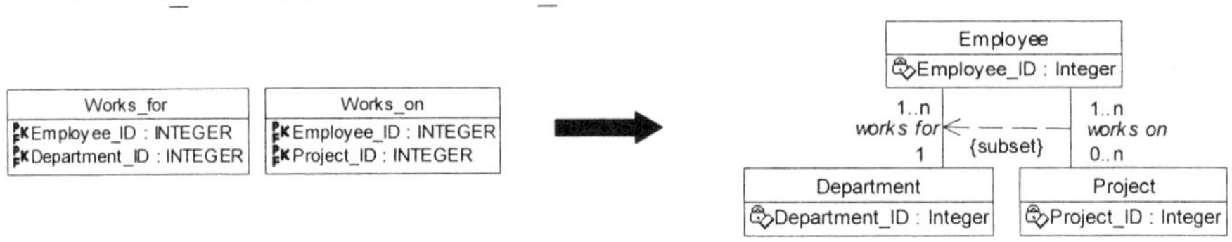

Figure 18. Mapping relationships: key overlap and data inclusion

- *When one composite relation represents a possibility, while another is the actual realization*. Indeed, we often use a constrained association to represent the fact that one entity can "do" something with another. For example, consider a relationship between `Can_be_assigned_to` and `Works_on` in Figure 19, when **key equality** and **data inclusion** hold on it. The former shows the mere possibility of instances of `Employee` being associated with instances of `Project`; i.e. an employee can be assigned to a project. But it does not depict the actual realization of this association, which is represented by a separate relation, `Works_on`, which captures what employees have been worked on projects[8]. Again, the relationship between `Can_be_assigned_to` and `Works_on` maps to a constrained association.

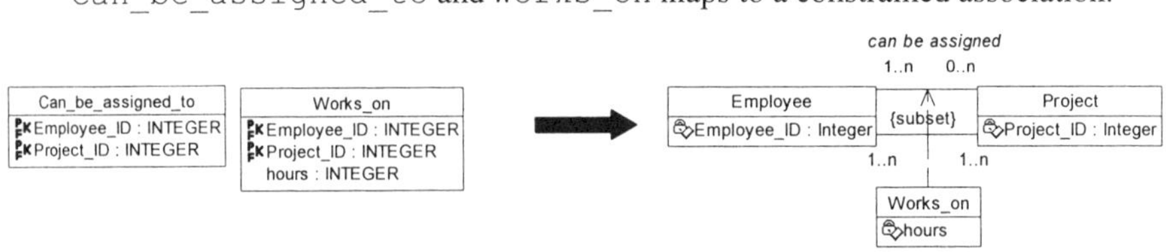

Figure 19. Mapping relationships: key equality and data inclusion

[8] The actual realization may involve additional attributes that are not part of the possibility. For example, in Figure 19 `Works_on` has an additional attribute `hours`.

6. Conclusions

We have proposed a novel approach to reverse engineering of relational databases to object databases. Based on an analysis of key, data and attribute correlations, as well as their combination, our approach has two important advantages over the existing approaches. First, it allows the user to extract more semantics from relational databases, which include inheritance and optimization structures. Second, it reduces user interaction, which is mainly required to confirm the extracted semantics and to give them more appropriate names[9].

Acknowledgements

This research is partly sponsored by Estonian Science Foundation (ESF) under the grant nr. 5766.

References

[1] J. Hainaut, J. Henrard, J. Hick, D. Roland and V. Englebert; Database Design Recovery, Proceedings of the 8th Conference on Advanced Information Systems Engineering, 1996, pp. 272 – 300

[2] J. Petit, F. Toumani, J. Boulicaut and J. Koulomdjian; Towards the Reverse Engineering of Denormalized Relational Databases, Proceedings of the 12th International Conference on Data Engineering, 1996, pp. 218 – 227

[3] W. Premerlani and M. Blaha; An Approach for Reverse Engineering of Relational Databases, Communications of the ACM, Vol. 37. No. 5, 1994, pp. 42 – 49

[4] A. Behm, A. Geppert and K. Dittrich; On the Migration of Relational Database Schemas and Data to Object-oriented Database Systems, Proceedings of the 5th International Conference on Re-technologies for Information Systems, 1997, pp. 13 – 33

[5] R. Chiang, T. Barron and V. Storey; A Framework for the Design and Evaluation of the Reverse Engineering Methods for Relational Databases, Data and Knowledge Engineering, Vol. 21. No. 1, 1996, pp. 57 – 77

[6] J. Albert, R. Ahmed, M. Ketabchi and M. Shan; Automatic Importation of Relational Database Schema in Pegasus, Proceedings of the 3rd International Workshop on Research Issues on Data Engineering: Interoperability in Multimedia Systems, 1993

[7] S. Ramanathan and J. Hodges; Reverse Engineering Relational Schemas to Object-oriented Schemas, Technical Report No. MSU-960701, Mississippi State University, 1996

[8] Z. Tari and J. Stokes; Designing the Reengineering Service for the DOK Federated Database System, Proceedings of the IEEE International Conference on Data Engineering, 1997, pp. 465 – 475

[9] Rational Rose; Using Data Modeler, version 1.3, 2001, http://www.rational.com

[10] Rational Rose; Unified Modeling Language (UML), version 1.3, 2001, http://www.rational.com

[11] D. Maier and S. Zdonik; Fundamentals of Object-oriented Databases, Readings in Object-oriented Databases, eds. D. Maier and S. Zdonik, 1990, pp. 1 – 32

Appendix: Schema Transformation Rules

Here we give a set of rules, with which the object database schema is incrementally extracted from the relational database schema. These rules can be divided into three groups: (1) rules for mapping relations, (2) rules for mapping attributes, and (3) rules for mapping relationships. Each rule is introduced with: (1) a list of preconditions, (2) a relational database construct (e.g. a relation, an attribute and a relationship), and (3) an object database construct, which is the result of applying the rule to the relational database construct, when the preconditions are met (e.g. a class, an attribute, an association, an

---

[9] The schema transformation process cannot be completely automated. User interaction is still necessary, when ambiguities occur and semantics cannot be inferred [4].

aggregation and a generalization). In the following, we write B, D, and P for base, dependent and composite relations, respectively: $R = B \cup D \cup P$ and $B \cap D \cap P = \varnothing$.

Table 3. Rules for mapping relations. Preconditions common to all the rules: $\neg\exists c \in C$: r=rel(c)

Rule	Preconditions	Relational database construct	Object database construct
RR1	$r \in B \cup D$	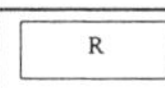	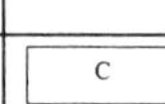
RR2	$r \in P$ key(r)=attr(r) $\forall i \in \{1\ldots n\}, n \geq 2$ $r_i \in B \cup D$ $\exists c_i \in C$: r_i=rel(c_i) K_i=key(r_i) $\cup K_i$=key(r) $\cap K_i = \varnothing$	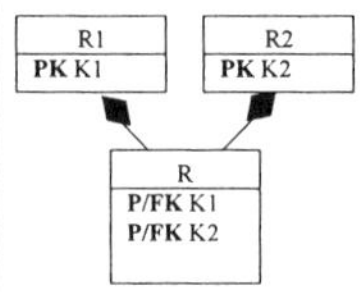	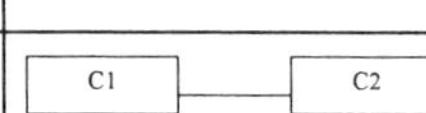
RR3[10]	$r \in P$ key(r)$\subset$attr(r) A$\subset$attr(r) A$\cap$key(r)=$\varnothing$ $\forall i \in \{1\ldots n\}, n \geq 2$ $r_i \in B \cup D$ $\exists c_i \in C$: r_i=rel(c_i) K_i=key(r_i) $\cup K_i$=key(r) $\cap K_i = \varnothing$	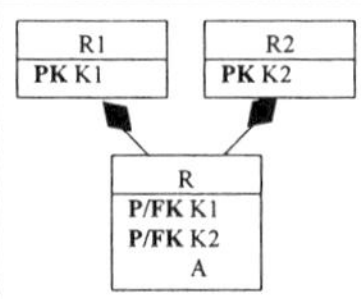	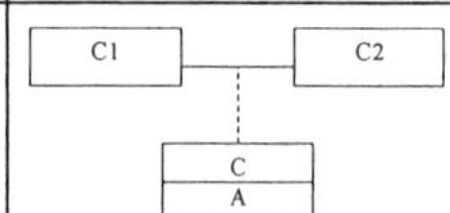
RR4	$r \in P$ K$\subset$key(r) $r_1, r_2 \in B \cup D$ $\exists c_1 \in C$: r_1=rel(c_1) $\exists c_2 \in C$: r_2=rel(c_2) $K_1 \cup K_2 \cup K$=key(r) $K_1 \cap K_2 \cap K = \varnothing$	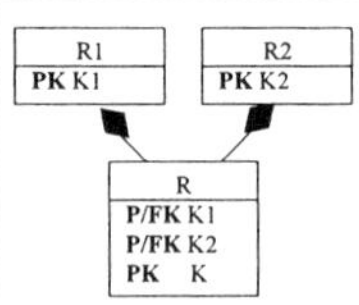	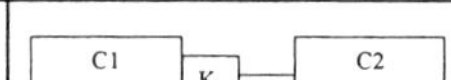

Table 4. Rules for mapping attributes

Rule	Preconditions	Relational database construct	Object database construct
RA1	$a \in$ attr(r) $r \in R$ $\exists c \in C$: r=rel(c) $\neg\exists r_1 \in R$: $a \in$ key(r_1) t=type(a)	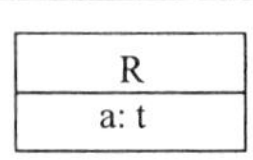	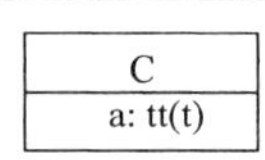

[10] For simplicity, in rules RR2 and RR3 only binary relationships (n=2) are shown.

Table 5. Rules for mapping relationships. Preconditions common to all the rules: $r_1, r_2 \in B \cup D$; $\exists c_1 \in C$: $r_1 = rel(c_1)$; $\exists c_2 \in C$: $r_2 = rel(c_2)$; $K_1 = key(r_1)$; $K_2 = key(r_2)$; $A_1 \subset attr(r_1)$; $A_2 \subset attr(r_2)$

Rule	Preconditions	Relational database construct	Object database construct
RK1	$K_1=K_2$, $\lvert K_1\rvert=\lvert K_2\rvert$ $r_1[K_1]=r_2[K_1]$	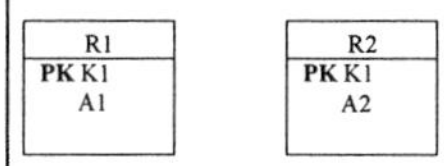	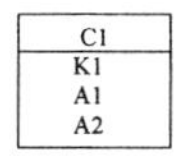
RK2	$K_1=K_2$, $\lvert K_1\rvert=\lvert K_2\rvert$ $r_2[K_1]\subset r_1[K_1]$	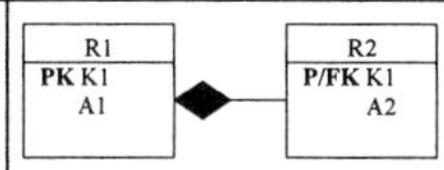	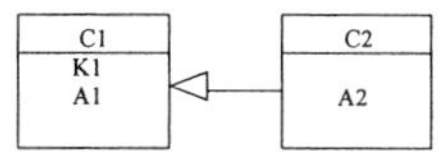
RK3	$K_1=K_2$, $\lvert K_1\rvert=\lvert K_2\rvert$ $r_1[K_1]\cap r_2[K_1]\neq\varnothing$ $r_1[K_1]-r_2[K_1]\neq\varnothing$ $r_2[K_1]-r_1[K_1]\neq\varnothing$ $A=A_1\cap A_2$ $\neg\exists c_{12}\in C$ $\neg\exists c\in C$	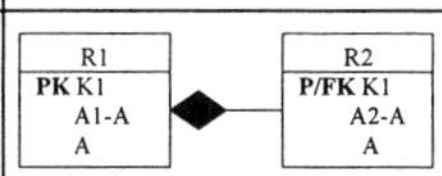	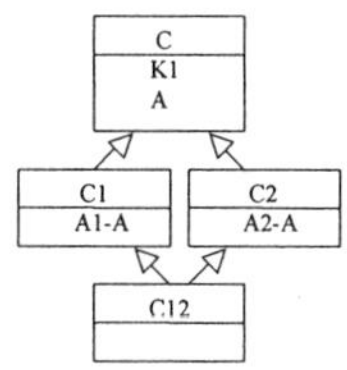
RK4	$K_1=K_2$, $\lvert K_1\rvert=\lvert K_2\rvert$ $r_1[K_1]\cap r_2[K_1]\neq\varnothing$ $r_1[K_1]-r_2[K_1]\neq\varnothing$ $r_2[K_1]-r_1[K_1]\neq\varnothing$ $A_1\cap A_2=\varnothing$ $\neg\exists c_{12}\in C$	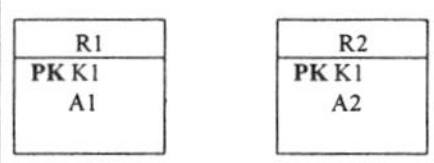	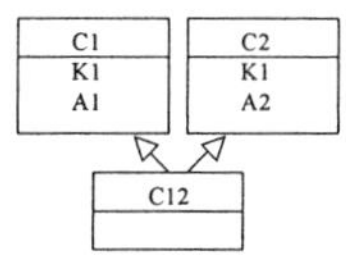
RK5	$K_1=K_2$, $\lvert K_1\rvert=\lvert K_2\rvert$ $r_1[K_1]\cap r_2[K_1]=\varnothing$ $A_1=A_2$, $\lvert A_1\rvert=\lvert A_2\rvert$	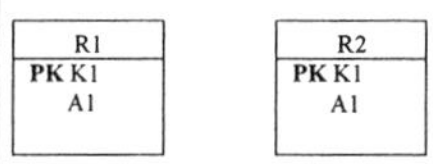	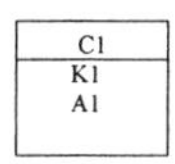
RK6	$K_1=K_2$, $\lvert K_1\rvert=\lvert K_2\rvert$ $r_1[K_1]\cap r_2[K_1]=\varnothing$ $A=A_1\cap A_2$ $\neg\exists c\in C$	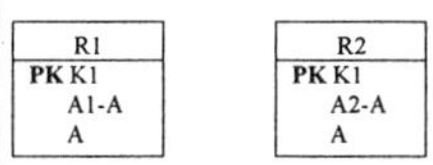	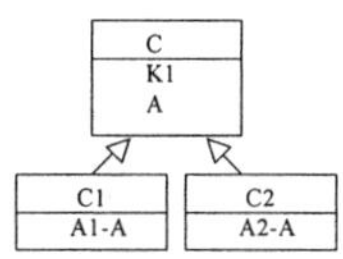
RK7	$K_1\subset K_2$ $r_1[K_1]\cap r_2[K_1]=\varnothing$	Not applicable	
RK8	RK7 is not applied $K_1\subset K_2$	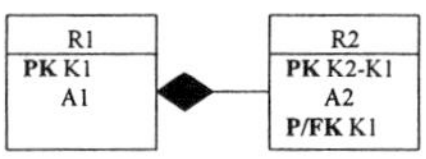	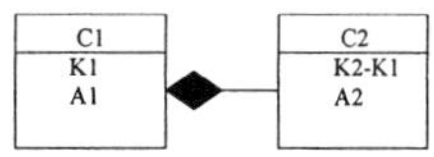
RK9	$K=K_1\cap K_2$ $K_1\cap K_2\neq\varnothing$ $K_1-K_2\neq\varnothing$ $K_1-K_2\neq\varnothing$ $r_1[K]=r_2[K]$ $\neg\exists c\in C$	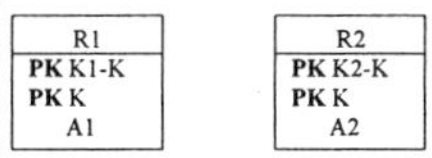	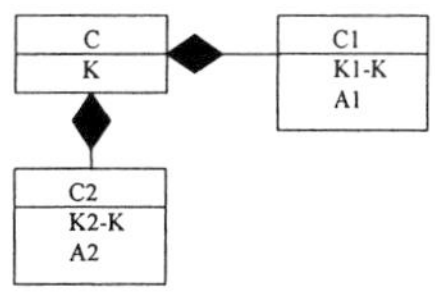

Databases and Information Systems
J. Barzdins and A. Caplinskas (Eds.)
IOS Press, 2005

Design Independent Modeling of Information Systems Using UML and OCL

Lina CEPONIENE, Lina NEMURAITE
Kaunas University of Technology, Faculty of Informatics,
Department of Information Systems,
Studentu 50-308, Kaunas, Lithuania
{kavalina, nemur}@soften.ktu.lt

Abstract. Design Independent Model (DIM) is proposed for definition of requirements for development of wide range of information systems conceived as processes or systems composed of services. Metamodel of DIM presents subset of UML 2.0 metamodel with additional constraints. The kernel for ensuring relative completeness and compatibility of requirements is a schema, in which interactions and state transitions related aspects of behavior are integrated with static structures of entities of the problem domain. Direct and reverse procedures for mapping between sequence diagrams and state charts are presented that help to reveal inconsistencies and incompleteness in definition of required behavior. Possibilities for implementation in CASE tools are investigated, case study and illustrating examples are presented.

Introduction

Information systems (IS) development is diverging from object-oriented to model-driven, where models and metamodels are playing central role instead of objects [1]. Also, the shift to separation of behavioral objects (services and processes) from informational objects (entities) is observed as opposed to object-oriented principles of tight binding structural and behavioral features together. This shift is raised by mismatch between business and software (separation of processes from entities), or between practical software paradigms and object-orientation (e.g. sending and retrieving messages from the queue instead of invoking operations on objects).

Today, two different processes in development of IS may be accentuated: creation of IS services (e-services or, more specifically, Web services [2]) and creation of e-business processes for management of e-services. Separation of process management from business functions carried out by different participants of process was the principle idea of Workflow Management Systems (WFMS). Today this idea has penetrated in Enterprise Application Integration (EAI), WFMS, Business Process Modeling and Execution languages (BPML, EBXML, BPEL, etc.) as general methodology for software development and exploitation. Another great methodological shift is Model Driven Architecture (MDA) [1], [3]. The essence of this approach is in idea that the foundation of software development is Platform-Independent Model (PIM) of software system. PIM is obtained from Business Model (formerly Computation Independent Model or CIM) and serves for definitions of multiple Platform Specific Models (PSM) from which implementations on different platforms may be generated.

In this paper Design Independent Model (DIM) is proposed for definition of requirements for development of wide range of information systems conceived as processes or systems composed of services. In our opinion, there is a gap between Business Model and Platform Independent Model. PIM, platform independent model, is dependent upon software architecture and design methodology. In contrast, DIM represents comprehensive definition of requirements – structure and behavior of IS independently of software architecture (e.g. object-oriented, component-based, service-oriented, Web) and design method (e.g. data-driven, event-driven, responsibility-driven etc.). Independency from design implies that operations are not allocated to control classes or components but defined rather as events triggering interface objects. The eventual purpose of DIM is formalization of process going from requirements to PIM, but it goes beyond limits of this paper. Introduction of DIM leads to transparency of design decisions that must be made during development of PIM.

DIM is defined in terms of actors, interfaces to target system (for development of e-services) or interfaces to process (for development of e-business processes), models of problem domains and sub-domains related with interfaces. In general case it may be represented visually by stereotyped UML [4] [5] class, sequence, state and activity diagrams with OCL constraints [6] [7] [8] [9] following principles of precise modeling [10] [11] and contract-based development [12]. These diagrams are partially overlapping views under DIM that actually defines schema of IS [13], fully representing state and behavior of IS under development in design-independent manner [14]. In current paper DIM is limited to creation of IS services and respectively usage of class, sequence and state diagrams. For development of stateless services there are no requirements for explicit management of process (in some cases, we may deal with services that are not joined by common process at all). In such case activity diagrams are not mandatory, though in being of common process they have explanatory importance and may be used in beginning of DIM development or generated from DIM when it is completed. In opposition, activity diagrams play the crucial role in development of IS where processes must be managed as they are able to integrate data flow and control aspects of behavior together with actors roles participating in process.

Particular attention is devoted to integration of interactions with state machines, which are specialized as state machines of entities and interfaces to behavior. Explicit separation and reconciliation of states of entities with states of services is similar to [15]. Interactions are not directly integrated with state machines in UML metamodel. Such integration is necessary for harmonization of IS model irrespective of development phase (requirements, design or implementation). In this work algorithm is proposed for direct and reverse transformations between sequence diagrams and state machines. There are many works on this topic (e.g., [16], [17], [18]), but bi-directional transformations were not possible. For this purpose some adjustments to UML metamodel must be made.

Models and algorithms presented in this paper, if implemented in UML CASE tools, would have effect on design practice, and not necessarily for development in MDA. Interactive generation of lacking diagrams from initial ones, caution about deficiencies would help designer to develop compact, unambiguous models and keep them in line instead of rapid sketching multiple uncoordinated views.

The paper is composed of 5 sections. In section 2, the structure of Design Independent Model of IS in general case is defined and metamodel integrating class, sequence diagrams and state charts is presented. Section 3 is devoted for derivation of state machines from sequence diagrams and vice versa. Section 4 introduces the case study illustrating the usage of DIM. Finally, section 5 concludes the paper and discusses possible future work.

1. Design Independent Model

Commonly use cases are used to capture behavioral requirements of software systems, but they often are textual descriptions in natural language and depend on individual interpretation, so are ambiguous [19] [20]. In this work model of use cases and their textual descriptions (initial requirements) are the input for creation of elaborated requirements model, i.e., DIM.

Model for comprehensive definition of requirements proposed in this work is based on many sources (e.g. [16], [21], [22], [23] [24] etc.). At conceptual level it was introduced in [14]. Elaborated model of requirements must define state and behavior of IS independently of future design. Behavior of IS has many aspects: interactions, state transitions, control flows and data flows. Different kinds of UML diagrams are devoted for representation of these different, often overlapping aspects, but all these diagrams are views under the same model of target IS. In requirements phase this model, named DIM, is specification, containing essential elements of IS model, which may be visually represented using class, sequence, state charts and activity diagrams (Fig.1).

In DIM, elaborated use cases are specified as interfaces to the target system together with actors – service users and external systems (service providers), whose interfaces must be accessed by the target system to provide requested services. Such form well conforms to requirements for development of components and Web services but it also may be applied for any IS or software system. Further, interfaces are specified by the set of operations identified from steps in use case specifications. Every operation is defined by its signature, pre and post conditions (e.g. [12], [25] etc.). Similar approach was proposed in [26], where use cases are supplemented with operational schemas, but grouping of operations is lost. In this work, relationships between use cases (<<include>>, <<extend>> and generalization) are presented as generalization relationships between interfaces strongly following Liskov substitution principle [27]. Other possible relationships between interfaces are dependencies with stereotype <<use>>. Usage dependencies [5] describe relationships between service users and interfaces, interfaces and external service providers, and different interfaces. The last kind of dependency may be discovered during use case analysis when sequence diagrams are created for modeling interactions between actors and interfaces.

Besides the interfaces, DIM includes entities and states of entities modeling the state of problem domain and sub-domains associated with particular interfaces. Possible relationships between entities are associations and generalizations, as proposed by authors of executable UML [10]. Entity and interface constraints (invariants) also are included in DIM.

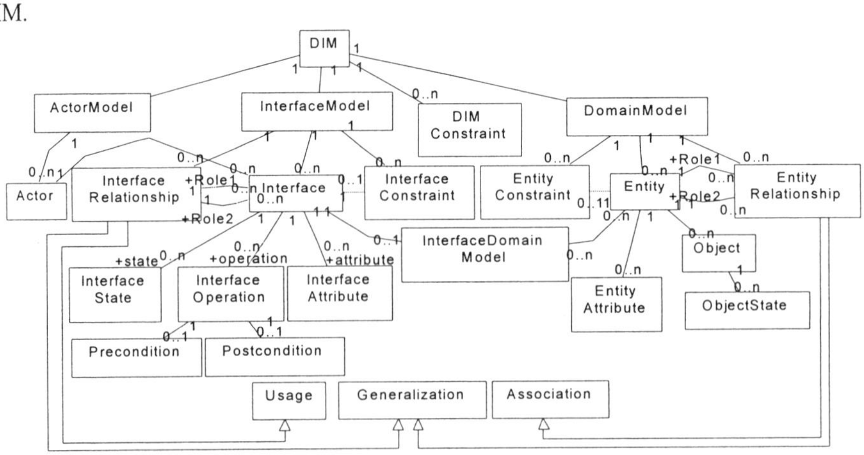

Figure 1. Metamodel of Design Independent Model

1.1 Sequence Diagram for Representation of Requirements

For development of DIM, one or more sequence diagrams for every use case are constructed from specifications of use case scenarios. These sequence diagrams must represent all desirable interactions between actors (service users), interfaces (i.e. use cases) of the system and possibly the interfaces to external systems, if they are needed to fulfill service requests. DIM deals with elaborated requirements; so initial use cases (or later interfaces) must be re-structured for reuse of behavior where advisable.

The main elements of the sequence diagram are *Lifelines* (used in UML 2.0 instead of classifier roles from previous versions of UML), *Messages* and *State Invariants*, representing states of the classifiers. Two types of *Messages* are distinguished: *Requests* and *Responses*, and two *Events* are connected to each message in sequence diagram: *Send Event* corresponds to one message end and *Receive Event* corresponds to the other one (Figure 2). *Received Request* may have attached *Constraint* that must be verified before fulfillment of *Request*.

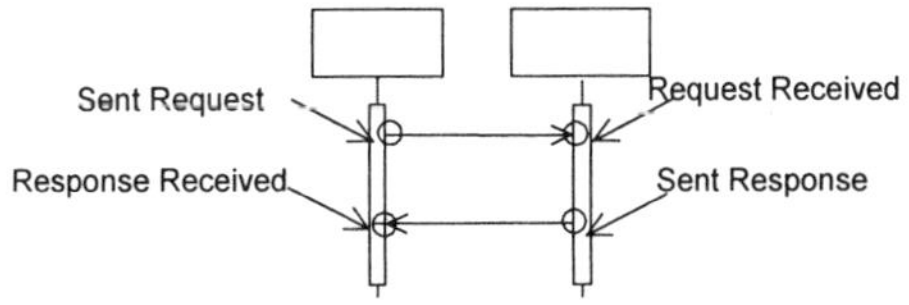

Figure 2. Event types

Sequence diagrams in DIM are based on UML 2.0 but have some additional adjustments. Metamodel of sequence diagrams, used for DIM, is presented in Figure 3 (there *Event* corresponds to *EventOccurence* in [5]). Messages are accepted to be synchronous as well asynchronous. For synchronous communication analysts often suppress *Response* messages in diagrams since *Response* implicitly follows after *Request*. For unambiguous representation of interactions *Response* messages are important and for synchronous messages must be included by default. Association connecting *Request* and *Response Messages* is added to UML metamodel, thus ensuring the possibility of tracking which *Response* is the answer to which *Request*.

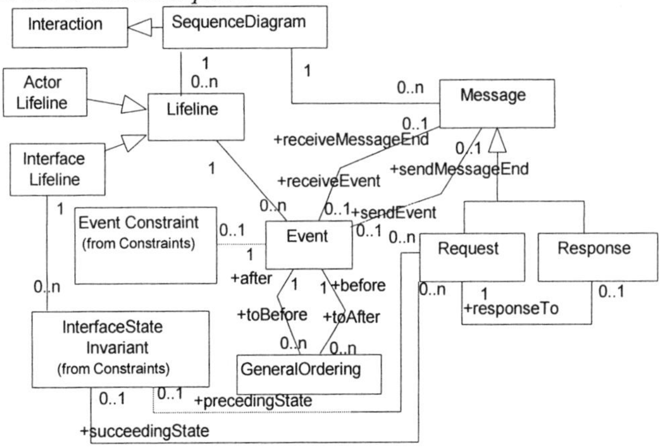

Figure 3. Sequence diagram metamodel for DIM

Request Messages received on *Lifelines* are associated with state invariants of this *Lifeline* by associations *precedingState* and *succedingState,* because *Received Request* messages correspond to the transitions between states of interface. *State Invariants* describing these states are associated with *Received* (preceding) and *Sent* (succeeding) *Request Messages*, and succeeding *Request Message* is associated with all following sequence of messages including *Response* to it.

Constraints for DIM sequence diagram metamodel:

– If there is a *Message* in the sequence diagram there must also be at least one *Lifeline* in this diagram too:

```
context SequenceDiagram
inv:self.Message->notEmpty() implies self.Lifeline->notEmpty()
```

– *Message* sender and receiver *Lifelines* belong to the same sequence diagram:

```
Context SequenceDiagram
inv:
self.Message->forAll(m|self.Lifeline->
exists(ls| ls = m.sendEvent.Lifeline) and self.Lifeline->
exists(lr| lr = m.receiveEvent.Lifeline)
```

– Every *Event* may be associated only with one *Message*, received or sent:

```
Context Event
inv:
self.sendEvent -> notEmpty() implies self.receiveEvent→isEmpty() or
self.receiveEvent -> notEmpty() implies self.sendEvent→isEmpty()
```

– Every synchronous *Request Message* must have *Response*:

```
Context Message
inv:
(self.MessageSort = synchCall or self.MessageSort = synchSignal)
and self.oclIsKindOf (Request) implies self.Response → notEmpty()
```

1.2 State Machines for Representation of Requirements

State charts define the dynamic behavior of objects of the corresponding classes during their entire lifetime. Protocol state machines are proposed for describing the behavior of an interface [5], but they do not describe events sent to other state machines. Capturing of sent events is important for definition of interactions with external systems. This possibility is included in state machines in DIM (Figure 4).

Relationships between *Event* class and its sender and receiver must be captured in state machines; otherwise it would be impossible to trace interactions from statechart diagrams. These relationships are established as roles of classifiers that correspond to lifelines on sequence diagram.

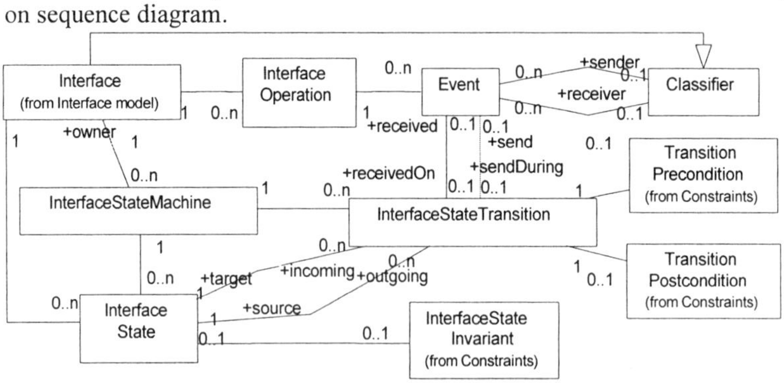

Figure 4. Interface state machine metamodel

Constraints of DIM state charts metamodel:

– If there is a transition in state machine there also must be at least one state in this state machine too:

```
context InterfaceStateMachine
inv:
self.InterfaceStateTransition -> notEmpty() implies
```

```
self.InterfaceState -> notEmpty()
```

– Event must have only one association with classifier – *Sender* or *Receiver*:

```
Context Event
inv:
self.sender->notEmpty() implies self.receiver->empty()
self.receiver->notEmpty() implies self.sender->empty()
```

– If transition has received *Event* associated with *Request Message*, it cannot have send event associated with *Response Message*:

```
Context InterfaceStateTransition
inv:
self.received.receiveMessageEnd.oclIsKindOf(Request) and
self.send.sendMessageEnd → notEmpty() implies
self.send.sendMessageEnd.oclIsKindOf(Request)
```

1.3 Relationships Among Sequence, Class and Statechart Diagrams

Elements of class, sequence diagrams and state charts are interlinked in UML metamodel, but relationships among them have a high degree of abstraction. Such flexibility supports different kinds of usage of UML models, but in our case it is necessary to tie models for assurance of consistency between different views of the same system.

Fragment of metamodel, integrating state charts, class and sequence diagrams, is presented in Figure 5. Here state charts are integrated with sequence and class diagrams via events triggering operations of interfaces from interface class diagram. Interface class diagram is integrated with domain model of the system as arguments of operations of the interface represent entities, attributes and data types from domain model (otherwise they are common built-in types). Operation constraints are written using states of entities from problem domain. Transitions of states of interfaces and entities are integrated via events associated with states and transitions. Transitions of states of entities occur in states of interfaces; states of entities constrain transitions of states of interfaces.

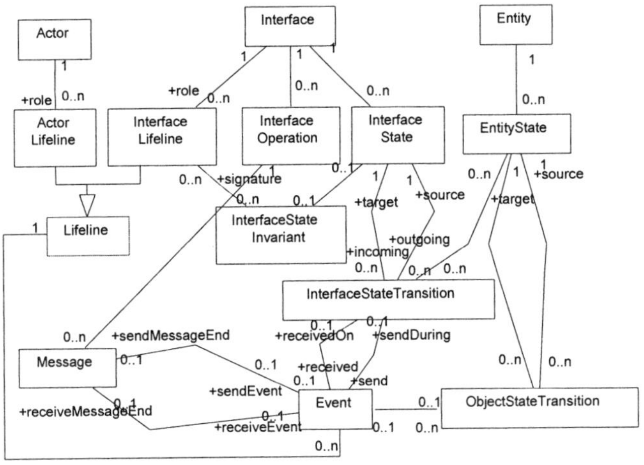

Figure 5. Fragment of metamodel integrating state charts, class and sequence diagrams

Part of constraints for DIM is represented on Figure 6. Using of these constraints removes ambiguity from relationships between representations of interactions and states. Similar constraints must be held between sequence diagrams and class diagrams, between state machines and classes; states of interfaces must be related with states of entities etc. Different kinds of constraints must conform too.

```
      Context DIM
      Inv:
      self.InterfaceModel.Interface →
-- For every interface default state (e.g.'WaitState') exists
      forAll(i|i.InterfaceState → exists(s0|s0.name = 'WaitState') and
-- For all Lifelines of Interface
      forAll(i|i.InterfaceLifeline → (l|
      forAll(e|l.Event→  --For all Lifeline Events
-- If Event is Received Request, Event State exists
       if e.receiveMessageEnd.oclIsKindOf(Request)then
        i.InterfaceState → exists(s|s.name = e.name.concat ('State') and
-- If Received Request Event is first event on Lifeline or it has
--preceding Sent Response, Transition exists from Wait State to event
-- State and Event is Received on this Transition (Fig.7)
        if e.toBefore.before → isEmpty() or e.toBefore.before →
        exists (e1|e1.sendMessageEnd.isKindOf(Response))
        then s.incoming →
           exists (t|t.source = s0 and t.target = s and t.received = e)
        endif and
-- If Sent Request succeeds Received Request Event, then Request is
-- sent during Transition to Event State (Fig.8)
        if e.toAfter.after.sendMessageEnd.isKindOf(Request)
        then t.send = e.toAfter.after
-- If Sent Response succeeds Received Request Event then transition
-- exists from Event State to Wait State
        else if e.toAfter.after.sendMessageEnd.isKindOf(Response)
          then s.outgoing →
           exists (t1|t1.received = e.toAfter.after and t1.source = s
           and t1.target.s0)
          endif
        endif
-- If Received Request is asynchronous type then transition exists
-- from Event State to Wait State (Fig.9)
       if (e.receivedMessageEnd.MessageSort = asynchCall or
           e.receivedMessageEnd.MessageSort = asynchSignal))
         then s.outgoing → exists (t1|t1.source = s and t1.target = s0)
       endIf )
      endIf
-- If Event e is Received Response then State exists having incoming
-- Transition during which Request for this Response was sent. This
-- State is the state of the last Received Request event before
-- Received Response event e (Fig.10)
   if e.receiveMessageEnd.IsKindOf(Response)
        then i.interfaceState → exists
         (s|s.name = e.toBefore →
       select(e1|e1.receiveMessageEnd.isKindOf(Request))
                        → last().name.concat('State') and
-- If Received Response is last Event on Lifeline or it has succeeding sent
-- Response then Transition exists from this State to Wait State and
-- Received Response is received on this Transition and its succeeding sent
      -- Response is sent (Fig.11)
      if e.toAfter.after → isEmpty() or
         e.toAfter.after.sendMessageEnd.isKindOf(Response)
      then s.outgoing → exists (t|t.source = s and
          t.target = s0 and t.received = e and
          if e.toAfter.after → notEmpty()
           then t.send=e.toAfter.After endif)
-- If Received Response has succeeding Sent Request then recursive
-- transition exists triggered by Response and Request is sent
-- during this Transition (Fig.10)
         else if e.toAfter.after.sendMessageEnd.isKindOf(Request))
          then s.outgoing → exists (t|t.source = s and t.target = s
          and t.received = e and t.send = e.toAfter.after)
          endIf
        endIf ) endIf endIf ))))
```

Figure 6. DIM constraints associated with sequence and state chart diagrams

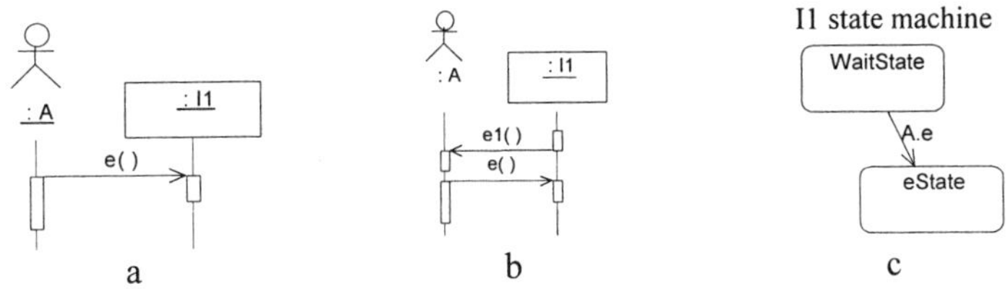

Figure 7. Representation of *Received Request Event*, first on *Lifeline* (a) or having preceding *Sent Response* (b), on sequence diagram (a, b) and on state machine (c)

Figure 8. Representation of *Received Request Event*, having succeeding *Sent Request*, on sequence diagram (a) and on state machine (b)

Figure 9. Representation of asynchronous *Received Request Event* on sequence diagram (a) and on state machine (b)

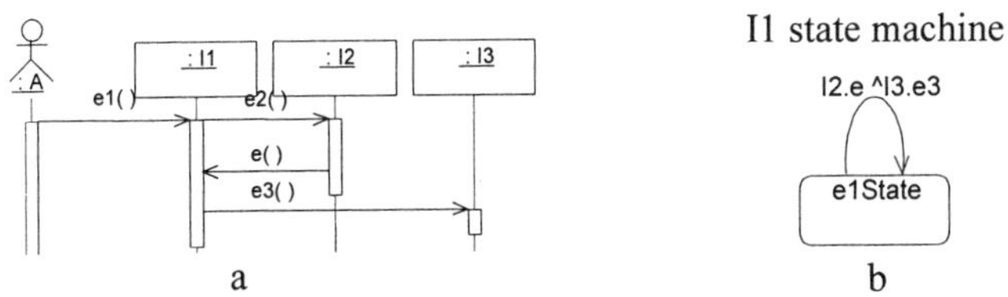

Figure 10. Representation of *Received Response*, having succeeding *Send Request Event*, on sequence diagram (a) and on state machine (b)

Figure 11. Representation of *Received Response Event* on sequence diagram (a) and on state machine (b)

In UML, it is possible to attach constraint to every element of model. In DIM, *Interface State Invariants*, *Event Constraints*, *TransitionPreconditions*, *Operation Preconditions* and *Post Conditions* are used. They have UML 2.0 semantics except *Event Constraints* that were used instead of *Interaction Constraints* [5]. Rules for consequent derivation of *Interface State Transitions Preconditions* and *Post Conditions* from *Event Constraints*, as soon as derivation of *Preconditions* and *Post Conditions* of *Operations* are elaborated, but they are not presented here because of limits of the paper.

Of course, pursuing of strict requirements loads with extra analytic work. CASE tool may help designer to keep different diagrams reconciled and to produce new diagrams from existing ones. Why is it needed? Because it is a way to see all peculiarities of the problem under investigation. If state machines would be derived from sequence diagrams, inconsistencies between interactions would be seen, as interactions of one classifier would be integrated from several sequence diagrams.

It is possible to proceed in reverse order – to make class diagrams or state charts and outspread them to interactions, then wrong traces would be clarified. More detail side of this kind of analysis is presented in the next section.

2. Reconciliation of Sequence and State Diagrams

The algorithm for derivation of state charts from sequence diagrams was developed that can be used for interactive generation of state chart designs from sequence diagrams or synchronization between state charts and sequence diagrams. The first version of this algorithm was presented in [28] and implemented using ArgoUML CASE tool. Developer, who manually creates state charts from sequence diagrams, can also use this process. Following the respective guidelines even in a case of non-automated derivation helps to maintain the consistency of the system description. The reverse process is proposed for obtaining sequence diagrams from state charts. It would help to test feasible scenarios of behavior defined by state machine, and to reveal missing ones.

2.1 From Sequence Diagrams to State Charts

The derivation of state machine from sequence diagrams consists of two stages: obtaining the event sequences from the lifelines from sequence diagrams and synthesizing the state charts from these sequences.

The sequence (starting from the top of lifeline) consists of pairs where the first member corresponds to the state and the second member corresponds to the transition in state chart according to constraints (Fig.6). For example, every *Received Request* e_i corresponds to transition in state chart diagram; if *Sent Request* succeeds e_i it is connected to e_i as a *Sent Event* of the *Transition*, etc.

Event constraints are added to corresponding transitions. Information about the senders and receivers of events is captured in state machine, thus ensuring possibility to re-create sequence diagrams from state machines (this is not employed in UML state machines, although provided in UML metamodel). Sender may be described similar to receiver: *Sender.Event.*

The sequence is created starting with *oclVoid* as the first element of the first pair (it corresponds to the *Wait State* in the state machine) then goes the first *Received Request* e_i as the second element of a pair. If there is succeeding *Sent Request* it is connected to e_i as a *Send Event*.

For the second pair, the first member is named as e_i with concatenated string “State”; the second member consists of at most two events succeeding e_i (the first – *Received Response,* if any; the second – *Sent Response* or *Sent Request*, if any).

The first member of the next pair is *oclVoid* (if the last analyzed event is *Sent Response)* or the name of event with concatenated string “State” (if the last analyzed event is *Sent Request*). The second member is constructed in similar manner as for the first pair.

If any element of the pair is missing, then *oclVoid* is used instead. The last element of the last pair is *oclVoid*.

The result of the first stage $\{\{e_1; e_2\};\{e_3; e_4\}\ldots\{e_{n-1};e_n\}\}$ is used in the second stage – generation of the corresponding state machine. The work of the algorithm is illustrated in Figures 12 and 13.

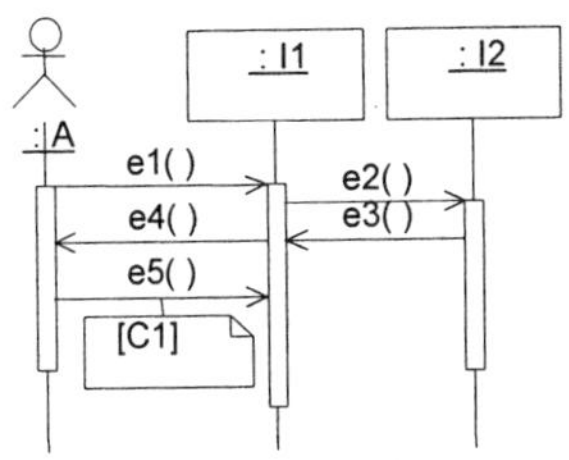

Sequence of events for I1

```
{{oclVoid; A.e1^I2.e2};
{e1State; I2.e3^A.e4};
{oclVoid; A.e5[C1]};
{e5State; oclVoid}}
```

Sequence of events for I2

```
{{oclVoid; I1.e2};
{e2State; ^I1.e3};
{oclVoid; oclVoid}}
```

Figure 12. Example of sequences of events obtained from sequence diagram

Figure 13. Resulting state machines for I1, I2

2.2 From State Charts to Sequence Diagrams

The algorithm for generation of sequence diagrams from state charts should produce a set of sequence diagrams where every sequence corresponds to one sequence of events from state chart. Sequence is created by walking through the state chart from *Wait State* to final state and remembering the path. In one sequence each transition can be taken not more than once (this helps to avoid endless traces). Transition may be included in sequence if its source state is included in sequence and it does not have received events dependent on events that are not included in preceding sequence (*Event e1* depends on event *e2*, if it is *Received Response* to *Sent Request e2*: *e2.Response.responseTo = e1).*

Thus the first stage of this algorithm is to obtain all sequences from a state chart. The sequences are described in the same manner as in generation state charts from sequence diagrams. Two traces obtained from the state chart diagram are represented on Figure 14.

For receiving sequence diagram pairs are analyzed according to their sequence:

– If *Lifeline* of the state machine owner, sender or receiver does not exist in the sequence diagram, it is added;

– If the first member of a pair is *Wait State*, the second member is mapped to *Request* message, received by state machine owner; *Sent Event* of this second member is mapped to *Request* message, sent by the state machine owner;

– If the first member of a pair is not *Wait State*, but the first member of subsequent pair is *Wait State*, the second member is mapped to *Response Message* and *Sent Event*, if any, is mapped to *Response Message*, sent by state machine owner to *Lifeline* of receiver of *Sent Event*;

– If the first member of a pair and the first member of subsequent pair is not *Wait State*, the second member is mapped to *Response Message*, sent to state machine owner from *Received Event* sender; *Sent Event*, if any, is mapped to *Request Message mq*, sent to *mq* receiver;

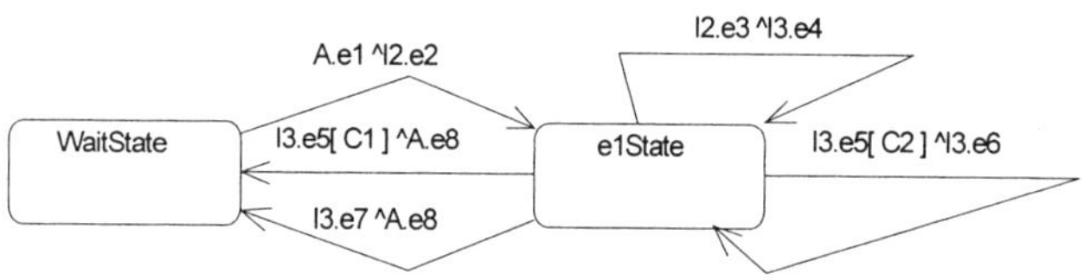


```
e3.Response.responseTo = e2 and e5.Response.responseTo = e4
and e7.Response.responseTo = e6
{{oclVoid;   A.e1^I2.e2}              {{oclVoid;   A.e1^I2.e2}
 {e1State;   I2.e3^I3.e4}              {e1State;   I2.e3^I3.e4}
 {e1State;   I3.e5[C1]^A.e8}           {e1State;   I3.e5[C2]^I3.e6}
 {oclVoid; oclVoid}}                   {e1State;   I3.e7^A.e8}
                                       {oclVoid; oclVoid}}
```

Figure 14. Example state chart and sequences obtained from this statechart

– Constraints are added as *Event Constraints* (Fig.15).

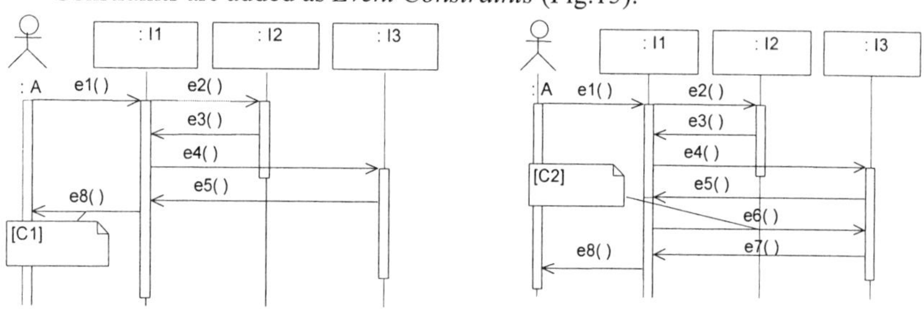

Figure 15. Sequence diagrams, derived from state machine on Figure 14

Reasoning about correctness of state machine gives possibility to improve behavior model, described by sequence diagrams. None of interactions or state related behavior definitions, presented on Figures 7–15 (except Fig.9), are complete. To ensure completeness, some additional constraints must be added to DIM metamodel, namely:

– For every synchronous *Sent Request* message at least two responsive transitions must be provided: one for the case, when response is received, and the other one, when response is not received, i.e., e1 and C = not(e1) (such condition may express timeout for waiting of event occurrence);

– For every *Received Response* several conditions may hold, and several transitions or several sent messages must be provided in correspondence with these conditions.

In the last example there are lacking messages and transitions for the cases, when *Responses* are not received from *I2* or *I3* (Fig.16).

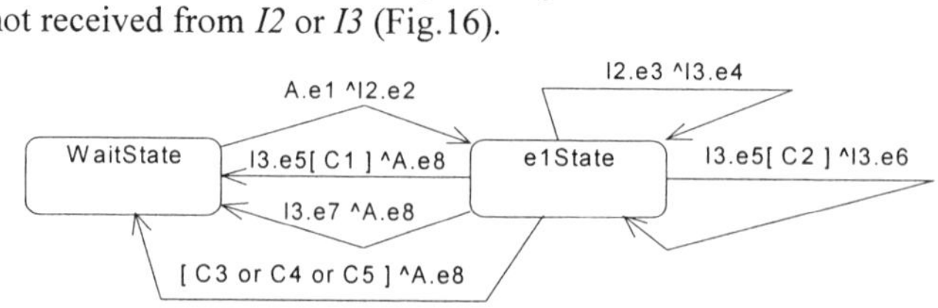


```
C1 = not (e3), C2 = not (e5), C3 = not (e7)
```

Figure 16. Improved state chart diagram

Additionally, different messages may emerge due to conditions originated from received responses. For example, after message *e3* received from *I2* at least two different messages may succeed: if *e3* conveys acceptance, request *e4* must be sent to *I2*, and if *e3* conveys rejection, response *e8* must be sent to *actor*. In general case, a set of transitions (with events and conditions) from one *Source State* must comprise tautology, and messages (with events and conditions) sent after one *Received Response* event also must comprise tautology.

In UML 2.0, the new elements – *Combined Fragments* – are proposed to use in sequence diagrams; these elements may ensure some canonical form of sequence diagram, in which all sequences for one initial *Received Request* may be represented. Also, the canonical form of state chart diagram for requirements definition may be set, where all above mentioned requirements should be ensured.

3. Design Independent Model Example

In this section the proposed method is illustrated by fragments of DIM for IS of Publications Agency. Use Case diagram of the system is presented in Figure 17. The resulting interface class diagram is presented in Figure 18. Relationships between states of interfaces and states of entities are illustrated in Figure 19.

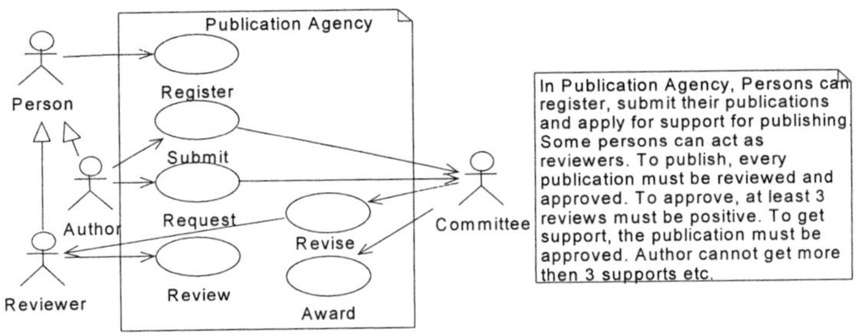

Figure 17. Use case diagram of Publication Agency

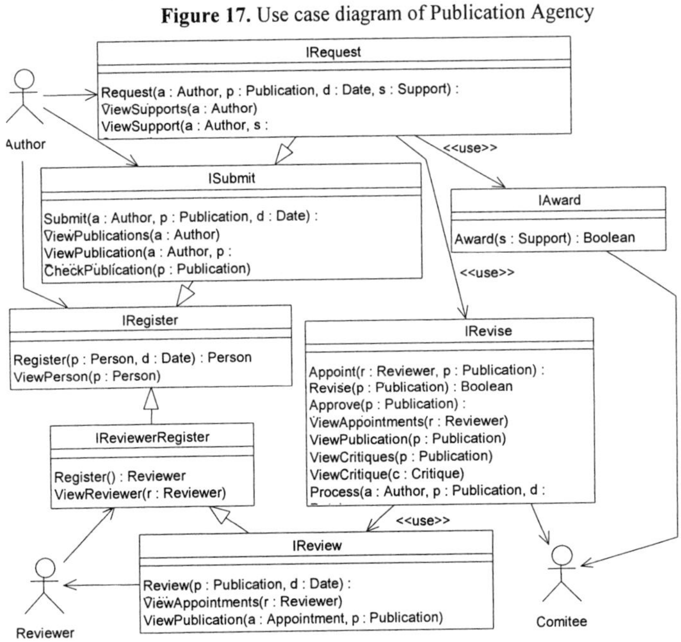

Figure 18. Representation of interfaces in class diagram

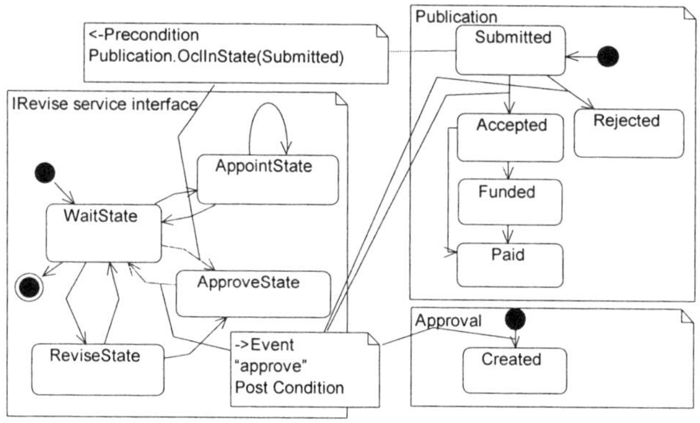

Figure 19. Relationships between states of interfaces and states of entities

4. Conclusion

Presented method for modeling of requirements to information systems consisting of services is devoted for making precise model in the sense that defined features of system should be understood unambiguously. Method is based on behavioral model where service by default is in wait state and can move to service state, during which service is provided, possibly in collaboration with other services or systems. How services are provided it is not considered in Design Independent Modeling – the main purpose is to define all actors (including external systems), interfaces, their operations, entities of problem domain, their relationships and constraints.

The idea of this work is in consequent derivation of model, fully representing design independent state and behavior of intended information system, from initial requirements. Use cases are formalized as interfaces; elaborated requirements may be presented in OCL or in visual form. Diagrams, supported with OCL constraints, are integrated and derivable from other ones. It is demonstrated, how direct and reverse mappings between sequence diagrams and state charts may be fulfilled and how analysis of state charts may reveal incomplete and inconsistent pieces.

When describing state charts, different authors prefer to use states of entities or states of actions. Really, both kinds of states are important and must be modeled in integrated manner. The form of state machines used in this paper has advantage against other kinds of state machines when service systems are modeled, as their semantics are close.

All reasoning is based on UML object model itself, on purpose to employ potential of this modeling language. In future work, it is intended to make the second step – from requirements to design, during which operations of interfaces must be allocated to control classes or components according rules of chosen design method.

References

[1] Frankel D. Model Driven Architecture: Applying MDA to Enterprise Computing. John Wiley & Sons, 2003.

[2] Web Services Architecture. Working Draft. W3 Consortium, 2002. http:// www.w3.org.

[3] Siegel J. Developing in OMG's Model-Driven Architecture. OMG document omg/01-12-01. http://www.omg.org.

[4] Jacobson I., Booch G., Rumbaugh J. The Unified Modeling Language User Guide. Addison Wesley, Boston (2000)

[5] Unified Modeling Language. Superstructure Specification. Version 2.0. OMG document ptc/03-08-02, 2003. http://www.omg.org.
[6] Warmer J.B., Kleppe A.G. The object constraint language: precise modeling with UML. Addison Wesley, 2000.
[7] Kleppe A., Warmer J. Extending OCL to Include Actions. Lecture Notes in Computer Science, 1939, pp. 440–450, 2000.
[8] Cook S. et all. The Amsterdam Manifesto on OCL. Lecture Notes in Computer Science, 2263, Springer-Verlag, 2002, pp.115–149.
[9] Unified Modeling Language: OCL. Version 2.0. OMG document ptc/03-08-08, 2003. http://www.omg.org
[10] Starr L. Executable UML. How to build class models. Prentice Hall, 2002.
[11] Mellor S.J., Balcer M.J. Executable UML. A foundation for model-driven architecture. Addison-Wesley, 2002.
[12] D'Souza D. F., Wills A. C. Objects, Components, and Frameworks with UML. The Catalysis approach. Addison Wesley, 1999.
[13] ISO/TC97/SC5/WG3. Concepts and Terminology for the Conceptual Schema and Information Base, 1982.
[14] Ceponiene L., Nemuraite L., Paradauskas B. Design of schemas of state and behaviour for emerging information systems. Emerging Database Research in Eastern Europe. Proceedings of the Pre-Conference Workshop of VLDB 2003. Brandenburg University of Technology at Cottbus, November 2003, p. 27-31
[15] UML Profile for Enterprise Distributed Object Computing Specification. OMG document ptc/o2-02-05. 2002. http://www.omg.org.
[16] Bordeleau F. and Corriveau J.-P.: From Scenarios to Hierarchical State Machines: A Pattern-Based Approach, in Proc. of OOPSLA 2000 Workshop: Scenario-based round-trip engineering, Tampere University of Technology, Software Systems Laboratory, Report 20, October 2000, 13-18.
[17] Mäkinen E. and Systä T.: An Interactive approach for synthesizing UML statechart diagrams from Sequence Diagrams, in Proc. of OOPSLA 2000 Workshop: Scenario-based round-trip engineering, Tampere University of Technology, Software Systems Laboratory, Report 20, October 2000, 7-12.
[18] Whittle J., Schumann J. Generating Statechart Designs From Scenarios. International Conference on Software Engineering, Ireland, 2000, pp.314-323.
[19] Kosters G., Six H.W., Winter M. Coupling Use Cases and Class Models as a Means for Validation and Verification of Requirements Specifications. Requirements Engineering, 6, Springer-Verlag, 2001, pp. 3-17.
[20] Rosenberg D., Scott K. Applying Use Case Driven Object Modeling with UML: An Annotated e-Commerce Example. – Boston: Addison Wesley, 2001. – 153 p.
[21] Astesiano E., Reggio G. Knowledge Structuring and Representation in Requirement Specification. Technical Report. DISI-Universita di Genova, Italy, 2002.
[22] Bock C. Three kinds of Behavior Model. Journal of Object-Oriented Programming 12(4), SIG Publications, 1999.
[23] Bock C. Goal-Driven Modeling. Journal of Object-Oriented Programming 13(5), SIG Publications, 2000
[24] Reggio G., Cerioli M., Astesiano E. An Algebraic Semantics of UML Supporting it's Multiview Approach. Lecture Notes in Computer Science, 2029, Springer Verlag, Berlin, 2001.
[25] Cheesman, J. Daniels J. UML Components. Adison Wesley, 2000
[26] Sendall S., Strohmeir A. From Use Cases to System Operation specifications. <<UML>> 2000, Lecture Notes in Computer Science, 1939, Springer-Verlag, 2000, pp. 1–15.
[27] Martin R.C. Agile Software Development, Principles, Patterns, and Practices. Prentice Hall, 2002.
[28] Nemuraite L.,Kavaliauskaite L., Ambrazevicius E.: Towards Ensuring Consistency to UML Models. Informacijos mokslai, 24, Vilniaus universitetas, 2002, pp. 85-97.

Databases and Information Systems
J. Barzdins and A. Caplinskas (Eds.)
IOS Press, 2005

UML + Spreadsheets as a Telecommunications Network Documentation and Analysis Tool

Guntis BARZDINS
Institute of Mathematics and Computer Science, University of Latvia, 29 Raina Boulevard, Riga, LV-1459, Latvia

Abstract. Although UML language and MDA (Model Driven Architecture) approach primarily are being developed for the needs of large software projects, these powerful ideas are fruitfully applicable also towards smaller projects, where having a precise model of a real-life system is essential. This case study illustrates how a complex telecommunications network can be precisely documented by means of an "executable model". Such executable model takes the form of a highly integrated spreadsheet, which is semi-automatically constructed from the OCL constrained UML description of the network structure. We also stipulate that this approach can be extended into generic MDA tool for creating complex spreadsheet applications from their OCL constrained UML models.

Introduction

Telecommunication systems and services are a notoriously complex domain consisting of a large number of subsystems. Industry-wide standards for dealing with this complexity in standardized manner are emerging, like eTOM (Enhanced Telecom Operations Map) [1], which provide a top-to-bottom view on the involved systems and interfaces.

Meanwhile the logical distance between such top-to-bottom view and the actual telecommunications transmission and switching network is so great, that in many cases an independent bottom-to-top approach is still appropriate. In this paper we focus on the bottom-to-top part of telecommunications operation IT support, which is heavily dependent on physical transmission technologies and architectures involved, and thus requires grassroots design of optimal IT support structure in each individual case. We claim in this paper that UML based MDA (Model Driven Architecture) is well suited for this kind of applications, and illustrate this by a case study from satellite communications domain.

On the methodological side, we claim that UML modeling and spreadsheet programming are a convenient and little explored technology match for implementing small, but logically advanced documentation projects.

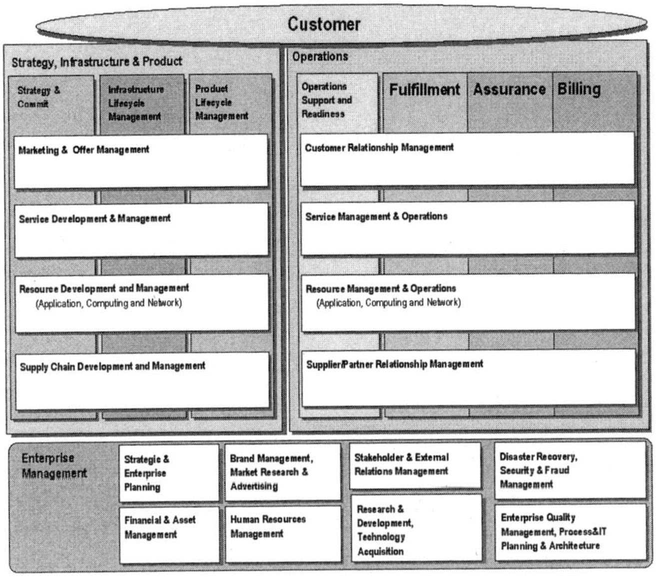

Figure 1. eTOM: enhanced Telecom Operations Map

1. Tools Used in the Case Study

Although theoretically any MDA tool could be used for such bottom-to-top cases, our observation is that the relatively small scale of these projects requires techniques optimized for quick (but not necessarily very efficient) implementations. We have found that UML modeling [2] and spreadsheet programming [3] provide a perfect match for these cases – powerful design techniques and simple implementation tools.

Despite their seemingly worlds-apart uses, the UML data modeling and spreadsheet programming have a common background – they both originate from the entity-relationship data representation model. It is obvious for UML class diagrams, but is true also for spreadsheet programming (although casual MS Excel spreadsheet user might not be aware of it). The full power of spreadsheet programming comes from lookup functions (LOOKUP, MATCH, INDEX, COUNTIF, SUMIF, etc.), which can be applied between worksheets (independent tables) of the workbook (database). We illustrate that lookup functions available in MS Excel cover the essential features of OCL – the Object Constraint Language [4], which is part of UML specification and in its expressive power is comparable to SQL for RDBMS. Therefore entity-relationship data representation model within this paper is used as a "glue" between more abstract UML modeling and more pragmatic spreadsheet programming. The effect of such integration is three-fold:

- It provides means for designing spreadsheets of unmatched complexity, which adequately capture the data structure of the real-world system to be modeled.
- It yields a clear UML-style documentation of the spreadsheet itself. UML model of the spreadsheet (which at the same time is also the UML model of the real-world system to be modeled) gives a new way of communicating spreadsheet internal logic between its users.
- Due to its clear structure, spreadsheet programming provides a simple transformation path from OCL constrained UML model to its executable implementation. This, besides being compliant with MDA (Model Driven Architecture) framework [5], also

suggests a useful extension to traditional spreadsheet software, like MS Excel, where spreadsheet application could be semi-automatically constructed from its OCL constrained UML description.

The rest of the paper is organized as follows: first we give a brief description of the real-world telecommunications system documentation problem, which triggered development of the described approach. Then we show in detail how this problem was solved using UML modeling and spreadsheet programming techniques, leading to an easy maintainable documentation (model) of a complex and frequently changing telecommunications system. The final part of the paper discusses the techniques used throughout the case study, and how they potentially could be extended into a more generic and automated tool.

2. Telecommunications System Documentation Problem

For telecommunications system engineers it is common to use diagrams like shown in Figure 2 to depict the structure of the particular telecommunications network setup. Such diagrams are sometimes accompanied with ad-hoc support documents listing telephone numbers, IP addresses, frequencies etc. configured on various elements of the telecommunications network. Only for very large and homogenous telecommunications networks (like public telephone network or large and homogenous ISP network) it is feasible to develop a specialized database application, where all essential configuration details are stored. For smaller and less homogenous networks it is usually left up to the telecommunications engineer to come up with the appropriate documentation, suitable for maintenance of the communications system throughout its lifetime. The result in this case is highly dependent on the presentation skills of the engineer.

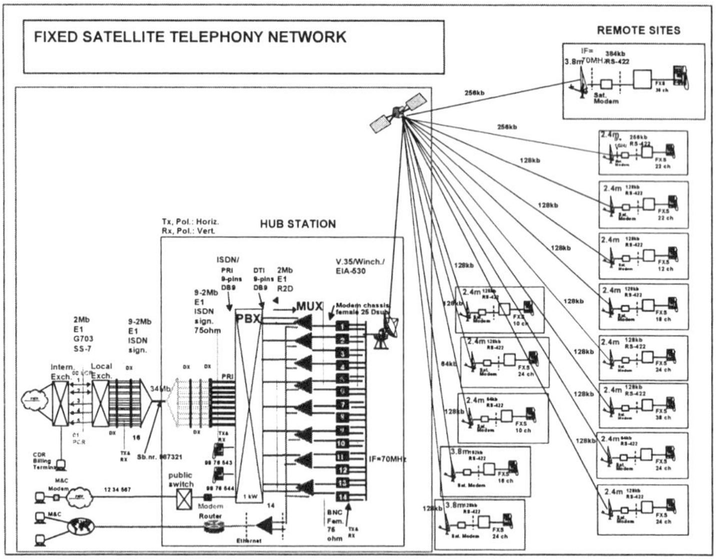

Figure 2. Traditional diagram of the satellite telephony network

But the situation with such diagrams and ad-hoc support documents easily gets out of hand, if the system is subject to frequent configuration changes – usually proper documentation of changes requires massive modifications at various parts of the documentation. And there are no any safeguards against human-error – like changes inconsistently reflected in various parts of the documentation. Result of this phenomenon is that complex and frequently changing telecommunications systems are extremely hard to maintain error-free.

Therefore a desired documentation method must avoid redundancies (so that the same network element does not need to be repeatedly described in several parts of the documentation – a common source of documentation errors), and must provide safeguards against human-error (for example, automatic crosschecks, which would reveal the logical inconsistencies within the documentation). Such semi-automatic documentation system would also make configuration changes virtually error-free – the documentation of the intended new configuration can be crosschecked for consistency even before it is implemented, thus avoiding system downtimes due to ill-planned changes.

A proper MS Excel spreadsheet seemed to be a natural solution to this documentation problem (development and ongoing support of the specialized software would be an overkill). But how such highly complicated spreadsheet can be designed? This is where we decided to turn to UML design methodology. Afterwards it came bit of a surprise, how well the two technologies actually complement each other.

3. System Description Phase

From object-modeling point of view, the system diagram depicted in the Figure 2 is effectively an instance diagram of a yet to-be-drawn object model of this system. Analysis of the system revealed the functionally essential elements, shown in the form of UML class diagram in Figure 3. This model looks very different from the drawing in Figure 2, because it is concerned primarily with the functionally essential aspects of the system, and not with the physical layout of the components. This transition is natural from the system-analyst point of view, but might seem quite unnatural from the communications engineer point of view.

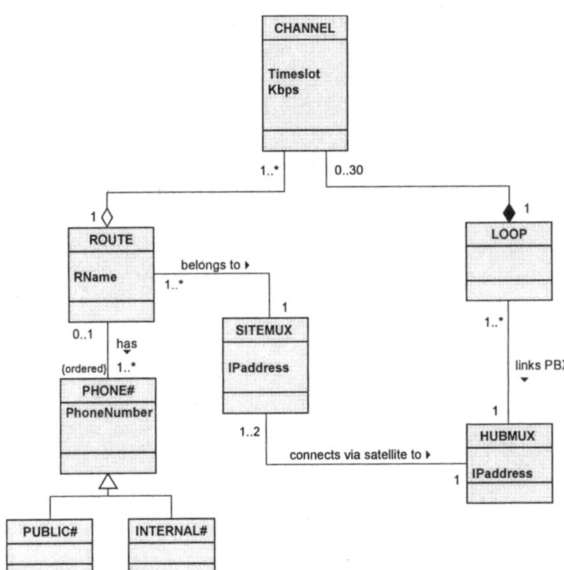

Figure 3. UML object model (class diagram) of the satellite telephony system.

Following comments should help to understand the object-model depicted in Figure 3:

- CHANNEL represents a single telephone line. It has two main attributes – compressed voice bandwidth in Kbps (when transmitted via satellite) and timeslot (when transmitted via LOOP).
- LOOP represents a bundle of up to 30 voice CHANNELs, which are physically transported over the same cable between the PBX (telephone switch) and HUBMUX (voice compression equipment).
- ROUTE represents a group of voice channels, which have the same telephone numbers associated in the PBX. Note that channels that belong to the same route can be scattered over several LOOPs.
- PHONE# represents the phone number assigned to the route. Several phone numbers may be assigned to the same route.
- There are two types of telephone numbers: PUBLIC numbers, which can be dialed from public network, and INTERNAL numbers, which can be reached only within the PBX.
- HUBMUX is the voice compression equipment used at the hub side of the network. It can serve several remote SITEMUXes.
- SITEMUX is the voice compression equipment used at the remote location connected via satellite link. Note that voice CHANNELs belonging to the same ROUTE can span several LOOPs, but they all must end up in the same SITEMUX.

The next step is to transform this purely conceptual UML class diagram into the spreadsheet. As the target environment we will use standard MS Excel spreadsheet program.

The transformation is based on viewing UML class diagram as an entity-relationship data model, and then creating the tables (MS Excel worksheets) with columns matching the attribute names of the classes. Figures 3 and 4 illustrate this transformation – first, additional attributes (primary keys, foreign keys, selectors) are added to the classes to uniquely encode class relationships (Figure 4). Then this modified class diagram is mechanically transformed into the Excel workbook: separate worksheets represent each class of the diagram, while columns in the worksheets represent attributes of the corresponding classes (Figure 5). Rows in the resulting worksheets represent individual class instances, with columns showing attribute values for each instance.

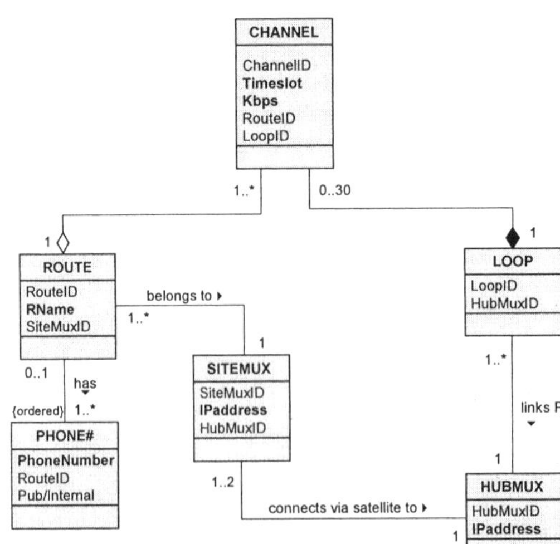

Figure 4. Additional attributes are added to encode relationships between the classes

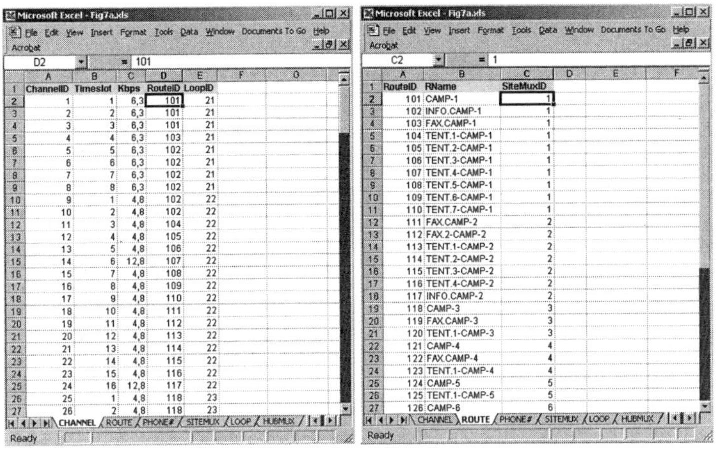

Figure 5. MS Excel workbook corresponding to the class diagram in Figure 4

By filling the newly created MS Excel workbook with data, we can encode there the actual configuration of the real-world telecommunications system. Note that this MS Excel workbook contains full information about the telecommunications system configuration and, importantly, – it avoids any information redundancy (no configuration information is being duplicated anywhere in the documentation).

But this workbook still has two drawbacks, which we will need to fix in the next section:

- It is not easy to find information in this workbook. For example, to find an answer to the question *"To which site a particular channel belongs to?"*, one would need to search through several tables.
- There are no any safeguards against human-error.

4. Report Generation and Constraint Checking Phase

If the previous chapter was dominated by UML design techniques to reveal the functional structure of the system and to encode it into MS Excel workbook in the non-redundant manner, then now is the time to unleash the power of spreadsheet programming. The essence of this phase is following: out of the system description columns created in the previous chapter (which fully and without redundancy describe the system), we can use spreadsheet formulas to pre-calculate any useful attribute of the system.

A "useful attribute" of the system can be specified either in natural language, or by means of OCL – Object Constraint Language [4], an add-on part of UML. Below are listed some (not all) useful attributes, part of which are depicted also in Figure 6. For comparison, attributes are defined in three different syntaxes – natural language, OCL, and MS Excel:

- Add a column to Channels worksheet, where formula calculates the SiteID to which this channel is routed.

  OCL notation: `Channel: self.Route.SiteMux.SiteMuxID`

  Excel formula: `=INDEX(ROUTE!C:C;MATCH(D2;ROUTE!A:A;);1)`
- Add a column to Channels worksheet, where formula calculates the HubMuxID through which this channel is routed.

  OCL notation: `Channel: self.Route.SiteMux.HubMux.HubMuxID`

  Excel formula: `=INDEX(SITEMUX!C:C;MATCH(INDEX(ROUTE!C:C; MATCH(D2;ROUTE!A:A;);1);SITEMUX!A:A;);1)`

- Add a column to Routes worksheet, where formula calculates one of phone numbers assigned to this Route.

  OCL notation: `Route: self.Phone#.PhoneNumber->first`

  Excel formula: `=INDEX('PHONE#'!A:A;MATCH(A2;'PHONE#'!B:B;);1)`

Besides that, we can create a new worksheet containing some crosscheck and overview values of the system, for example:

- Total number of channels.

  OCL notation: `Channel.allInstances->size`

  Excel formula: `=COUNTIF(CHANNEL!A:A;">0")`

- Total number of phone numbers in use.

  OCL notation: `Phone#.allInstances->select(RouteID>0)->size`

  Excel formula: `=COUNTIF('PHONE#'!B:B;">0")`

- Total number of misrouted channels: the Loop, to which Channel C belongs, is connected to the HubMuxID=A, but the Route associated with channel C belongs to SiteMux, which is connected to HubMuxID=B, and A <>B.

  OCL notation: `Channel.allInstances -> select(Route.SiteMux.HubMuh <> Loop.HubMux)->size`

  Excel formula: first mark misrouted channels in a separate column K on Channels worksheet by formula

  ```
  =(INDEX(LOOP!B:B;MATCH(E2;LOOP!A:A;);1)=INDEX(SITEMUX!C:C;
  MATCH(INDEX(ROUTE!C:C;MATCH(D2;ROUTE!A:A;);1);SITEMUX!A:A;);1))
  ```

  and then calculate `=COUNTIF(CHANNEL!K:K;FALSE)`

Such overview values are specifically oriented towards crosschecking the system description integrity – for example, if our intention was to change only telephone number assigned to some route, then it should not change the overview value for the number of phone numbers in use. And the number of misrouted channels, of course, must be 0 at all times.

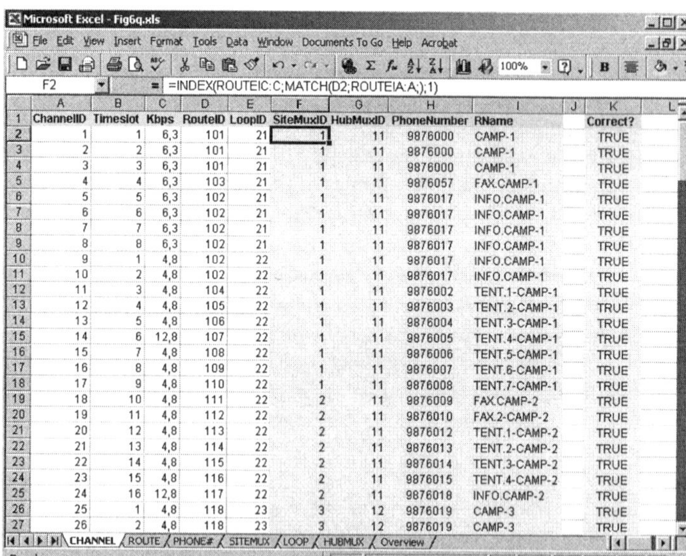

	A	B	C	D	E	F	G	H	I	J	K
1	ChannelID	Timeslot	Kbps	RouteID	LoopID	SiteMuxID	HubMuxID	PhoneNumber	RName		Correct?
2	1	1	6,3	101	21	1	11	9876000	CAMP-1		TRUE
3	2	2	6,3	101	21	1	11	9876000	CAMP-1		TRUE
4	3	3	6,3	101	21	1	11	9876000	CAMP-1		TRUE
5	4	4	6,3	103	21	1	11	9876057	FAX.CAMP-1		TRUE
6	5	5	6,3	102	21	1	11	9876017	INFO.CAMP-1		TRUE
7	6	6	6,3	102	21	1	11	9876017	INFO.CAMP-1		TRUE
8	7	7	6,3	102	21	1	11	9876017	INFO.CAMP-1		TRUE
9	8	8	6,3	102	21	1	11	9876017	INFO.CAMP-1		TRUE
10	9	1	4,8	102	22	1	11	9876017	INFO.CAMP-1		TRUE
11	10	2	4,8	102	22	1	11	9876017	INFO.CAMP-1		TRUE
12	11	3	4,8	104	22	1	11	9876002	TENT.1-CAMP-1		TRUE
13	12	4	4,8	105	22	1	11	9876003	TENT.2-CAMP-1		TRUE
14	13	5	4,8	106	22	1	11	9876004	TENT.3-CAMP-1		TRUE
15	14	6	12,8	107	22	1	11	9876005	TENT.4-CAMP-1		TRUE
16	15	7	4,8	108	22	1	11	9876006	TENT.5-CAMP-1		TRUE
17	16	8	4,8	109	22	1	11	9876007	TENT.6-CAMP-1		TRUE
18	17	9	4,8	110	22	1	11	9876008	TENT.7-CAMP-1		TRUE
19	18	10	4,8	111	22	2	11	9876009	FAX.CAMP-2		TRUE
20	19	11	4,8	112	22	2	11	9876010	FAX.2-CAMP-2		TRUE
21	20	12	4,8	113	22	2	11	9876012	TENT.1-CAMP-2		TRUE
22	21	13	4,8	114	22	2	11	9876013	TENT.2-CAMP-2		TRUE
23	22	14	4,8	115	22	2	11	9876014	TENT.3-CAMP-2		TRUE
24	23	15	4,8	116	22	2	11	9876015	TENT.4-CAMP-2		TRUE
25	24	16	12,8	117	22	2	11	9876018	INFO.CAMP-2		TRUE
26	25	1	4,8	118	23	3	12	9876019	CAMP-3		TRUE
27	26	2	4,8	118	23	3	12	9876019	CAMP-3		TRUE

Figure 6. MS Excel workbook from Figure 5, populated with additional calculated attributes

MS Excel workbook shown in Figure 6 is the final result of our design efforts (full example is available at `http://www.ltn.lv/~guntis/mdaex.xls`). The resulting spreadsheet has following important properties:

- Now we can find any useful information about the system easy. For example, answer to the question *"To which site a particular channel belongs to?"* is automatically pre-calculated and visible in the spreadsheet line describing the channel.
- There is no information redundancy – each system parameter is stored (and modified) only in one place. From there, all calculated cells are updated with correct values automatically.
- Overview parameters of the system can be used to spot description inconsistencies.

Note that this spreadsheet is a true model of the real-world communications system itself – the telecommunications engineer can spend more time designing (and crosschecking) changes to the system in this virtual model, rather than in the physical system. Only when the changes seem OK in the model, engineer proceeds with their physical implementation. Thus documentation is always up-to-date – it is actually updated already in the change design phase, before the physical change is even implemented.

5. Generic MDA Framework for Spreadsheet Applications

The key remaining question is – how universally applicable is the well-working approach illustrated in this case study? For reasons mentioned below, we suggest that it is generally applicable to tasks, where OCL constrained UML static structure model captures majority of system logic. Development of semi-automatic tools supporting this approach could potentially spark their rather massive use due to ubiquity of MS Excel spreadsheet users.

From data representation point of view spreadsheets are similar to relational databases – they both support lookup functions and can contain embedded procedures, where bulk of data-driven business logic can be embedded. In both cases external code should be used to implement interactivity (note, that also spreadsheet program can interact with external code through `SetCellValue(...)` and `GetCellValue(...)` methods – see details in [3] about server-based spreadsheet engines).

From MDA methodology point of view transformations used in the case study must be fairly easy to encode in a formal transformation description language, like the one proposed in [6]. In this case MS Excel workbook shown in Figure6 could be generated by automatic (or semi-automatic) transformation from UML class diagram in Figure 3 and OCL constraints defined in Section 4. This would result in a clean MDA framework for creating complex Excel spreadsheets from UML and OCL models. Although we are not aware of any tool being developed along these ideas, the implementation appears to be rather straightforward and would result in substantially enhanced complex spreadsheet development tool.

Our conclusion is that spreadsheet programming based implementation of UML models should not be neglected, as this might be adequate for smaller projects, fast prototyping, or learning purposes. The case study shows that it works quite well at least in telecommunications operation environment.

References

[1] TMFORUM: NGOSS and eTOM. http://www.tmforum.org/TMFC1859 eTOM Overview.pdf (2003)

[2] G. Booch, I. Jackobson, J. Rumbaugh. The Unified Modeling Language. Reference Manual, Addison-Wesley, 1999.

[3] M. Smialek. Spreadsheet Programming: MS Excel as Component Development Environment, http://www.devx.com/enterprise/Article/11686 (2003)

[4] J. Warmer, A. Kleppe. The Object Constraint Language: Precise Modeling with UML, Addison-Wesley, 1999

[5] OMG: MDA Guide Version 1.0.1, http://www.omg.org/docs/omg/03-06-01.pdf (2003)

[6] E. D. Willink. A concrete UML-based graphical transformation syntax – The UML to RDBMS example in UMLX. Workshop on Metamodeling for MDA, University of York, England, November 2003

Ontologies

Databases and Information Systems
J. Barzdins and A. Caplinskas (Eds.)
IOS Press, 2005

Merging DAML+OIL Ontologies

Patrick LAMBRIX and He TAN
Department of Computer and Information Science
Linköpings universitet, Sweden

Abstract. Ontologies are being used nowadays in many areas. Within each area there are a number of ontologies, each with their own focus, that contain overlapping information. In applications using multiple ontologies it is therefore of interest to be able to merge these ontologies. In this paper we describe a prototype implementation of SAMBO, an ontology merge tool for DAML+OIL ontologies. The tool generates suggestions for merging concepts and relations and for creating is-a relationships between concepts. We evaluate different strategies for the generation of suggestions. We also compare our tool with the ontology merge tools Protégé-2000 with PROMPT and Chimaera in terms of the quality of suggestions and the time it takes to merge ontologies using these tools.

1. Introduction

Intuitively, ontologies can be seen as defining the basic terms and relations of a domain of interest, as well as the rules for combining these terms and relations. Ontologies are being used nowadays in many areas for communication between people and organizations, as the basis for interoperability between systems, and as query models and indexes to repositories of information.

Within an area there are always a number of ontologies, each with their own focus. For instance, in bioinformatics (e.g. [8]), the ontologies cover different aspects in molecular biology such as molecular function and cell signaling. Many of these ontologies contain overlapping information. For instance, a protein can be involved in both cell signaling and other biological processes. In applications using ontologies it is therefore of interest to be able to use multiple ontologies. However, to obtain the best results, we need to know how the ontologies overlap and align them or merge them into a new ontology. Another reason for merging ontologies is that it allows for the creation of ontologies that can later be composed into larger ontologies. Also, companies may want to use de facto standard ontologies and merge them with company-specific ontologies.

In this paper we describe a prototype of the ontology merge tool SAMBO. In its current implementation it helps users merge DAML+OIL ontologies and performs description logic reasoning. We describe DAML+OIL in section 2 and SAMBO in section 4. Further, we describe a number of evaluations in section 5. SAMBO can generate suggestions for merging, but allows the user to choose different algorithms for the generation of these suggestions. In the first evaluation we compare the different algorithms. In the second evaluation we compare SAMBO with two well-known ontology merge tools, Protégé-2000 with PROMPT and Chimaera (described in section 3), regarding the quality of the suggestions and the time it takes to merge two ontologies using these tools.

2. DAML+OIL

DAML+OIL [4] is an ontology language built on Web standards such as XML and RDF. It takes a frame-based approach, and inherits the expressiveness and reasoning power from description logics. DAML+OIL can be seen as a very expressive description logic. More precisely, DAML+OIL is equivalent to the SHIQ description logic with the addition of existentially defined classes and data types [7]. This equivalence allows DAML+OIL to exploit implemented description logic systems to provide highly optimized reasoning services. DAML+OIL has already been widely adopted, although it may be superseded later by its successor OWL.

From a knowledge representational point of view ontologies can have the following components (e.g. [19]): concepts, relations and axioms. DAML+OIL supports the representation of all these kinds of components. Further, it has concept and relation constructors (e.g. boolean operators and quantifiers) that allow for the definition of new concepts and relations based on already existing concepts and relations.

3. Related Work

Protégé-2000 is software for creating, editing and browsing ontologies. The design and development of Protégé-2000 has been driven by two goals: to be compatible with other systems for knowledge representation and to be an easy to use and configurable tool for knowledge extraction. Protégé-2000 is available as free software and should be installed locally. It also has a number of plug-ins, among others PROMPT, which is an algorithm for merging and aligning ontologies [18, 13]. When merging two ontologies PROMPT creates a list of suggested operations. An operation can, for instance, be to merge two terms or to copy a term to the new ontology. The user can then perform an operation by choosing one of the suggestions or by specifying an operation directly. PROMPT performs the chosen operation and additional changes that follow from that operation. The list of suggestions is then updated and a list of conflicts and possible solutions to these conflicts is created. This is repeated until the new ontology is ready. PROMPT was previously called SMART and a high-level description of the algorithm is given in [12]. In [15] evaluation criteria for mapping or merging tools were proposed. First, an evaluation should be driven by pragmatic considerations: input requirements, level of user interaction, type of output and content of output. Tools that satisfy a user's pragmatic requirements can then be compared with respect to a performance criterion based on precision and recall. Protégé-2000 with PROMPT was evaluated according to these criteria.

The initial goal for developing Chimaera [3] was to provide a tool that could give substantial assistance for the task of merging knowledge bases produced by different users for different purposes with different assumptions and different vocabulary. Later the goals of supporting testing and diagnosing ontologies arose as well. The user interacts with Chimaera through a web browser [11]. The two main tasks when merging two ontologies in Chimaera are to merge two semantically identical terms from different ontologies so that they are referred to by the same name in the resulting ontology, and to identify terms that should be related via is-a, disjointness or instance relationships and provide support for introducing those relationships. Chimaera also supports the identification of the locations for editing and performing the edits. Today, Chimaera has support for merging taxonomies of concepts and for merging attributes. To assist the user Chimaera generates name resolution lists that suggest concepts that are candidates to be merged or to have taxonomic relationships not yet included in the merged ontology. Chimaera also generates a taxonomy resolution list where it suggests taxonomy areas that are candidates for reorganization. On the basis of these lists the user decides what should be done.

In [9] we evaluated how well PROMPT and Chimaera work for merging bio-ontologies. The ontologies we used for testing were Gene Ontology ontologies and Signal-Ontology. A larger survey of ontology tools and methodologies was performed by the OntoWeb Consortium [17].

4. SAMBO

The current implementation of SAMBO is a web-based system that helps a user merge two DAML+OIL ontologies into a new DAML+OIL ontology with unique names for terms (currently concepts or relations). The system separates the merging process into three steps: merging relations, merging concepts and introducing is-a relationships. Each step should be finished before the next step is started. In each step, the user can choose to manually merge terms or introduce is-a relationships in the ontologies or to have the system propose suggestions. The user can choose to accept or reject the suggestions. Upon acceptance of a suggestion, the system performs the actual merging or includes the is-a relationship, computes the consequences and makes the additional changes that follow from the operation. Upon rejection of a suggestion, it is checked that we do not have two different terms with the same name. If so, one of the terms needs to be renamed. In each case the suggestion list is updated. At each point in time the user receives information on which operations have been performed and how many suggestions there are left for the step. After all suggestions are processed or the user decides that no more merging should be performed or no more relationships between the ontologies should be added, the system copies the terms of the original ontologies that are not merged to the new ontology.

Regarding the generation of the suggestions we use the following strategies. For the first step (relations) the system generates a suggestion when the names of the terms in the different ontologies are the same or when one is an affix of the other. In steps two and three the SAMBO system gives the user the choice to use one of several underlying algorithms. This allows the user to experiment with different strategies and obtain better results. Currently, we have implemented the following strategies.

1. N-gram string matching
2. Edit distance string matching
3. Linguistic matching + Porter stemming
4. Linguistic matching + Porter stemming + WordNet
5. Structure-based strategy

N-gram and edit distance matching are simple string matching algorithms. N-gram matching computes a similarity between strings based on n-grams. An n-gram is a set of n consecutive characters extracted from a string. Similar strings will have a high proportion of n-grams in common. In the second algorithm the similarity between strings is based on the edit distance. This is defined as the number of deletions, insertions, or substitutions required to transform one to the other. The greater the edit distance, the more different the strings are. Algorithms three and four are linguistic matchers. We assume that the name of a term is represented as a string of words. The algorithm computes the similarity of the strings based on the similarity of the pairs of words. Within the algorithms we employ the Porter stemming algorithm. Algorithm four also uses WordNet to compute the similarity of the pairs of words. The matching thresholds, e.g. the similarity value in the linguistic matcher, can be modified by experts. In the structured-based strategy we use the structure of the ontologies and already merged concepts to propose new suggestions. In our implementation a path between two concepts is a composition of one or more is-a links in the is-a hierarchy. The user receives a list of already merged concept pairs and can ask the system to generate new suggestions based on the paths between two merged concepts. Thus, the strategy is based on the intuition that concepts between two given merged concepts in the is-a hierarchy have a good chance of being similar to each other.

In the second step (merging of concepts) the user can choose any of the strategies. A suggestion is generated based on the chosen strategy. We also make explicit use of the fact that several ontologies define a 'synonym' relation and thus check whether synonyms match the concept names.

In the third step (is-a relationships between concepts) the user can choose one of the linguistic matchers. The system generates a suggestion when a concept name in one ontology contains all the words of a concept name in the other ontology based on the linguistic matcher.

SAMBO uses further the FaCT [6] system to provide a number of reasoning services. (Our implementation is partly based on OilEd [2] ideas and implementation.) The user can ask the system to check whether the new ontology is consistent. She can receive information about unsatisfiable concepts and cycles in the ontology. The user can also receive an updated DAML+OIL ontology that is generated from the FaCT representation and therefore it contains less redundancy and it contains explicit statements for derived relationships.

5. Evaluation

We have performed two different evaluations. In the first evaluation we compare the quality of the suggestions that are generated by the different algorithms and strategies in our system. In the second evaluation we compare SAMBO with two well-known ontology merge tools regarding quality of the suggestions and the time it takes to merge two ontologies.

5.1 Comparison of algorithms

In this evaluation, we compare the quality of the suggestions generated by the different matching algorithms and strategies in our system.

Test Ontologies

As test ontologies we have chosen two bio-ontologies that are available from Open Biological Ontologies (OBO) [16]. OBO is an umbrella address for structured shared controlled vocabularies and ontologies for use within the genomics and proteomics domains. The ontologies we chose are the Medical Subject Headings (MeSH) and the Anatomical Dictionary for the Adult Mouse (MA). MeSH is a controlled vocabulary produced by the American National Library of Medicine and used for indexing, cataloging, and searching for biomedical and health-related information and documents. It consists of sets of terms naming descriptors in a hierarchical structure. These descriptors are organized in 15 categories, such as the category for anatomic terms and the category for organisms. We used the MeSH category for anatomic terms, including approximately 1400 terms. MA is cooperating with the Anatomical Dictionary for Mouse Development (EMAP), to generate an anatomy ontology covering the entire lifespan of the laboratory mouse. The adult mouse anatomy ontology we used describes anatomical structures for the postnatal mouse (Theiler stage 28), including approximately 2350 terms. The ontology is represented as a directed acyclic graph. It organizes anatomical structures spatially and functionally, using is-a and part-of relationships. The two ontologies cover a similar subject domain, anatomy, and are developed independently. We translated the two ontologies from the GO flat file format to DAML+OIL retaining identifiers, names, synonyms, definitions, and is-a and part-of relationships.

Evaluation Results

Using the first linguistic matcher (using the Porter stemming algorithm, but not WordNet), our system found 377 suggestions. When also WordNet was used, 82 more suggestions are found. (WordNet has a good coverage of anatomy [1], so it was likely that new suggestions would be found.) These suggestions can be divided into four types. In the first type one term is the plural of the other (e.g. `ganglion` and `ganglia`). In the second type the terms are synonyms (e.g. `midbrain` and `mesencephalon`). These suggestions are useful. Further, there is the case where there is a semantic relationship between the terms. For instance, our system suggested `bile canaliculi` and `bile duct`. According to WordNet canaliculi is a kind of duct. This is a useful suggestion for is-a relationships. We found that for every suggestion in this category, there always was an is-a or part-of relationship in the original ontologies. Finally, 12.5% of the suggestions from WordNet were due to different senses of words. For instance, `nerve` and `cheek` was suggested. These are synonyms in the sense of impudent aggressiveness, but not in the anatomical sense.

Using the N-gram matcher we obtained 661 suggestions (of which 370 were also found by the first linguistic matcher). As it is a flexible string matcher, it is able to find suggestions based on small differences in the spelling of words. For instance, it suggests `brain stem` and `brainstem`. It is able to find useful suggestions that are not found using WordNet, for instance, `striatum` and `neostriatum`. At the same time, 24.3% of the generated suggestions contain unrelated terms. For example, the suggestion `coeliac artery` and `iliac artery` refers to arteries but coeliac is of or relating to the abdominal cavity and iliac is of, relating to or located near the ilium. This matcher has also the characteristic that for a term in one ontology often many similar terms in the other ontology are suggested for merging or is-a relationship. We found that for 53 concepts there were multiple suggestions. (For the 53 concepts together there were 165 suggestions.) For instance, for `hippocampus` the following were suggested: `hippocampus`, `hippocampus CA1` and `hippocampus CA2`. SAMBO displays these suggestions simultaneously.

Edit distance matching is a good approximate string matching algorithm that allows to eliminate spelling errors. In this test, however, it found the smallest number of correct suggestions.

When making use of the fact that these ontologies also contain information about synonyms, 9 more suggestions are found (e.g. `cerebellum lobule IX` and `uvula`).

We also experimented with the structure-based matcher with different path lengths. The maximum length of an is-a path in the MeSH anatomy ontology is 10, while for the MA ontology this is only 4. In this experiment we achieve the best result with maximum path length of 3. In this case we receive two new correct suggestions. A higher path length does not give more new results. One reason for the fact that only two correct new suggestions were found, is that, in our tests, most concepts in the paths were already merged. However, it may also be that the is-a relation alone does not define a good enough neighborhood of a concept to find useful suggestions. Further testing will make this clear.

5.2 Comparison of tools

In this evaluation we compare SAMBO with Protégé-2000 with PROMPT and Chimaera regarding the quality of suggestions and the time it takes to merge ontologies using these tools. We note that for Protégé-2000 with PROMPT and Chimaera we have used the systems as they are provided in the basic distribution. Extensions to the basic algorithms are being made. For instance, Anchor-PROMPT [14], an extension to PROMPT, uses paths between merged concepts in the two ontologies to find new suggestions.

Test Ontologies

We tested the tools using three 'cases' taken from two different domains: biological ontologies and ontologies about academia.

For the first two cases we used a part of an ontology from the Gene Ontology Consortium (GO) [5] together with a part from Signal-Ontology (SigO). The Gene Ontology Consortium is a joint project which goal is to produce a structured, precisely defined, common and dynamic controlled vocabulary that describes the roles of genes and proteins in all organisms. Currently, there are three independent ontologies publicly available over the Internet: biological process, molecular function and cellular component. The GO ontologies are becoming a de facto standard and many different bio-databases are today annotated with GO terms. The terms in GO are arranged as nodes in a directed acyclic graph, where multiple inheritance is allowed. The purpose of the SigO project is to extract common features of cell signaling in the model organisms, try to understand what cell signaling is and how cell signaling systems can be modeled. SigO is a publicly available controlled vocabulary of the cell signaling system. It is based on the knowledge of the Cell Signaling Networks databank [20] and treats complex knowledge of living cells such as pathways, networks and causal relationships among molecules. The ontology consists of a flow diagram of signal transduction and a conceptual hierarchy of biochemical attributes of signaling molecules. In our tests two cases were created. Each case consists of one part of SigO and one part of GO. Each case was chosen in such a way that there was an overlap between the GO part and the SigO part. The first case, *behavior* (B), contains approximately 60 terms from GO and approximately 10 terms from SigO. The second case, *immune defense* (ID), contains approximately 70 terms from GO and approximately 15 terms from SigO. We used more terms from GO than from SigO because the granularity of GO is higher than the granularity of SigO for these topics.

We also created a case, *academia* (AC), using two ontologies from the DAML ontology library. The first ontology describes an employment hierarchy in academic institutes. The ontology was developed at Carnegie Mellon. It contains 31 terms. The other ontology describes employees, publications and academic activities in a computer science department and was developed at the University of Maryland. It contains 83 terms. The overlap between these two ontologies is about employees in academic departments of a university.

Evaluation Results

In table 1 we show the results regarding the quality of the generated suggestions in terms of precision and recall for the cases *behavior* (B), *immune defense* (ID) and *academia* (AC), respectively. (The numbers for PROMPT and Chimaera for the B and ID cases are taken from [9].) Precision measures how many of the suggestions are relevant while recall measures how many of the relevant suggestions the system actually proposed. SAMBO performs perfect with respect to quality of suggestions on the *behavior* case and good on the *immune defense* case. However, it performs worse than PROMPT in the *academia* case.

Table 1. Quality of suggestions

Tool	Case	Sug.	Correct	Missing	Recall	Precision
PROMPT	B	3	3	2	0.6	1
Chimaera	B	16	4	1	0.8	0.25
SAMBO	B	5	5	0	1	1
PROMPT	ID	4	4	5	0.444	1
Chimaera	ID	10	4	5	0.444	0.4
SAMBO (N-gram)	ID	9	7	2	0.777	0.777
SAMBO (edit distance)	ID	9	7	2	0.777	0.777
SAMBO (ling. matcher)	ID	8	7	2	0.777	0.875
SAMBO (ling. matcher, WordNet)	ID	9	7	2	0.777	0.777
PROMPT	AC	7	5	6	0.454	0.714
Chimaera	AC	1008	8	3	0.727	0.008
SAMBO (N-gram)	AC	10	5	6	0.454	0.5
SAMBO (edit distance)	AC	10	5	6	0.454	0.5
SAMBO (ling. matcher)	AC	9	4	7	0.363	0.444
SAMBO (ling. matcher, WordNet)	AC	14	4	7	0.363	0.285

We can have a look at the experiments for SAMBO in more detail. Tables 2 and 3 show the suggestions created by SAMBO (N-gram) for the *immune defens*e case and the *academia* case, respectively. The first column describes whether the suggestion was for a merge (M) or is-a relationship (R). The last column describes whether the suggestion was correct (C), wrong (W) or not found by the system (NF). In the *immune defense* case, the two missing suggestions are for is-a relationships between concepts. They might have been found by using more flexible string matching. However, more flexible string matching may also lead to more wrong suggestions. The wrong suggestions in *immune defense* have to do with similar concepts where the matching was too flexible. As we mentioned before, if a concept from the first ontology is involved in multiple suggestions, the user receives these multiple suggestions simultaneously. Further, when the user has decided on merging, the suggestion list is updated and therefore not all the wrong suggestions will actually be shown to the user. In the *academia* case there are two missing suggestions for merging and the system did not find any correct suggestions for the is-a relationships. The names of the terms in the two ontologies differed too much for the algorithm to find them.

Based on these tests the quality of SAMBO's merging suggestions is good. However, the system behaves from good to poor when suggesting is-a relationships. This can be explained by the fact that we use linguistic and string matching. Linguistic matching and string matching find concepts that are similar in name. This works for some examples (e.g. `Antigen Processing and Presentation` is-a `antigen presentation`), but there are also many examples for which this strategy does not work (e.g. `fever` is-a `inflammation`).

Table 2. Suggestions for *immune defense* with N-gram

	Gene Ontology	Signal Ontology	
M	Synonym	synonym	C
M	immune response	Immune Response	C
M	B-cell activation (synonym: B-cell proliferation)	B Cell Activation	C
M	B-cell activation (synonym: T-cell proliferation)	T Cell Activation	W
M	T-cell activation (synonym: T-cell proliferation)	T Cell Activation	C
M	T-cell activation (synonym: T-cell proliferation)	B Cell Activation	W
M	complement activation	Complement Signaling (synonym: complement activation)	C
R	antigen processing	Antigen Processing and Presentation	C
R	antigen presentation	Antigen Processing and Presentation	C
R	inflammatory response	Inflammation	NF
R	activation of natural killer cell activity	Natural Killer Cell Response	NF

Table 3. Suggestions for *academia* with N-gram

	Academic Ontology of UM	Academic Ontology of CMU	
M	Student	Student	C
M	Administrative Staff	AdministrativeStaff	C
M	Director	Director	C
M	Faculty	Faculty	C
M	Professor	Professor	C
M	Assistant (includes research and teaching assistants)	Assistant (is a kind of administrative staff)	W
M	Postdoctoral Fellow	PostDoc	NF
M	Technical Staff	Systemstaff	NF
R	Research Programmer	Research	W
R	Research Staff	Research	W
R	Research Scientist	Research	W
R	Research Engineer	Research	W
R	Masters Student	GraduateStudent	NF
R	PhD Student	GraduateStudent	NF
R	Research Staff	ResearchAssistant	NF
R	Visiting Staff	VisitingProfessor	NF

Table 4 illustrates the results regarding the time it took to merge the test ontologies. The total work time is computed as the work time based on the suggestions plus the time for merging and adding the relationships that the systems missed to propose. Due to the good quality of suggestions compared to the other systems as well as the easy to use user interface and the automatic copy operations, working with SAMBO was much faster than working with the other two systems.

Table 4. Time of merging process

Tool	Case	Work based on suggestions	Total work time
PROMPT	B		15
Chimaera	B	8	8
SAMBO	B	3	3
PROMPT	ID		28
Chimaera	ID	10	15
SAMBO	ID	4	6
PROMPT	AC		20
Chimaera	AC	18	22
SAMBO	AC	3	7

6. Conclusion

We have described SAMBO, a merging tool for DAML+OIL ontologies, evaluated the suggestions generated by its different algorithms and strategies and compared the system in terms of quality of suggestions and work time with PROMPT and Chimaera. Regarding the quality of the suggestions SAMBO did well compared to the others and outperformed the other tools regarding time. We note that our test ontologies did not require the full expressivity of DAML+OIL. Therefore, the full description logic reasoning capabilities of the tool have not been exploited yet in this test. We will also investigate further on more advanced strategies for the generation of suggestions.

The system still needs to be evaluated with respect to other factors as described in e.g. [9] and [10]. For instance, we already know that the user interfaces of Protégé-2000 and Chimaera have a more elaborate visualization of the ontologies than our current implementation. Also, these systems allow for many different input and output formats and are more full-fledged ontology engineering tools than our system.

Acknowledgements

We thank Vaida Jakoniene and Bo Servenius for comments on the system. We also acknowledge the financial support of the Center for Industrial Information Technology and of the EU Network of Excellence REWERSE (Sixth Framework Programme project 506779).

References

[1] Bodenreider, O., Burgun, A. Characterizing the Definitions of Anatomical Concepts in WordNet and Specialized Sources. Proceedings of the First Global WordNet Conference, pp 223-230, 2002.

[2] Bechhofer, S., Horrocks, I., Goble, C., Stevens, R. OilEd: a Reason-able Ontology Editor for the Semantic Web. Proceedings of the Joint German Austrian Conference on Artificial Intelligence, LNAI 2174, pp 396-408, Vienna, Austria, 2001.

[3] Chimaera. http://www.ksl.stanford.edu/software/chimaera/

[4] DAML+OIL. http://www.w3.org/TR/daml+oil-reference

[5] The Gene Ontology Consortium. Gene Ontology: tool for the unification of biology. Nature Genetics, 25(1):25-29, 2000. http://www.geneontology.org/.

[6] Horrocks, I. FaCT and iFaCT. Proceedings of the International Workshop on Description Logics, pp 133-135, Linköping, Sweden, 1999.

[7] Horrocks, I. DAML+OIL: a Reason-able Web Ontology Language. Proceedings of the International Conference on Extending Database Technology - EDBT02, LNCS 2287, pp 2-13, Prague, Czech Republic, 2002.

[8] Lambrix, P. Ontologies in Bioinformatics. Chapter in Dubitzky, W. and Azuaje, F. (eds.) Artificial Intelligence Methods and Tools for Systems Biology, Kluwer, to appear, 2004.

[9] Lambrix, P., Edberg, A. Evaluation of ontology merging tools in bioinformatics. Proceedings of the Pacific Symposium on Biocomputing - PSB03, pp 589-600, Kauai, HI, USA, 2003.

[10] Lambrix, P., Habbouche, M., Pérez, M. Evaluation of ontology engineering tools for bioinformatics. Bioinformatics, 19(12):1564-1571, 2003.

[11] McGuinness, D., Fikes, R., Rice, J., Wilder, S. An Environment for Merging and Testing Large Ontologies. Proceedings of the Seventh International Conference on Principles of Knowledge Representation and Reasoning - KR2000, pp 483-493, Breckenridge, Colorado, USA, 2000.

[12] Noy, N.F., Musen, M. SMART: Automated Support for Ontology Merging and Alignment. Proceedings of the 12th Workshop on Knowledge Acquisition, Modeling and Management, Banff, Canada, 1999.

[13] Noy, N.F., Musen, M. PROMPT: Algorithm and Tool for Automated Ontology Merging and Alignment. Proceedings of Seventeenth National Conference on Artificial Intelligence, pp 450-455, Austin, TX, USA, 2000.

[14] Noy, N.F., Musen, M. Anchor-PROMPT: Using Non-Local Context for Semantic Matching. Proceedings of the IJCAI01 Workshop on Ontologies and Information Sharing, Seattle, WA, USA, 2001.

[15] Noy, N.F., Musen, M. Evaluating Ontology-Mapping Tools: Requirements and Experience. Proceedings of the EKAW Workshop on Evaluation of Ontology Tools, Siguenza, Spain, 2002.

[16] Open Biological Ontologies. http://obo.sourceforge.net

[17] OntoWeb Consortium. Deliverable 1.3: A survey on ontology tools. 2002. http://www.ontoweb.org

[18] Protégé with PROMPT. http://protege.stanford.edu/index.html

[19] Stevens, R., Goble, C., Bechhofer, S. Ontology-based knowledge representation for bioinformatics. Briefings in Bioinformatics, 1(4):398-414, 2000.

[20] Takai-Igarashi, T., Nadaoka, Y., Kaminuma, T. A Database for Cell Signaling Networks. Journal of Computational Biology, 5(4):747-754, 1998.

Databases and Information Systems
J. Barzdins and A. Caplinskas (Eds.)
IOS Press, 2005

Combining FCA and a Logic Language for Ontology Representation

Hele-Mai HAAV
Institute of Cybernetics at Tallinn University of Technology,
Akadeemia 21, 12618 Tallinn, Estonia

Abstract. The most important problem of ontology design is to guarantee accuracy, transparence and consistency of ontology representation. This paper provides automatic way of transforming lattice based ontology representation to logical expression by defining first order logic model of concept lattice based ontology expression. As the latter represents only taxonomic relationships between concepts, then ontology designer is given possibility to add additional concepts, properties, and ontological relationships using a rule language based on first order logic. Validation of ontology, reasoning about ontology, and search is done using logical inference.

1. Introduction

Ontology has its origin in philosophy as a science of being. In the philosophical sense, ontology is a particular system of categories accounting for a certain vision of the world (a domain of interest [1].

Formal ontologies provide machine-processable semantics of information that is shared between different agents (humans or software). Ontologies play important role in many application domains including semantic web services, business systems, industrial systems, intelligent information retrieval, etc.

As analysis of existing ontologies and ontology design methodologies shows, it is very difficult for a designer to develop accurate, transparent and consistent ontology [2, 3, 4]. Domain ontology design usually involves work of domain experts, software engineers and ontology designers. All parties need to reach to design agreement.

In order to help experts in ontology design process, an initial domain ontology can be automatically or semi-automatically built from domain-specific knowledge captured in domain-specific texts, documents, or data [5, 6, 7]. Ontology designer can do further development of the initial ontology.

Next step after extraction of the initial ontology is to formally express ontology in some logic language for sharing knowledge by software agents; ontology based reasoning purposes and search. This can also be done automatically or semi-automatically.

The main goal of this paper is to present a model of concept lattice based ontology expression in the form of first order language in order to make it possible to represent in addition to taxonomic relationships also non-taxonomic relationships between the concepts. This work is continuation and extension of our previous work on semi-automatic extraction and expression of domain-specific ontologies using Formal Concept Analysis (FCA) [7]. In our previous approach, an initial representation of domain ontology is extracted from a set of domain-specific texts as a concept lattice using FCA and NLP [see 7]. As an extension to this work, our current approach presented in this paper, takes concept lattice based ontology expression and maps it automatically to a set of rules (and facts) in first order language. This is done according to the first order language model of concept lattice based ontology description. As concept lattice based ontology description represents only taxonomic relationships between concepts, then ontology designer is given possibility to add additional concepts and relationships (part-of, related to, etc) using a rule language based on first order logic. Validation of ontology, reasoning about ontology and search can be done using logical inference. As our approach uses first order language, then it is possible to attach different ontology inference engines for practical applications by translating ontology expression to any inference engine rule language.

Our approach is applicable in many application domains, where domain specific ontologies can be extracted form web catalogs, product catalogs, domain specific dictionaries and texts etc.

The rest of the paper is structured as follows. Section 2 gives a motivation for the approach and refers to the related works. General framework of proposed approach is presented in section 3. First order language model of concept lattice based ontology expression is given in section 4. Section 5 shows how a rule language can be used to present non-taxonomic relationships. Section 6 concludes the work.

2. Motivation and Related Work

Recent study of DAML ontology library by Tempich and Volz [2] shows that semantic web ontologies are designed in rather heterogeneous way and many semantic web ontologies fail in being usable for inference. This indicates that quality of ontology representation is not high enough.

Also analysis and evaluation of ontology development methodologies [3] result in conclusion that high level design methods do not support well nontrivial ontology-based reasoning. The latter is an important feature of any ontology development methodology, because an ontology expression created by human experts can easily be inconsistent.

Our motivation in this paper is to provide assistance to ontology designer in ontology design process in order to guarantee accuracy, transparence and consistency of ontology representation by automatic or semiautomatic methods of ontology extraction and expression.

Related work on this topic can be grouped as follows:

- Work on extraction of taxonomic relationships and concept hierarchies from given text or data
- Work on formally representing intended semantics of ontology description in some logic language

We have been inspired by Formal Concept Analysis (FCA) [8] as one of the methods of learning concept taxonomies. FCA algorithmically constructs concept lattice from binary relationship between objects and their attributes. FCA is used in ontology engineering for merging ontologies in FCA-MERGE method [9]. There are other methods available for extraction of taxonomic relationships as hierarchical clustering techniques [10] or pattern-based approaches [11, 4]. Also works on extraction of concept hierarchies are of interest. Some related works can be found in [12, 13].

There are many approaches developed within Semantic Web community in order to define mapping of RDF schemas to some of logical languages for providing formal semantics [4]. Most widely used are description logic [14], first-order logic and F-logic [15]. Grüninger and Fox [16] provide ontology development methodology (TOVE project), which allows to manually transform informal natural language specifications into computable model expressed in first order logic. $(KA)^2$ ontology uses F-logic, also Ontobroker [17] and Text-To-Onto [4] systems use F-logic for inference.

To the best of our knowledge, we do not know works, which report about direct automatic or semi-automatic extraction of logical descriptions of a domain ontology.

In the approach described in this paper, we also do not extract logical expressions of domain ontology but instead we rely on initial ontological structure in the form of a lattice learned from domain-specific texts. We provide automatic way of transforming lattice based ontology expression to logical expression by defining first order logic model of concept lattice based ontology expression.

3. Concept Lattice Based Ontology Expression

3.1 Formal Concept Analysis

In this section, Formal Concept Analysis (FCA) developed by Ganter and Wille [8] is very briefly introduced in order to give basis for understanding our approach.

FCA is a result of an attempt to restructure mathematical order theory and lattice theory and as such gives a new interpretation of complete lattices as concept lattices. Currently, FCA can be seen as a field of applied mathematics used in a number of applications of conceptual analysis [18], knowledge representation [9], data analysis, etc.

FCA allows extraction of similar groups of objects from a set O of objects described by a set of attributes C using binary relationship R on $O \times C$. For $o \in O$ and $c \in C$ we note $R(o,c)=1$ if oRc (the object o has the attribute c, the relationship R holds) and $R(o,c)=0$ if $o \neg Rc$ respectively.

The table K(O,C,R) is called formal context that can be used to extract groupings and relationships between objects and attributes. In the framework of FCA the discovered groups represent the closed set of the Galois connection induced by R on the pair O and C. Two mappings are defined as follows.

If X is an arbitrary part of O, then we may define mapping f that maps X onto the set of all elements of C that are related to all elements of X as follows:

$$f(X) = \{y \in C \mid \forall x \in X: xRy\}$$

Similarly, if Y is an arbitrary part of C, then we may define the following mapping:

$$g(Y) = \{x \in O \mid \forall y \in Y: xRy\}$$

These mappings mean that f(X) is the set of all attributes shared by all objects in X and g(Y) is the set of all objects that have all attributes in Y.

The mappings f and g are monotonously decreasing in the following sense:

$$X \subseteq X' \Rightarrow f(X') \subseteq f(X)$$

$$Y \subseteq Y' \Rightarrow g(Y') \subseteq g(Y)$$

The pair of mappings (f, g) meets the criteria of Galois connection between the powerset P(O) and the powerset P(C). The closed subsets of both O and C form two lattices wrt the set inclusion. Hence, we have two lattices L_O and L_C that are isomorphic. Consider the set L of all pairs of corresponding parts of lattices L_O and L_C so that each element of this set is the Cartesian product of closed parts of O and C (i.e., $X \times f(X)$ or $g(Y) \times Y$ wrt R).

These pairs are called *formal concepts* by Wille [8]. In FCA, X is referred to as the concept extent and Y as the concept intent. Thus, a formal concept is a pair of a set of objects that have common attributes (the extent), and the defining set of attributes that they have in common (the intent). Each concept is uniquely determined by either its extent or intent. The partial order relation on the set of all formal concepts of the given context K(O,C,R) is determined by the set inclusion between the extensions or equivalently by the reverse inclusion between intensions of concepts.

In this paper, we define the partial order relation on L based on the set inclusion between the extensions of concepts as follows,

$$X \times Y \leq X' \times Y' \text{ iff } X \subseteq X' \text{ and } Y' \subseteq Y.$$

This lattice L is called the Galois lattice or formal concept lattice of the relationship R on O × C. This partial order defines subconcept and superconcept relationships. Concept lattice is a complete lattice [8] meaning that for each set of formal concepts, there is always a greatest lower bound (glb or greatest common subconcept) and a least upper bound (lub or least common superconcept). Given an object o, there is always a most specific concept, whose extent contains o, this concept is called *object concept*. Dually, for each attribute c, there is a most general concept whose intent includes c, this is called *attribute concept*.

FCA is an algorithmic method for construction of concept lattices for a given context. There are several algorithms for generating the set of all formal concepts and construction of line diagrams of concept lattices. Excellent comparison of performance of those algorithms can be found in [19].

3.2 Extraction of a Concept Lattice

In this section we briefly overview our method [7] used to extract a concept lattice from domain-specific texts. This is a basis of our new approach presented in this paper.

According to the method, first task is to produce a formal context K(O,C,R) for extractable domain ontology. As a set of objects O, domain-specific text sources are considered. A set of noun-phrases from the texts is taken as a set of attributes C. It is assumed that noun-phrases denote/indicate concepts used in application domain, as text sources are domain specific and use specific vocabulary. Binary relationship R between descriptions (texts) of domain entities and noun phrases is discovered during the NLP process of text sources.

Next step is to perform FCA on the context. As a result, formal concept lattice that corresponds to the domain ontology is constructed.

For our current approach presented in this paper, the resulting lattice is reduced and certain naming procedure is performed in order to use it as concept lattice based ontology expression (see section 3.4 below). In the following sections, small examples from real estate domain are provided in order to illustrate FCA and its usage in our approach.

3.3 A Real Estate Domain Example

As an example, consider real estate catalogs. For example, Table 1 represents formal context for real estate domain. For each description of a real estate item, there is a corresponding text source (e.g. A1 corresponds to text about real-estate item one, etc.). Noun-phrases are extracted from each catalog entry by using NLP tools. In the table below, existence of a relationship between a catalogue entry (text) and a noun-phrase is denoted by 1. We feel free to display only some noun-phrases chunked from real estate catalog entries. This is in order to obtain a small sample lattice in the example. In this paper, we are not interested in how noun-phrases are extracted from the texts but we concentrate to the formal context obtained from the texts and presented as in the Table 1 below.

Table 1. Real estate domain context

Objects	**Attributes (Noun-phrases)**					
	Real estate	Family house	Country house	Summer house	Blockhouse	Apartment
A1	1	1				
A2	1		1			
A3	1		1	1		
A4	1				1	1

After applying FCA to this context we get a set of formal concepts of this given context as follows.

1. <{A1}, {real estate, family house}>
2. <{A2, A3}, {real estate, country house}>
3. <{A3}, {real estate, country house, summerhouse}>
4. <{A4}, {real estate, blockhouse, apartment}>
5. <{A1, A2, A3, A4}, {real estate}>
6. <{Ø}, {real estate, family house, country house, summerhouse, blockhouse, apartment}>

The graphical representation of the concept lattice is depicted in the form of Hasse diagram as following Fig. 1.

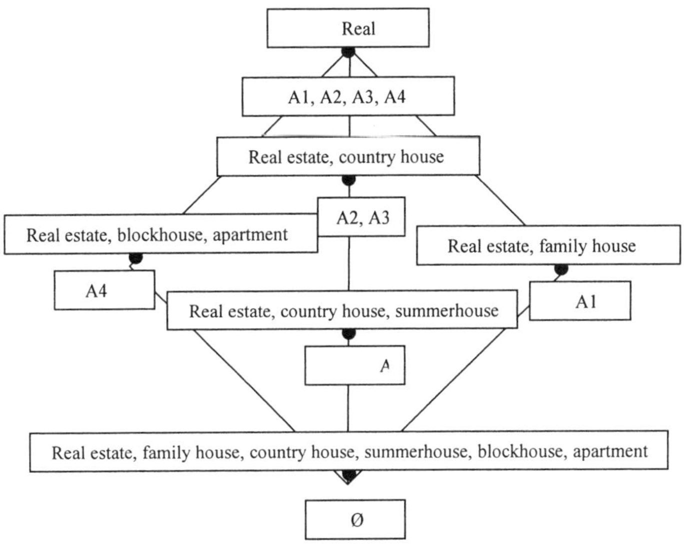

Figure 1. Concept lattice of real estate domain

Each node in this lattice (denoted by black circle) is a formal concept. For example, one of the formal concepts of the context described in Table 1 is as follows:
{A1, A2, A3, A4,} × *{Real estate}*, where the set *{A1, A2, A3, A4,}* is the extent of the concept and the set *{Real estate}* is its intent. This is top element of the lattice. This is most general concept; it has one attribute that is shared by all real estate catalog entries. The bottom element of the lattice (lattice bottom) is the least general concept that is defined by all the attributes at the same time. In our example, there are no objects that are defined by all the attributes, so the extent of the bottom concept is empty.

Sub and super-concept relationships between the formal concepts are represented by edges in the Hasse diagram in Fig. 1. For example, the formal concept *{A3}* × *{Real estate, Country house, Summerhouse*} is a sub-concept of the concept *{A2, A3}* × *{Real estate, Country house}*.

Inheritance of attributes is also present in the concept lattice according to sub and super-concept relationships.

3.4 Concept Lattice Based Ontology Expression

In this section, we define concept lattice based ontology expression. There might arise confusion in using of the word concept in ontology research comparing to FCA. There is no direct notion in ontology field to denote formal concepts. Ontology concepts can be compared to FCA attributes, as both can be considered as unary predicates on the set of objects.

Our idea in creating concept lattice based ontology expression is to use duality feature of concept lattice, i.e. lattice of intensions and lattice of extensions of concepts are connected via Galois connection. In principle, in our case we consider only lattice of intensions of concepts as a useful structure for ontology expression learned during the concept lattice construction process.

There is redundant information in concept lattice. For a formal concept C= (X, Y), X will be present in every ancestor of C and symmetrically, Y will appear in every descendant. The two kinds of redundancy can be eliminated from concept lattice without losing any information as shown in [18]: redundant attributes in formal concepts intents and redundant objects in formal concepts extents.

Our reduction procedure has 2 steps: elimination of redundant attributes from full concept lattice and elimination of lattice of extents.

Elimination of redundant attributes. Let Y′ be the set of elements of Y (intent) that do not appear in any descendant of Y. To eliminate redundant attributes we define a pair (X, Y) as a pair (X, Y'), where Y' $= \{y \in C \mid g(y) = X\}$. Taking union of the Y' sets for the ancestors of a pair (X, Y'), including this pair itself recovers the initial pair (X, Y).

Elimination of lattice of extents. From the reduced lattice above, we eliminate lattice of extents L_O and get reduced lattice of intents L_{CR} of formal concepts.

We call the resulting lattice L_{CR} of reduction procedure as concept lattice based ontology expression. Fig. 2 shows the lattice L_{CR} of our example.

There are 3 aspects that need to be clarified as follows:

1. The resulting lattice is dependent on the context. Whenever we change formal context, we get a new lattice structure. For learning initial domain lattice we need to go through concept lattice construction procedure several times using different formal contexts (e.g. Real Estate Catalog entries in our example).
2. After reducing concept lattice to its intentional part, we need to give formal concepts the names. Naming in our case can be done as follows:
   a) A concept gets a unique name that is the name of the attribute(s) of formal concepts, which are left after reduction procedure. For each attribute c, there is a most general concept whose intent includes c, this is called *attribute concept* and the name of this concept is the name of corresponding attribute(s). Let us recall that attributes of formal concepts indicate domain specific concepts in our approach.
   b) After the previous naming procedure, there might be nodes that do not get names. In principle, the names for these nodes need to be provided by domain expert or ontology designer. It is possible automatically generate formal names (e.g. c1, c2…) for those nodes and then ask advice from human expert.
3. Human expert should manually add concept descriptions in natural language to each concept to express meaning of a concept in application domain.

In order to illustrate what was said above the Fig. 2 depicts real estate domain ontology produced from concept lattice shown in Fig.1 using reduction and naming procedures.

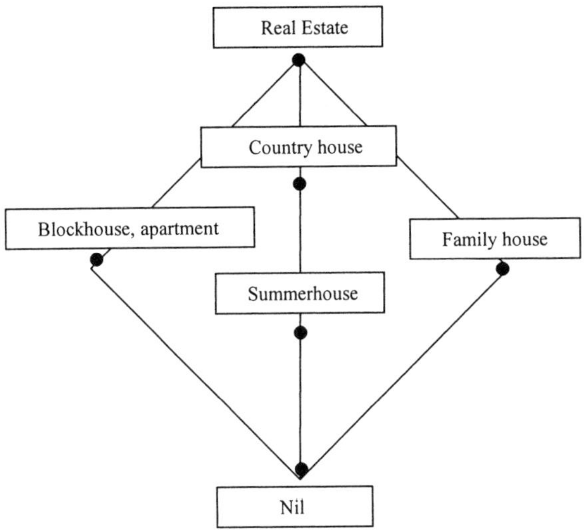

Figure 2. Reduced concept lattice and naming of concepts

This lattice displays names of concepts according to reduction and naming procedure. One concept has combined name: blockhouse-apartment. This indicates that there might be non-taxonomic relationship between the concepts or these could be separate concepts but a given formal context was not complete enough to enable to extract them. The bottom element of the lattice is empty, which is denoted by generated name Nil.

The lattice does not refer to objects forming an extent of a certain concept but these can be recovered if necessary.

The following is concept lattice based ontology expression (see Fig. 3 below) of the same domain but another formal context is used. It illustrates the naming of concepts that do not get name by the naming procedure.

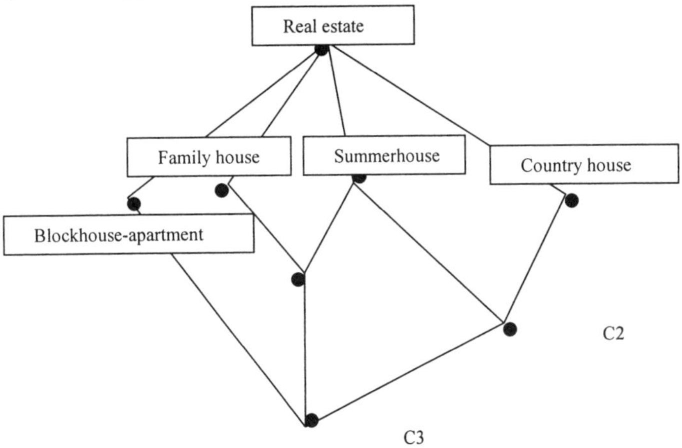

Figure 3. Naming of concepts

The nodes denoted by generated concept names C1, C2, C3 do not have labels in the lattice. Ontology designer may analyse the lattice above and find out that concept C1 is sub-concept of both concepts: Family house and Summerhouse. C1 then denotes the concept that is a family summerhouse and C2 denotes a country summerhouse respectively. The naming is up to ontology designer in this case.

In conclusion, our approach is about learning domain specific ontology from extensionally defined collections of instances contained in data (e.g. real estate catalog). It is assumed that the instances represent relevant knowledge for inductive extraction of intensional descriptions for a given domain.

Our next goal is to define first-order logic model for concept lattice based ontology expression in order to provide richer language for representation of ontological relationships and specify intended meaning of the descriptions.

4. First Order Logic Model of Concept Lattice Based Ontology Expression

In the previous section, we have defined domain ontology as a concept lattice reduced to its intensional part L_{CR} (see for example Fig. 3). In this section, we provide first order logic model for L_{CR}. At the moment we are not interested in extensions, which gave birth to full concept lattice for a formal context of a domain. In principle, it might be interesting to take extensions also into account when building a logic model, but it is out of scope of this work as we are concentrated to reasoning about concepts rather than their extensions. Nevertheless, existence of full concept lattice gives always an opportunity to find extent of a given concept.

In order to build first order logic model for concept lattice based ontology expression we need to define mappings from lattice structure to a first order language.

4.1 Language Constructs

We use standard syntax for first order logic and define a simple rule language based on Horn clauses as follows.

An alphabet of the rule-language is defined as follows:

1. Set of constants **N** that consists of the set of concept names **C**, names of properties A, and special names **any** (lattice top, empty top is always True) and **nil** (lattice bottom, empty bottom is always False).
2. Set of variable names **V**. Uppercase letters denote the variables in **V**.
3. Set of predicate symbols **P.**

Terms are either constants or variables. An atom (atomic formula) is a formula of the form $p(t_1,...,t_n)$, where p is a predicate symbol and $t_1,...,t_n$ are terms. A formula is called ground if it contains no variables.

Horn rules (clauses) have at most one atom on its head and they are formulas of the following form:

$$A \leftarrow B_1, B_2,...,B_n ,$$

where B_1, $B_2,...,B_n$ is conjunction of atoms B_i, i=0,...,n and universal quantification of variables is assumed.

A definite clause has exactly one atom in the head. A ground rule contains no variables. A rule with an empty head is considered to be a query or constraint. Ground rule with an empty body is called a fact.

Rules are used to define conceptual relationships and operations on lattice. Inference rules based on concept lattice are predefined.

In addition, users are also provided to specify constraints or any other ontological relationships in the form of the rules.

An interpretation for the rule-language is defined as a set of ground atoms constructed from predicate names in **P** and constants in **N.** As the language is general, then different inference engines can be used. For example, Java Expert Systems Shell (Jess) [20] or Gandalf [21] can be used as inference engines.

4.2 Mappings of Concept Lattice to Rule Language

The mappings from lattice to rule language are defined as follows.

Mapping concepts
Concepts are represented using their names in ontology as constants in rule language. For example, house, summerhouse etc. If we like to refer to extents, then concepts can also be represented by predicates like house(X).

Mapping taxonomic relationships
Predicate isa is used to represent partial order relationship between concepts. For example, the predicate isa(summerhouse, real-estate) defines partial order relationship between the concepts summerhouse and real-estate stating that *summerhouse* isa *real-estate* (i.e. summerhouse is subconcept of realestate). The predicate subconcept(Concept1,Concept2) is used to denote that *Concept1* is an immediate subconcept of *Concept2*. Subconcept predicates are automatically generated according to the given lattice L_{CR}.
Rules for lattice axioms
As L_{CR} is complete lattice, then the rules for lattice axioms are as follows:

Reflexivity:
isa(Concept, Concept)

Transitivity:
isa(Concept1, Concept2)←subconcept(Concept1, Concept2)
isa(Concept1, Concept2) ← isa(Concept1, Concept3), isa(Concept3, Concept2)

Predicate subconcept denotes that Concept1 is an immediate subconcept of Concept2.

Antisymmetry:
equal(Concept1,Concept2) ← isa(Concept1,Concept2), isa(Concept2,Concept1)

Rules for lattice operations
As L_{CR} is a complete lattice, then for each set of concepts, there exists always a greatest lower bound (glb or greatest common subconcept) and a least upper bound (lub or a least common superconcept). Lattice meet is used to calculate glb and join is operation to calculate lub. We define these operations using the following set of rules.

Meet operation
common_subconcept(C, C_1,…,C_k) ← isa(C, C_1),…, isa(C, C_k)
greatest_common_subconcept(C, C_1,…,C_k) ← common_subconcept(C, C_1,…,C_k), common_subconcept(T, C_1,…,C_k), isa(T, C)

The predicate common_subconcept(C, C_1,…,C_k) means that the concept C is a common subconcept of the set of concepts {C_1,…,C_k}

The predicate greatest_common_subconcept(C, $C_1,\ldots,C_k$) means that the concept C is the greatest common subconcept of the set of concepts $\{C_1,\ldots,C_k\}$.
Symmetrically, we define predicates and rules for join operation as follows:

Join operation
common_superconcept(C, $C_1,\ldots,C_k$) ← isa(C_1, C),..., isa(C_k, C)
least_common_superconcept(C, $C_1,\ldots,C_k$) ← common_superconcept(C, $C_1,\ldots,C_k$), common_superconcept(T, $C_1,\ldots,C_k$) isa(C, T)

Logical model of a given lattice L_{CR} can automatically be generated on the basis of mappings presented above. This process is demonstrated in the following example.

4.3 Examples About Mappings and Inference

On the basis of concept lattice based ontology expression shown in Fig. 3 the following set of ground subconcept atoms is generated.

subconcept(familyhouse, real-estate)
subconcept(summerhouse, real-estate)
subconcept(countryhouse, real-estate)
subconcept(blockhouse-apartment, real-estate)
subconcept(c1,familyhouse)
subconcept(c1,summerhouse)
subconcept(c2,summerhouse)
subconcept(c2,countryhouse)
subconcept(c3,blockhouse-apartment)
subconcept(c3,c1)
subconcept(c3,c2)

Inference rules for lattice axioms and operations can be used to decide taxonomic relationships between concepts as well as perform lattice operations.

For example, to find the least common superconcept of the set of concepts *{summerhouse; countryhouse}*, we define the following query: least_common_superconcept(X, summerhouse, countryhouse).

The answer is the concept *real-estate*. If we are interested in finding the greatest common subconcept of the concepts *familyhouse* and *summerhouse*, then the corresponding query is as follows: greatest_common_subconcept(X, familyhouse, summerhouse). The answer is the concept *c1*.

We may be interested in all the superconcepts of the concept *familyhouse*, for example. The query isa(familyhouse, X) gives the list of ground atoms as an answer. In our example, this is one atom isa(familyhouse, real-estate).

5. Representing Non-Taxonomic Relationships

As we have seen, taxonomic relationships between concepts can be automatically generated from a given lattice based ontology expression L_{CR}. It is easy to add more rules and facts to the ontology expression above. In order to define non-taxonomic relationships the corresponding groups of predicates and rules should be defined.

Properties of concepts
For defining properties of concepts, the following predicate can be used:
hasproperty(Conceptname, Propertyname).

For example, the following ground atom can be added to the sample ontology description above: hasproperty(real-estate, location).

Inheritance of properties
Inheritance of properties can be represented by the following rule:
hasproperty(C1, X)←isa(C1,C2), hasproperty(C2, X)

We may ask the query: hasproperty(countryhouse, X) and receive the answer that it has also property *location* as the concept *real-estate.*

Ontological relationships
Ontological relationships like part-of, related-to etc can be easily represented via predicates. The following predicates demonstrate opportunities adding other ontological relationships:

partof(C1,C2)
relatedto(C1,C2)
synonyms(C1,C2), etc.

For example, according to predicates above, ontology designer can add the following ground atoms to the ontology model:

partof(apartment, blockhouse)
relatedto(blockhouse, city)
synonyms(familyhouse, familyhome)

Additional specific inference rules can easily be added to the set of predefined lattice based inference rules.

In the end, non-taxonomic relationships give additional possibilities for ontology representation and reasoning about ontology.

6. Conclusion

We have shown how concept lattices can be used for ontology representation and we have stated notion of concept lattice based ontology expression.

We defined a Horn clause model of concept lattice based ontology expression in order to enhance ontology expression with descriptions of non-taxonomic relationships between concepts. According to this model, a given lattice based ontology expression can automatically be transformed to logical expression in first-order language.

Validation of ontology, reasoning about ontology and search can be done using logical inference. As our approach uses first-order language, then it is possible to attach different ontology inference engines for practical applications by translating ontology representation to any inference engine rule language.

Our future work is concentrated to experimental application of the provided logical model using the first-order theorem prover Gandalf [21].

Acknowledgements

This research was partially funded by Estonian Research Foundation by the Grant no 5766.

References

[1] Guarino N., Formal ontology and information systems. In Proceedings of FOIS'98-Formal Ontology in Information Systems, Trento, Italy, 6-8 June 1998, IOS Press, 1998

[2] Tempich C. and Volz R., Towards a benchmark for Semantic Web reasoners-an analysis of the DAML ontology library, Sure Y (ed) Proceedings of Workshop of Evaluation of Ontology-based Tools (EON 2003) at 2nd Int. Semantic Web Conference (ISWC 2003), USA, 2003

[3] Fernandez-Lopez M. and Gomez-Perez A., Overview and Analysis of Methodologies for Building Ontologies, The Knowledge Engineering Review, Vol. 17:2, 129-156, Cambridge University Press, 2002

[4] Maedche A., Ontology Learning for the Semantic Web, Kluwer Academic Publishers, 2002

[5] Gangemi A, et al, An Overview of the ONIONS Project, Data and Knowledge Engineering, 31, 1999

[6] Maedche, A. and Staab, S., Discovering Conceptual Relations from Text, Proceedings of the 14th European Conference on Artificial Intelligence, IOS Press, Amsterdam, 2000

[7] Haav H-M., Learning Ontologies for Domain-specific Information Retrieval, W. Abramowicz (Ed), Knowledge-Based Information Retrieval and Filtering from the Web, Kluwer Academic Publishers, 2003, ch 15, pp 285-300

[8] Ganter B. and Wille R., Formal Concept Analysis, Mathematical Foundations, Springer, 1999.

[9] Stumme G. and Maedche A., FCA-Merge: Buttom-up Merging of Ontologies, Proceedings of the 17th International Joint Conference on Artificial Intelligence, Seattle, USA, Morgen Kaufmann

[10] Manning C. and Schuetze H., Foundations of Statistical Natural Langauge Processing. MIT Press, Cambridge, 1999

[11] Hobbs J., The Generic Information Extraction System, Proceedings of the 5th Message Understanding Conference (MUC-5), Morgan-Kaufmann, 1993

[12] Assadi H., Construction of a regional ontology from text and its use within documentary system, N. Guarino (Ed), Formal Ontology in Information Systems, Proc. Of FOIS-98, Trento, Italy, 1999, pp 236-249

[13] Hofmann T., The Cluster-Abstraction Model: Unsupervised Learning of Topic Hierarchies from Text Data, Proceedings of 16th International Conference on Artificial Intelligence (IJCAI-99), Stockholm, Sweden, 1999, pp 682-587

[14] Baader F. et al., (Eds), Desription Logic Handbook: Theory, Implementation and Applications, Cambridge University Press, 2002

[15] Kifer M., Lausen G., and Wu J., Logical Foundations of Object-Oriented and Frame-Based Languages. Journal of ACM, 1995, 42(4), pp 741-843

[16] Grüninger M. and Fox M. S., Methodology for Design and Evaluation of Ontologies, Workshop on Basic Ontological Issues in Knowledge Sharing, IJCAI-95, Montreal (available at www.eil.utoronto.ca)

[17] Decker S., Erdmann M., Fensel D., and Studer R., Ontobroker: Ontology Based Access to Distributed and Semi-Structured Information, Meersman R. et al. (eds.): Semantic Issues in Multimedia Systems. Proceedings of DS-8. Kluwer Academic Publisher, Boston, 1999, 351-369.

[18] Godin R., Missaoui R., and Alaoui H., Learning Algorithms Using a Galois Lattice Structure, Proc. of the Third Int. Conference on Tools for Artificial Intelligence, IEEE Computer Society Press, CA, 1991, pp. 22-29.

[19] Kuznetsov S. O. and Objedkov S. A., Comparing performance of Algorithms for Generating Concept Lattices, In: Proceedings of Internatinal Workshop on Concept Lattices-based KDD, ICCS'2001

[20] Jess Home Page http://herzberg.ca.sandia.gov/jess/

[21] Tammet T., Gandalf, Reference Manual, University of Göteborg, Sweden, October 1997

Information Systems

Databases and Information Systems
J. Barzdins and A. Caplinskas (Eds.)
IOS Press, 2005

A Contribution to Web Digital Archive Integration from the Parliamentary Management System 'SIAP'

Carmen COSTILLA[1], M.ª José RODRIGUEZ[2], Juan P. PALACIOS[2], José CREMADES[3], Antonio CALLEJA[3] and Raúl FERNANDEZ[3]
[1] *Titular Professor, Information Systems and Databases (SINBAD) Research Group Dept. Ingeniería de Sistemas Telemáticos (DIT), ETS Ingenieros de Telecomunicación (ETSIT), Technical University of Madrid (UPM), www.sinbad.dit.upm.es, costilla@dit.upm.es*
[2] *Ph. Researchers at the SINBAD- DIT-ETSIT-UPM*
www.sinbad.dit.upm.es, mjrodriguez@sinbad.dit.upm.es; jppalacios@sinbad.dit.upm.es
[3] *Ph. Researchers at the SINBAD-DIT-ETSIT-UPM and Founder Partners of CRC Information Technologies (www.crcit.es), {jcremades, acalleja, rfernandez}@crcit.es*

Abstract. Nowadays, millions of users choose the web as their main communication channel. As a consequence, everyday we find more activities (government, commerce, finance, digital libraries...) developed through the web. Archives are an example of this trend. They are being digitized and there is already an important initiative to standardize web access to them. Firstly, this paper introduces the SIAP system we have built; it has been running at the Asamblea of Madrid since 1999. Secondly, we present DAWIS-UPM[1] project (Digital Archive Web Information System) whose main goal is to design a web architecture providing integrated access to different digital archives. This architecture is based on mediators and wrappers making the virtual, flexible and dynamic 'on-the-fly' integration of digital archives possible. Finally, a semantic integration -based on ontologies, according to current international archival standards- is here proposed.

1. Introduction and Motivation

The web is a universal information space that every day freely incorporates new server sites that may contain very different Web Information Systems (**WIS**). Web data management is a relevant topic for governments, museums, libraries, businesses and other institutions. As the current web is inherently heterogeneous in data formats and data semantics, the semantic web is very necessary.

The web is anarchic with regard to the capabilities of linking semantic affinity information contents of a specific domain. However, the fundamental is that – on the web – these domains could have a universal scope, and this is one of the most important communications and computer science goals you can get. Today, the semantic web

---

[1] DAWIS-UPM stands for **D**igital **A**rchives **W**eb **I**nformation **S**ystem, the TIC2002-04050-C02-02 national research project (2002-2005), funded by the Spanish Ministry of Science and Technology within Communications and Information Technologies National Programme, developed at the DIT-ETSIT-UPM.

constitutes an important challenge for future web intelligence [7, 49]. Current research proceeds quickly in this direction.

Research into analyzing web sites faces two challenges due to the huge amount of on-line published data and the complex structures found in them [3]. We think it is absolutely essential to improve the virtual integration of heterogeneous and distributed **WIS** [46]. Focused on a specific domain, the web will allow the dissemination of unified and useful information to the user [27].

We have today new web technologies related to interoperable web, such as XML and web services [4]. However, web data integration has to improve considerably. **WIS** are being still integrated by hand – with tedious processes and ad hoc work – that produce highly vulnerable results. To cover current needs – when many different sources are offering a wide variety of semantics – it is not enough. Besides, the related software applications also present a high degree of diversity.

In this context, archives are a very interesting and representative WIS investigation area. From a technical point of view, archives hold huge amounts of documentary information, where the multimedia, the diversity of material and format aspects is the summit of all its splendour. The (semi-) automatic integration of many web Digital Archives (**DA**) permits user easily access to global information.

This paper describes some aspects of web DA integration, thanks to the experience acquired from our recent lines of research. A just finished project is due to a Parliamentary Integrated Management System, – **SIAP**[2] –, that we have built. **SIAP** was funded by the Parliament of Madrid, whose first product (called SGP) has been running successfully at the Asamblea of Madrid since 1999. Nowadays, our ongoing research is focused on heterogeneous web data source integration and semantic web, applied to web DA virtual integration.

We have investigated DA conceptual modelling in order to define a virtual web integrated architecture through a unified semantic **mediator layer** (made up of ontologies, mappings and data repositories) and **wrappers** (coupling heterogeneity between data sources, using XML). We are starting this current investigation, and our final (and future) idea is the implementation of our results as different web service levels (Java libraries) of a web DA integration architecture.

We think that building a web DA integrated architecture of distributed heterogeneous data integration, with dispersed components and services, running very different web functions is a paradigmatic application [44]. As well as the services, the overall functionality depends on the ability of the specification and implementation of semantics that each web query requires.

Section 1.1 briefly introduces the international standard regulating archive content description, and section 1.2 summarizes the DA concept. Section 2 introduces the **SIAP** system containing a **Parliamentary Digital Archive Management System**. In Section 3, we describe the **Web Integrated Architecture** proposed for the **DA**. We remark here on the ontological aspects that we have considered, applied to some archival standards (DCMI, ISAD, ISAAR) and SIAP system. Finally, the conclusions are given.

1.1 Archives

Archives constitute a worthy and representative application area. They contain a huge amount of documentary information on human activity. Archives hold different contents

---

[2] SIAP *Parliamentary Information Management System*, a CRC Information Technologies product (www.crcit.es)

such as culture, business, living and relaxing. However, their main and common goal is: *the propitiation of an easy access to the information content, guaranteeing the safety and custody of what they hold.* Consequently, when you want people to access archive content (such as a catalyst of cultural and historical heritage), undoubtedly the web is the optimal media for a universal safe propagation of content.

The General International Standard Archival Description ISAD(G) (2nd ed.) [24] is the most important one for archivists. Its main goal is the identification and description of the content and context of archive material, in order to promote its management and accessibility. ISAD(G) is developed by the International Council on Archives (ICA) [24], a worldwide organization of archival community professionals. ISAD(G) defines a set of elements (metadata terms) for archive description that may be applied regardless of the format of the archive material. Some combinations of these elements constitute the unit of description.

Figure 1 represents the ISAD hierarchical unit of description levels. Every level has different degrees of detail. Additionally, the International Standard Archival Authority Record for Corporate Bodies, Persons and Families ISAAR(CPF) is a second one. Section 3 describes how to use these two standard concepts.

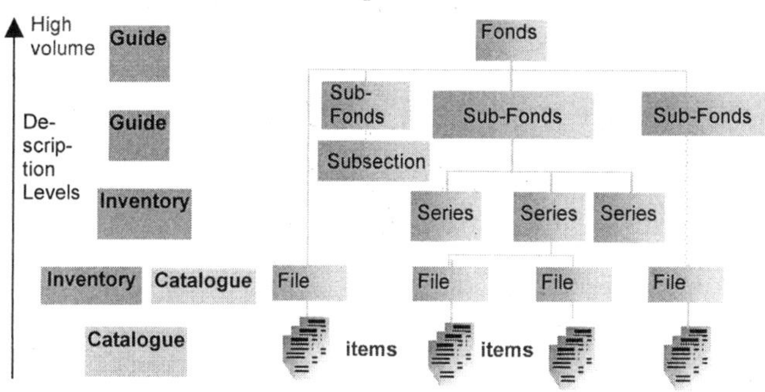

Figure 1. ISAD(G) hierarchical description levels of archive organization

1.2 Digital Archives

The **DA** holds a collection of **documentary data** in digital format joined to **descriptive data** on the organization, its goal and the content of these documents. The descriptive data are **metadata** for archive description at any level that ISAD(G) proposes, as shown in Figure 1.

Any DA can store many different contents (corporate databases, data warehouse, thesaurus for information retrieval, web pages, e-mails, documents, maps, photos, etc.). It is true that, in some archives, the process of digitizing could be difficult, slow and may be impossible obtain completely. However, currently more and more institutions are trying to do it in order to offer a higher degree of accessibility. Hence, the web DA is disseminating certain contents that, generally, consist of metadata such as: descriptive data for indexing thesaurus of Information Retrieval, summaries, index, keywords, synonymous and antonyms lists (broader terms, narrow terms, etc.) and other related terms [1]. All this information is stored in what we call DAs, as digital data source repositories. Nowadays, this information is what -through web sites- users are accessing in order to get certain relevant aspects on any fixed unit of description combination (item or collection of items, file or collection of files, Sub-Series, Series, Sub-Series and/or Series collections, Sub-Fonds, collection of Sub-Fonds and Fonds).

2. 'Asamblea of Madrid' Parliamentary Archive

The *Asamblea of Madrid* parliamentary archive has totally digitized its documentary fonds, since the beginning of our democracy. This DA management system is powerful and ideally advanced for the web.

The DA management system runs at the Asamblea de Madrid as a part of a broader system called SGP (Parliamentary Management System). SGP was funded by the Asamblea de Madrid (1997-2000) and was built at our research group together with people from Spanish businesses CRC Information Technologies (CRC IT). Nowadays, CRC IT commercializes SGP as SIAP and this product is sponsored by our Technical University of Madrid (UPM), by our Technical High School of Telecommunication Engineering (ETSIT) and by Oracle and Cronos Ibérica enterprises.

SIAP manages and controls the workflow of overall parliamentary documentary information and automatically creates the many types of parliamentary document produced by political activity. For the ***SIAP*** design we used methodologies [13, 26] and CASE tools; in addition, the main design guidelines were: the proper political activity, institutional norms, Political Initiative typology and the nature of political documentation. In order to provide *useful information*, as intelligent as possible, we follow the guidelines in [3, 12, 45, 46] and our experience from previous systems [10, 11] among others.

The ***SIAP*** system structures Political Initiative and Parliamentary jobs (Plenary sessions, Committees, etc.). It has many applications (www.crcit.es/siap). ***SIAP*** workflow sends the document to be considered established by the regulation, according to each type of political initiative. Moreover, it is controlled when a document is sent to be published into the Official Bulletin, the Sessions Journal and on dynamic web pages automatically created for a specific user type. ***SIAP*** associates the document to the respective File (*Expediente* in Spanish), to the corresponding Daily Order, to the pertinent Official Bulletin, to the previous scripted political sessions, etc. Besides, ***SIAP*** also knows where the original document is stored and where the multiple document copies are held.

Section 2.1 summarizes the more general aspects of ***SIAP***, successfully running at the *Asamblea of Madrid* since 1999, and section 2.2 describes its **Archive Management System.**

2.1 SIAP system: Objectives and Architecture

SIAP models Parliament's overall structure and organization. It has demonstrated a high degree of security and effectiveness for those procedural steps in every political initiative. The main goal of ***SIAP*** is to obtain a perfect integration of the parliamentary information with the proper activity of the Parliament.

SIAP controls input documents and produces a huge amount of output documents, some of them with a public nature. It makes the entire Parliament of Madrid Official Bulletin, and the greatest part of Sessions Journal. Besides, SIAP applies documentary searching to other external Official Bulletins (from foreign institutions). If it were convenient, all this information would be dynamically published on the web.

SIAP has an open architecture to be implemented into the Institution network. It has an underlying object-relational database running into the Database Server and its applications could also work on its own (Law Budget Project, Archive Management, Registers, etc.).

SIAP runs on Oracle in Client/Server (C/S, two ties) and in Internet/Intranet (three or four ties). In C/S, for providing services to civil servants that update the information system (with a high degree of protection and security). In C/S, protocols SQL*Net8 and

OCA and, in Web OAS (Oracle Application Server and IAS of Oracle 9i) are used. For both, IntermediaText for information retrieval (before Oracle Context) is used. Figure 2 represents the documentary workflow and ***SIAP*** open architecture; where BOAM means Asamblea of Madrid Official Bulletin and DDSS means Sessions Journal.

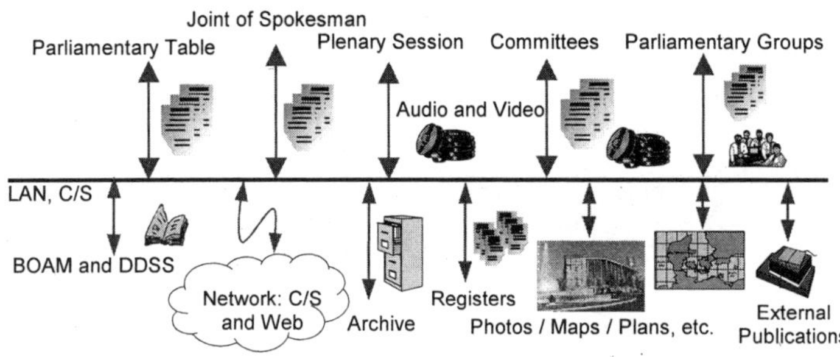

Figure 2. Documentary workflow and ***SIAP*** open architecture (C/S and Web)

2.2 SIAP Digital Archive Management System

Parliament activity produces documents tailored to a specific type of Parliamentary Initiative. ***SIAP* Archive Management System** holds especially relevant information organized in **files** (descriptive folders) according to the initiative typology and objective. The **file** collects related information on political activity and marks the procedure type to schedule with all documents that are held inside it (initiatives, photographs, maps, graphics, audio, video, etc.).

Additionally, the file incorporates other descriptive information. Concisely, it includes the following: 1) **Identification**, 2) **Censure or judgment of the contained documents**, 3) **Schedule**, 4) **File Classification and its respective documents**, 5) **Allocation**: topographical, informatics and address of Institution, departments, dependencies, etc., 6) **File prosecution** made up of: a) **Workflow** considering all states that must or should be adopted by the file and any of the related documents annexed to the file, b) **Current status**, c) **History** and, finally, 7) **Relationships** between files by subject, type, date, state, descriptors, etc.

This **DA** handles queries against the underlying object-relational database (**SQL**) and/or to the thesaurus with a powerful **information retrieval**. It searches thousands of publications and locates the subject of interest in a few seconds; locates the annexed document to the file in an almost instantaneous response, independently of the Legislature where it is stored [1]; and, finally, creates a lot of report results [9] (www.asambleamadrid.org).

SIAP controls the all of the information from any Legislature and knows the topographical key signatures of its complete archive fonds. For example, The Madrilenian Court can instantaneously locate all files and documents dealing with *anorexia,* since the beginning of the Spanish democracy and can reproduce video fragments of the political sessions (Plenary and Commissions) in which this illness was discussed.

Through web applications, the parliament can negotiate the information exchange between similar institutions at any level, in an integrated and intelligent way. ***SIAP*** can offer **global information** produced by many others Institutions, as well as its own [20, 36, 40].

The ***SIAP*** **Digital Archive** was tested for three hours in a ***Parliamentary Archives and Political Parties Conference*** (Nov. 1999) held at the Asamblea of Madrid, and it was considered a powerful and complete system. Figure 3 shows its main screen.

If ***SIAP*** runs in other institutions, web information capabilities will grow in a spectacular way. We want to remark on how easy is to get dynamic interoperability on the web when data sources have the same WIS (same design and semantic control, similar software, etc.) (See Fig. 4). On the other hand, in heterogeneous WIS environments it could be hard to provide easy and powerful access to global information.

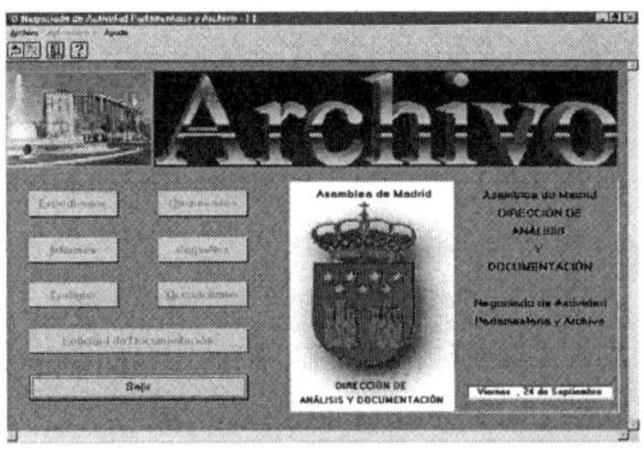

Figure 3. The SIAP digital archive at the asamblea of Madrid

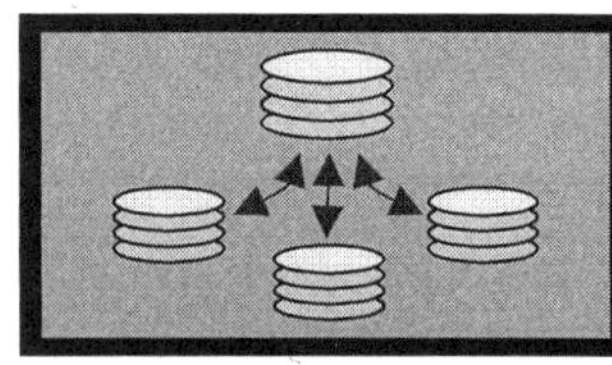

Figure 4. Web multi-parliamentary

3. Integration of Web Digital Archives

As the current web is inherently heterogeneous, semantic web will be absolutely necessary, mainly where you need a certain conceptual organization. Besides, public DA integration should allow access to inherited cultural content without distances, languages and cultures barriers.

There are many important initiatives based on DA (OAI, DCMI, EAD, etc.) in order to make them accessible to the web user. The systematic and (semi-) automatic generation of web DA allows the administration of the huge documentary legacy held into the Archives to be improved, as well as facilitating the creation of new DA available to any web user.

However, we still do not have a good integrated web solution for making it possible for any DA in the world to be '*added*' to a web environment (integrating several DAs), independently of DA content and description level. A web user is still unable to access a **virtual, global, integrated DA,** whose semantics will be universally guaranteed by a web-integrated system.

Due to the large amount of non-digitized, nor computerized Archives; an *ad hoc* solution to develop a concise DA or to integrate '*n*' certain pre-existing archives is not enough. For this reason, we are seeking a generic solution allowing any DA be published on the web, as well as their generic integration. This kind of integration sends the user queries to 'n' DA just as if they were a single virtual archive.

3.1 Virtual Integration of Web Heterogeneous Data Sources

Research on the dynamic and virtual web integration of multiple heterogeneous DA represents a very different challenge to the classic database static schema integration. The latter is obtained by using a fixed federation of local participant databases schemata in a global static schema [40].

In the virtual integration of web heterogeneous data sources there is neither centralized information management, nor federated information, nor integration information systems dealing with materialized data [15]. In the virtual integrated systems, data sources remain as autonomous data inside every locality. Integration is only virtual (not materialized), as we describe later.

In spite of having many standards in databases and static schema integration reference models, a systematic solution for web dynamic integration, requires a standard that does not yet exist (solutions are often static and ad hoc).

However, the question of the non-existence of a standard web is old [20]. Since the first attempts to regularize data exchange on the Internet (EDI) until the proliferation propitiated by the web (http, html, xml, obi, cxml, etc), it has still not been possible to model the interactions among a system components properly [33]. We are dealing with many independent WIS, whose first requirement for becoming part of any integration is to guarantee total autonomy to each local participant in the dynamic and virtual integration (DAs, in our case). Therefore, the DA does not federate in an integrated static global schema. On the contrary, it is produced by a Web Mediator System [47]. By doing so, data remains in local data sources, and integration takes place on the fly, during the query pre-processing in the mediator.

3.2 Introduction to DAWIS-UPM

DAWIS-UPM [6, 43, 18] is a project which global objective is the definition of a web-integrated architecture, providing a virtual and dynamic access to many different DA. DAWIS-UPM is the successor of SIAP, now looking for the web integration of multiple DA. In this sense, we find the acquired experience in the previous project we have carried out for the Parliament of Madrid very valuable.

So, heterogeneous web data source integration and semantic web, applied to DA is the focus of our current research. We are investigating DA conceptual modeling in order to define a virtual web integrated architecture through a unified semantic mediator layer (made up of ontologies, mappings and data repositories) and wrappers (coupling heterogeneity between data sources). Communication and interoperability are based, among others, on web services and XML respectively.

As well as other aspects, DAWIS-UPM needs to provide a systematic and (where possible) automatic definition of the underlying reference architecture [6], through which the web user could access a lot of heterogeneous DA [17] transparently. This means the consideration of the following aspects: web architecture, DA model specification, query processing, wrappers, and "mediators".

The generic goal is mainly to provide a semantic unification and web knowledge enrichment. To obtain this, we do not need to start from scratch. There are interesting proposals such as FEDORA and OAI, that offer current interoperable architectures limited to providing remote access and C/S, to a determined DA [19, 38, 39, 42]. Undoubtedly, they are an important first step.

Figure 5 represents the proposed architecture that will be service-based and compliant to current web-centric architectures (J2EE).

We are aware of the achievement of a **Dynamic, Global and Virtual Web DA Integration**. It constitutes an important challenge for current WIS. To give an example, we are researching in order to make it possible for a web user to query something like this: *"Find all documents dealing with equestrian statues of the King Carlos V"*, avoiding the consideration of which physical archive he/she needs to refer to for this query. If it were necessary, the system would be in charge of informing about the origin of each document that makes up the overall answer.

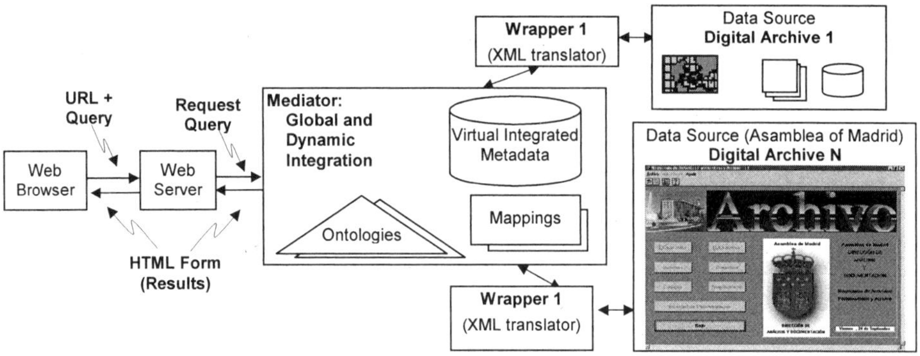

Figure 5. Web digital archive integrated architecture through mediator and wrappers

The final idea is the implementation of this architecture as Java libraries based on service-levels [18], where "mediator" and "wrappers" will be dynamically generated starting from the web user query. In order to achieve these goals, it is necessary to consider the following architectural topics: flexibility, integration and their related semantic problem (considered in section 3.3), interoperability and cooperation, and efficiency, as we discussed in [6].

We want to remark that a web services architecture does not solve the semantic integration problem [4, 29]. Web services allow data between remote applications to be exchanged, but they do not guarantee that the receiver application will be able to understand them. As pointed out in [44], the problem is the lack of a global ontology.

Since the Web Application term appeared (WebApp) [33], its capabilities have been growing continuously. For example, initially some WebApp supported only static pages, while now, in J2EE mark, a combination of Servlets and JSPs allows to generate dynamic pages.

In ***DAWIS-UPM***, Servlets are of special interest, because they have a certain semantic orientation. Servlets are Java classes for request processing and corresponding answers generation, supporting part of the application logic (although it will be only in an explicit way) [29]. Moreover, CORBA is another standard that needs a wide range of services providing design and development of current distributed systems [4]. However, the puzzle is not yet complete. Again, the weakest part is the *semantics of the systems*, business logic rules of any participant application. Therefore, it is not surprising that it were in the server applications environment where the alarm first sounded. Thus, in order to solve the dynamic and virtual web integrated access problem, ***semantics of the systems*** is an important and preliminary characteristic, which must be considered.

Thus, we consider ontologies that contribute to reinforcing these semantic necessities. This consideration is different to current distributed architecture approaches, which only consider defined static interactions between software components (during the compilation) and between programs and users [41].

As well as the semantic problem, it is also necessary to provide mediation services [47] for web query processing [32]. Considering Figure 6 and coming first from the front-end browser to the back-end DA, the next main services are needed:

a) From the portal web, the global query arrives at the mediator where the query concepts and relationships are matched to a certain ontology (coded in OWL) and the information that repository holds (pre-processing).

b) The mediator, taking into account the related mappings with terminological semantic unification, is able to inference knowledge at this virtual integrated level.

c) The mediator, once the user query is pre-processed, and making use of the mappings between mediator concepts and concepts coming from each DA; knows which local queries (as web query components) need to be invoked –via wrapper- for any DA manager execution. Each locality returns the answer to the mediator, in a well-known format, previously specified by the mediator.

Now, coming from any back-end DA to the front-end browser, ***DAWIS*** should have, at least, services offering the following functionalities:

d) Any DA establishes the query types it wants to offer to the mediator, in order to be a specific dynamic integration participant.

e) In each DA, every XML schema contains the extracted concepts that make its query capabilities possible (tailored to the measurement of each possible local participant integration). These schemata determine the way any DA is taking part of any global architecture.

f) These XML schemas must to be compliant to those "mappings" defined between the "wrappers" and mediator ontology.

Finally, we want to remark here that we are making only a preliminary requirement specification on the necessary web architecture. This justifies our contribution to the semantic web focused on DA application that we discuss in the rest of this paper.

3.3 Ontologies in semantic web

According to [2], the web will evolve unavoidably toward a semantic web. Then, computers will find the meaning of data following the *links* to the definitions of the terms and key rules. With this, designing automatic services will be easier; because the main bottleneck, for getting the semantic web, is the modelling of the information. To solve this problem and to be able to code the necessary information, we have XML and ontologies as better options [21].

An ontology describes shared knowledge on a specific domain in order to establish a communication via between humans and programs [1, 8]. Each ontology provides a precise definition of the terms (concepts) of a specific domain; that is to say, it facilitates a common language for sharing and to re-using knowledge of some phenomenon in the domain of interest [21, 49] between applications and groups.

Consequently, applications can speak to each other and interpret, without ambiguity, the information they exchange. For this reason, they have been accepted as powerful description tools that allow making explicit the semantic web. Since the early 1990s, a lot of research [2, 5, 7, 22, 23, 49] deals with these semantic problems, where ontologies play a fundamental role.

Moreover, XML is a more mature technology than ontological languages (DAML+OIL, OWL) because of the amount of user communities and available tools (http://sws.mcm.unisg.ch/links.html#daml). Kim affirms that ontologies should be adopted where the capability to represent semantics will be as necessary as forgetting the maturity advantages that XML already offers (http://www.xml.org/) [27].

We need to build ontologies [48] because DA domain is not simple and it needs to be used for a wide audience. We adopt ontologies as a paradigm that reduces complexity, and XML is the technology used for the interoperability of archive data.

3.4 Ontologies in DAWIS

As with the aforementioned architectural aspects, the archival conceptual modeling does not start from scratch either. On the one hand, we have the well-known DCMI (Dublin Core Metadata Initiative) ontology. On the other hand, ISAD(G) and ISAAR(CPF) are international standards specifying how archival content should be described. These standards are widely followed by archivists. Finally, a fourth special Parliamentary ontology - coming from the SIAP Parliamentary DA - is also being undertaken.

All these items will be considered for achieving the wide ontological unification of DA domain that we are looking for.

DCMI Ontology

The Dublin Core Metadata Initiative (DCMI) [14] promotes metadata standards and develops specialized vocabularies for describing resources. According to the RFC 2396, a resource is "something that has identity".

The DCMI proposal was came about in 1995 inside the Online Computer Library Center (OCLC). Since then, DCMI collaborates with the normalization and the development of technologies that allow a bigger efficiency of the use of metadata. Today, DCMI is a pillar inside a semantic web, an obligatory reference to all ontology groups that want to adopt a common standard for the semantic description of any very general resource. DCMI ontology was formally adopted by the European Committee of Normalization. In 2001 it was accepted as the ANSI/ISO Z39.85 norm and, since February of 2003, it is the ISO 15836-2003 standard.

Although DCMI is not a specific ontology for describing archival material, it could be perfectly adapted to describe any kind of resources. Since 2000, DCMI is already a free ontology that we have taken through the Protégé tool.

ISAD(G) & ISAAR Ontologies

Since 1999, ISAD(G) has been widely used as unquestionable classification base of national archives [31]. Standardization models play a very important role for the future success of accessibility. ISAD(G) defines twenty-six descriptive elements (metadata terms) specific to archival description. Only six of these elements are considered essential and must be used in any archival unit of description.

ISAD(G) specifies the identification, content and context of archival information units. It also provides links in order to fix any combination of archival information elements. One of the main archivist activities is the description of record creators, a difficult job in documentation and maintenance. In this sense, it is good practice to independently record the creator(s) of a unit of description and then link this to the related unit of description. This practice enables the creator description to be shared between different archival documents, or repositories, only adding linking information.

Moreover, ISAAR(CPF) [25] can be considered as a complement to ISAD(G) to specifically manage the description of authority records. It defines twenty-seven descriptive elements; where the concept “Authority record” is similar to the concept of Unit of Description in ISAD(G). In this sense, an Authority Record describes an archival authority (Corporate Bodies, Persons and Families) as well as the Unit of Description describing an archival unit of information.

Regarding ISAAR(CPF) and ISAD(G) elements, in [43] we have defined the UML respective models. Starting from these, we have developed ISAD(G) and ISAAR(CPF) ontologies. Figure 6 shows a partial graph of this ISAD(G) Ontology.

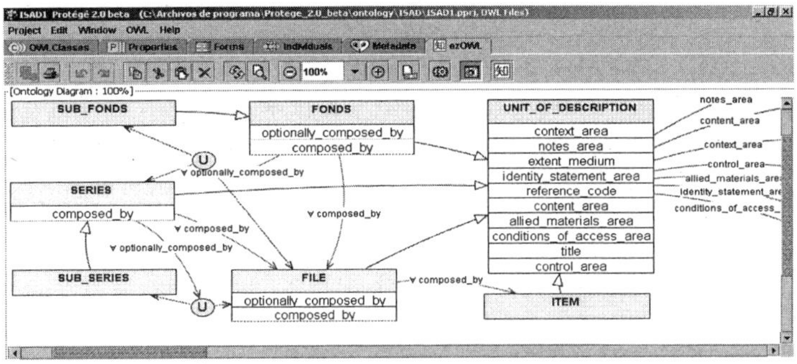

Figure 6. ISAD ontology partial graph

SIAP Ontology

Both, ISAD(G) and ISAAR(CPF), propose general rules that can be adapted to describing any archive; nevertheless, these rules are not considered for the peculiar description of special archives. In our case, Parliamentary archives have a part of their concept that are particular, due especially to the workflow that has to follow each Parliamentary file and peculiar parliamentary organization. In this sense, we have developed a SIAP ontology that comprises particular Parliamentary concepts.

In order to define this specific ontology, we start from the ***SIAP*** DA conceptual model (see Figure 7). Parliamentary DA type has certain particular information, and also specific description rules, where the semantics of the terms and the elements can be different from other kinds of archives.

But the interest extends further on, because -in the future- the integration way, here considered, will allow us to conceive a system able to offer rules to integrate these specialized concepts of a diverse nature (police, government, etc.). Ontologies are complex to build all in one go. So, building them bit a bit, testing each new (kind of) addition empirically and developing appropriate learning techniques for each bit, you may automate the process; and, next time, you may build a new one in a more systematic way.

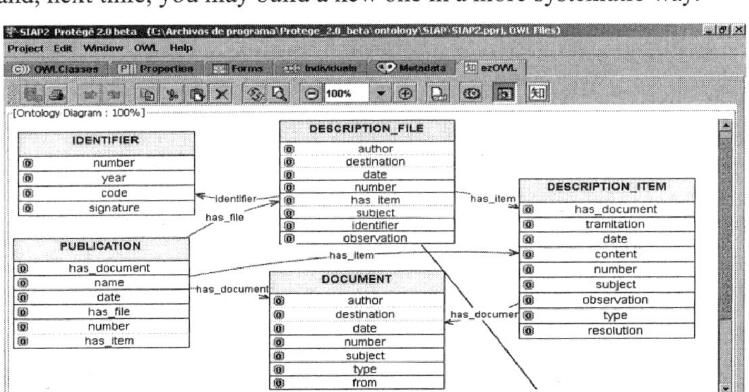

Figure 7. SIAP ontology partial graph

Summarizing, we have the DCMI ontology (free on the web) and three original ontologies based on ISAD(G), ISAAR(CPF) and SIAP that we have developed [21, 28]. The majority of current semantic web is built in this way. In fact, ontologies are now singular web "*hot-spots*" whose semantics are bigger than usual. Today, that web has several hundred of these isolated ontologies [49]. We consider this kind of semantics as a worthy first step that we call the ***single-ontologycal-centric*** approach.

Generally speaking, an important requirement of all different ontologies -closely related to a specific domain- should converge in only one semantic unification and, where possible, with a universal scope. In our DA case, it will allow a semantic, powerful and really open integrated web ontology inside the mediator layer to be guaranteed [22].

3.5 Ontological Integration Kernel

According to the previous semantic unification we have stated, this section discusses our preliminary approach concerning to the definition and development of a global ontology [49], unifying the semantics of the diverse existent ontologies related to DA domain [5].

Regarding this, we consider two important previous works: *Observer* [34, 35, 37] (http://sol1.cps.unizar.es:5080/OBSERVER) and *Harmony* [16, 30] (http://metadata.net/harmony). They constitute an important contribution to our current research.

Observer considers ontologies as if they were on only one horizontal plane. Then, inter-ontological mappings could be also considered in a parallel plane. For this reason, we call this kind of relationships "***inter-ontological-horizontal-mappings***". As well as this, *Observer* provides the capability to query many specific ontologies from different domains. Only if the user wants, the system allows a new query formulation referring other ontology concepts. In this sense, the answer enrichment is provided. We want to recognize the *Observer's* great improvement regarding query processing against many ontologies.

On the other hand, *Harmony* -a digital library project- defines a common "*ABC metamodel*" for metadata integration coming from many other ontologies. *Harmony* considers the *ABC model* as if it were on a higher plane, acting as an umbrella (containing very general concepts), allows other more specific ontologies to be attached on a lower plane (CIDOC/CRM, MPEG7, IMS,). Then, inter-ontological mappings are always provided on a vertical plane. For this reason, we call this kind of relationship "***inter-ontological-vertical-mappings***". Again, we want to appreciate this valuable contribution.

Considering these two project results as pillars, we propose to develop specialized techniques and methodologies that can guarantee a correct **shared and common semantic integrity between current DA ontologies**. Our ontological integration kernel implementation will be carried out on the definition of an ontological unification approach, operating as a kernel in which current (more or less) specialized ontologies could be semantically connected, as Figure 8 represents.

The distinctive feature ***DAWIS*** presents is the semi-automatic ontological kernel generation for a global integration that does not have to be defined in either a manual or static way. Thus, this ontological kernel will be in charge of providing a common understanding for fundamental DA concepts. In order to obtain a flexible and scalable integration, "bridges or steps" between ontologies - such as an intelligent and generic "crosswalk"- need to be provided. This allows knowledge inference from any ontology to another.

We are aware that inter-ontological semi-automatic mapping definition can turn out to be a difficult and expensive task, caused by the semantic heterogeneity. So, today, our research is continuing in this direction applying the bit-by-bit empirical methodology we have previously mentioned.

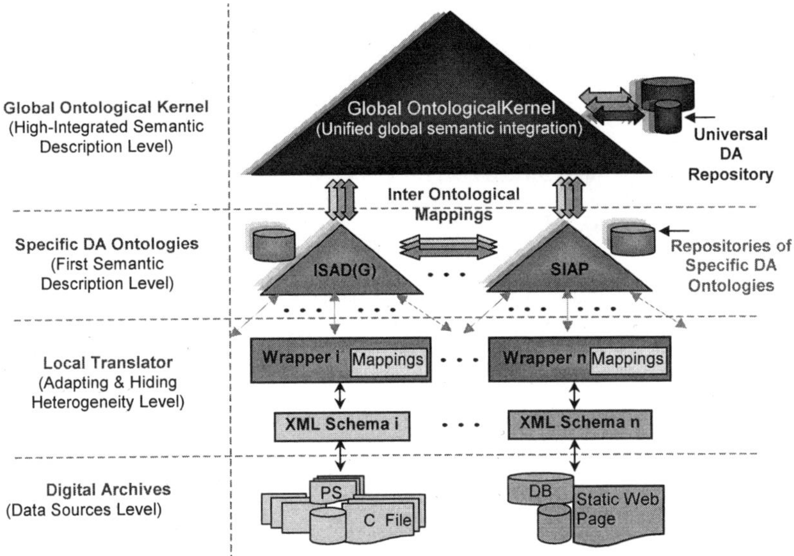

Figure 8. Integration model based on the ontological unification of the mediator

4. Conclusions

This paper focuses on current Web Information Systems (WIS) applied to Digital Archives (DA) looking for a virtual, flexible and dynamic integration by means of a Mediator and Wrapper architecture. We start from the Parliamentary Integrated Management System that has been running at the *Asamblea of Madrid* since 1999. After this, the paper presents four current ontologies closely related to DA domain. Applying an empirical step-by-step methodology, a high-integrated semantic description level is proposed by means of a global ontological kernel inside the Mediator layer.

This research is still in its early stages. However, we consider the adopted empirical method as a preliminary corner stone, to establish a solid foundation for the achievement that this web architecture is looking for (Java libraries and web services).

Acknowledgments

The authors would like to thank the anonymous reviewers. Their valuable comments and suggestions greatly improved the quality of this paper.

This work is partially granted by the Ministry of Science and Technology (MCYT-TIC2002-04050-C02-02, DAWIS-UPM project), by Community of Madrid (07T/0056/2003/3, EDAD-UPM project) and by Spanish NoE Databases, RedBD, MCYT-TIC2000-3250-E.

References

[1] Baeza y Berthier, *Modern Information Retrieval*. Addison Wesley, 1999.

[2] Berners-Lee T, Hendler J, Lassila O, *The semantic Web*. Scientific American 284, 5,pp.34-43, May 2001.

[3] P. Bernstein, A. Halevy and R. Pottinger, *A Vision of Management of Complex Models*. ACM Sigmod Record, pp. 55-63, Vol. 29, N. 4, Dec. 2000.

[4] Brittenham P, *Web Services Development Concepts (WSDC 1.0)*, IBM Software Group, May 2001

[5] Calvenese D, Giacomo de G and Lenzerini M, *A Framework for Ontology Integration*. In Proc.of the First Semantic Web Working Symposium, pp 303-316 – 2001.

[6] Costilla C, Eibe S, Menasalvas E, Sáenz J, Marcos E, Cavero J y Vela B, *DAWIS: Enfoques Preliminares sobre la Arquitectura de Referencia para la Integración de Archivos Digitales en Web*, Taller de la Red de Excelencia de Bases de Datos en España (RedBD) VII JISBD'02, El Escorial, Madrid, Nov. 2002.

[7] Chandrasekaran B, Josepheson J, Benjamins VR, *Ontologies: What are they? Why do we need them?* IEEE Intelligent Systems, 14(1):20-26, 1999.

[8] Clyde W, Hossapple, Joshi KD, *A collaborative approach to Ontology design*. Communications of the ACM. Vol. 45 n. 2, pp.42-47, Feb. 2002.

[9] Costilla C, Calleja A, Cremades J, *SIAP: Sistema de Información para Ayuntamientos y Parlamentos*, Revista Círculo de Usuarios de Oracle, CUORE, Sección *'Vivat Academia'*, 8 pages, Oct. 2003.

[10] Costilla C, Bas MJ, Villamor J, *SIRIO: A Distributed Information System over a Heterogeneous Computer Network*. ACM SIGMOD RECORD, Vol. 22, N. 1, pp. 28-34, March 1993.

[11] Costilla C, Bas MJ, Villamor J, *The Object Data Model and its Graphical User Interface for an Object Database System*. Proc. of Data Management System' Biwit'95, pp. 161-172, IEEE Computer Science Press, 1995.

[12] Costilla C, *A Contribution to Knowledge Communication in Distributed Knowledge-Based Systems*, in *Research into Networks and Distributed Applications* (R.Speth ed.), North-Holland, pp.1153-1162, 1988.

[13] Costilla C, *Sistemas de Bases de Datos. Conceptos, Técnicas y Lenguajes*, ed. Serv. Publicaciones ETSI Telecomunicación-UPM, ISBN: 84-7402-271-1, Madrid, 1999.

[14] DCMI. *Dublin Core Metadata Element Set, V. 1.1: Ref. Desc.* Feb. 2003 http://dublincore.org/documents/2003/02/04/ dces/

[15] Domenig R, Dittrich KR, *An Overview and Classification of Mediated Query Systems*. ACM SIGMOD RECORD, Vol. 28, N. 3, pp. 63-72, Sep.1999.

[16] Doerr M, Hunter J,Lagoze C,*Towards a Core Ontology for Information Integration*,JODI 4(1),Apr.2003

[17] Doan A and McCann R, *Building Data Integration Systems: A Mass Collaboration Approach*, 18th Int. Joint Conf. on Artificial Intelligence (IJCAI), 2003. Aug. 2003, http://www.isi.edu/info-agents/workshops/ijcai03/papers

[18] Eibe S, Costilla C, Menasalvas E y Acuña C, *DAWIS: Una Arquitectura de Integración Web para el Acceso Integrado a Archivos Digitales*. VIII Jornadas de Ingeniería del Software y Bases de Datos, JISBD'03, pp. 583-591, Alicante, Nov. 2003.

[19] *The Mellon Fedora Project: Digital Library Architecture Meets XML and Web Services*, Forthcoming European Conf. on Research and Advanced Technology for Digital Libraries, Rome, Sept. 2002.

[20] Florescu D, Levy A and Medelzon A, *Database Techniques for the World Wide Web: A Survey*. ACM Sigmod Record, Sept. 1998.

[21] Gruber, T. R. *Towards principles for the design of ontologies used for knowledge sharing*. Int. Workshop on Formal Ontology, 1993.

[22] Guarino N and Welty Ch., *Evaluating Ontological Decision with ONTOCLEAN*. Communications of the ACM. Vol. 45 n. 2, pp. 42-47, Feb. 2002.

[23] J. Hammer, J. McHugh, H. Garcia-Molina, *Semistructured Data: The TSIMMIS Experience*, Proc. I East-European WS on *Advances in Database and Information Systems*, ADBIS'97, St.Petersburg, Russia, Sept. 1997

[24] International Council on Archives, *ISAD(G) General International Standardization Archival Description*, 2nd ed, ISBN 0-9696035-5-X, Ottawa 2000. http://www.ica.org/biblio/cds/isad_g_2e.pdf

[25] *International Standard Archival Authority Record for Corporate Bodies, Persons and Families*, 2nd ed. Feb. 2003 http://www.hmc.gov.uk/icacds/eng/ISAAR(CPF)2. pdf

[26] Jacobson, Booch y Rumbaugh, *The Unified Software Development Process*. Adison Wesley, 1999.

[27] Kim, H. *Predicting how ontologies for the semantic web will evolve*. Communications of the ACM. Vol. 45, n. 2, pp. 47-51, Feb. 2002.

[28] Knoblock C, Minton S, Ambite JL, Ashish N, Modi P, Muslea I, Philpot A, and Tejada S, *Modeling web sources for information integration,* Proc. 15th National Conf. on AI and Tenth Innovative Applications of AI Conf. (AAAI-98), pp. 211-218, Wisconsin USA 1998

[29] Kreger H, *Web Services: Conceptual Architecture (WSCA 1.0)*, IBM Software Group, May 2001.

[30] Lagoze C and Hunter J, *The ABC ontology and model*, Journal of Digital Information, JODI, Nov. 2001, 2(2), http://jodi.ecs.soton.ac.uk/Articles/v02/i02/Lagoze/

[31] Landis B, *What is ISAD(G)?,* Description Section, Society of American Archivists, 2000.

[32] LeFurgy, *Levels of Service for Digital Repositories*, D-Lib Magazine, 8(5), May 2002.

[33] Leymann F, *Web Services Flow Language (WSFL 1.0)*, IBM Software Group, May 2001.

[34] Mena E, and Illarramendi A, *Ontology-Based Query Processing for Global Information Systems.* Kluwer Academic Publishers, July 2001.

[35] Mena E, *Observer: An Approach for Query Processing in Global Information Systems based on Interoperation across Pre-existing Ontologies.* Tesis Doctoral, Universidad de Zaragoza, 1998.

[36] Mendelzon A, Mihaila G and Milo T, *Querying the World Wide Web.* Journal of Digital Libraries, (1,1) '97.

[37] Mena E, Illarramendi A, Kashyap V, and Sheth A, *OBSERVER: An Approach for Query Processing In Global Information Systems Based on Interoperation across Pre-Existing Ontologies*, Distributed and Parallel Databases, vol. 8, pp. 223-271, 2000

[38] Open Archives Initiative. *Implementation Guidelines for the Open Archives Initiative Protocol for Metadata Harvesting Protocol* Version 2.0 of 2002-06-14 Document Vers 2002/06/13T19:43:00Z

[39] OAI, *The OAI Protocol for Metadata Harvesting Protocol* Vers. 2.0., Feb. 2003 http://www.openarchives.org/OAI/2.0/openarchivesprotocol.htm

[40] Özsu MT, Valduriez P, *Principles of Distributed Database Systems*, 2nd ed, Prentice-Hall, 1999.

[41] Payette S and Lagoze C, *Flexible and Extensible Digital Object and Repository Architecture,* Second European Conference on Research and Advanced Technology for Digital Libraries, Heraklion, Crete, Greece, Sept. 21-23, LNCS, Vol. 1513, Springer, 1998

[42] Prom CJ and Habing TG, *Using the Open Archives Initiative Protocols with EAD*, JDCL'02, July 13-17, 2002, Portland, Oregon, USA, 2002.

[43] Sáenz J, Costilla C, Marcos E y Cavero J, *Una Representación en UML del Metamodelo Estándar ISAD(G) e ISAAR(CPF) para la Descripción de Archivos Digitales.* VIII Jornadas de Ingeniería del Software y Bases de Datos, JISBD'03, pp. 519-528, Alicante, Nov. 2003. http://sinbad.dit.upm.es

[44] Shirky C, *Web Services and Context Horizons.* IEEE Computer, pp. 98-100, September 2002.

[45] Wand Y, Storey VC and Weber R, *An Ontological Analysis of the Relationship Construct in Conceptual Modelling.* ACM Transaction on Database Systems, pp. 494-528, Vol. 24, N. 4, December 1999.

[46] Wiederhold G,*Weaving Data into Information.Database Programming & Design*, Freeman pubs,Sep'98

[47] Wiederhold G and Genesereth M, *The conceptual basis for mediation services.* IEEE Expert, (12,5), pp.38–47, Sep.-Oct. 1997

[48] Wache H, Vögele T, Visser U, et all., *Ontology-Based Integration of Information –A survey of Existing Approaches.* Proc. IJCAI-01 Workshop: Ontologies and Information Sharing. Seattle, WA, 2001, pp 108-117. http://www.isi.edu/info-agents/ workshops/ijcai01/ papers

[49] Zhong N, Liu J, Yao Y (eds.), *Web Intelligence*, Springer Verlag, 2003

Databases and Information Systems
J. Barzdins and A. Caplinskas (Eds.)
IOS Press, 2005

e-EDU – An Information System for e-Learning Services

Tarmo ROBAL, Ahto KALJA
Department of Computer Engineering, Tallinn University of Technology
Raja 15, 12618 Tallinn, Estonia
tarmo@pld.ttu.ee, ahto@cs.ioc.ee

Abstract. There have been many different educational and learning support systems, some of them have failed, and some have been more or less successful. The general practice among lecturers has been to develop their own systems to manage tasks, study results, materials. The e-EDU system provides a general approach to solve these problems, taking into account the real needs of the lecturers of TUT. The e-EDU system proposes a new and localized approach as a service to enliven the daily studies at the university as well as offers means for distance learning.

1. In Need for New Approach

Over the years, universities have developed several different kinds of information systems, including systems that were supposed to support learning processes via providing materials, links and other information valuable during an educational course.

The majority of universities' information systems concentrate on the management of study process results, students and the subjects they have performed. These kinds of systems actually deal with the students' information at the beginning of the course, when students have to declare courses in the inception of a new semester. The next phase, when these systems deal actively with students' information is the end of the term, when the results need to be entered into the system. As we see, usually there is no support service for the course itself. The general practice has been that each lecturer has his or her own system, how to manage the course: students, their tasks and results, course materials. It is rather a lecturer-central system than an online service.

With the WebCT, Lotus LearningSpace [1,2] and other systems alike, these problems can be revealed. However, in most cases they are rather difficult to use and do not satisfy the real needs of the academic personnel of a particular university, country, region or an educational system.

The e-EDU stands for an online learning support service being developed at the Department of Computer Engineering at Tallinn University of Technology (TUT). Its purpose is to provide a supporting service for students and lecturers. Two lecturers, who had similar systems on their own, invoked the e-EDU development. Today, the system, which grew out of these two early versions, is being used as an active learning support system in seven subjects taught by the department. The subjects are Informatics I and II, Computers I and II, Digital Systems Design and Test, Testing of digital systems, Fault-tolerant systems. These courses are being taught to more than 200 students each semester. The system has been under development from the year 2002. The e-EDU takes into account the real needs of the lecturers of the department and in the nearest future will compete with WebCT as the basic learning system for distance learning. It's not yet another WebCT! We have to take an advantage of tomorrow's technologies already today!

2. Access from Everywhere

One of the advantages of the e-EDU system is that it is a web-based system with many interfaces. Therefore, it can be accessed from almost everywhere (Fig. 1.), regardless of whether you are using a PC, Workstation, PDA, or Laptop. The system is built up on the assumption of service-based architecture in a way that several interfaces are supported. For example, lecturers do not need to have any student information on paper, instead they have their PDA, Laptop or other portable device and can have an access to the IS via LAN or WiFi. The majority of premises of Tallinn University of Technology are covered with WiFi network, which is a good prerequisite for such a system development and exploitation.

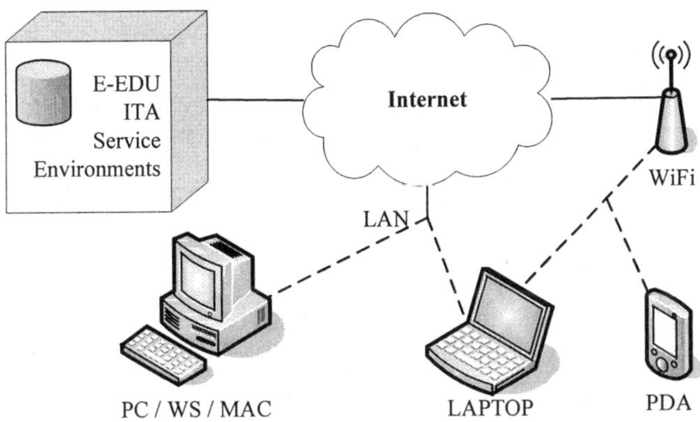

Figure 1. Ways to access the service

3. System Functionality

The functionality of the e-EDU system is based on the needs of students and the academic personnel. The main goal of the service is to decrease and mitigate the information processing of student data performed by the academic personnel. On the other hand, as technological potential improves, we need to consider today's and tomorrow's means for making the study process more effective.

As far as the system development is concerned, the concepts of web applications development discussed in the author's earlier paper: The Rational Unified Process with the "4+1" view model of Software Architecture – a Way for Modelling Web Applications [3], are being applied. The system's architectural design is described using the Unified Modelling Language, the graphical user interface as mock-up pictures.

The e-EDU system has three major views, according to the users, that are implemented via different web services. The views are as follows:

- The students' view, as a web application edu.pld.ttu.ee;
- The teachers' view, as a part of the department's intranet application ITA;
- The administrator's view, as a part of the ITA intranet application;

The functionality provided by the e-EDU can be grouped as follows (Fig. 2.):

- User authentication via a general authentication system or using the Estonian Citizen ID-card;
- Course registration management;
- Registered students management;
- Course material management;
- Assignments and students' results management;

- Communication between academic personnel and students;
- Tests generation and crediting;
- General course management – administrators only;
- Archived data management – administrators only;
- System back-up management – administrators only.

As we are dealing with a system that is rather laborious and risky to implement fully, system elaboration has been divided into development cycles by functional modules according to the demand for certain functionality. The development is based on the RUP technology. The system development started off with the most critical part: assignments and results management. The next development cycle added the management system for course materials. Today, a third cycle of development of this module has been finished. In the nearest future, other functional modules will be developed in the following order: announcements subsystem, test generation and evaluation subsystem, communication subsystem for data exchange with other e-learning systems. Further development of already existing modules will continue, in order to add functionality step-by-step, improve the systems functional behaviour and to make the applications user-friendlier.

Both, the e-EDU web portal and the ITA intranet application are represented over the Internet using the HTTPS protocol with applied security certificates to attain more secure connection than the regular HTTP protocol would provide. Also, organizational measures are applied to enforce the security. As we see later in this article, the lecturer can perform the majority of activities with the course; however, there are operations, which can be carried into effect only by an administrator. System development and testing are rigorously separated from the operative "LIVE" system. As security descriptions are not the aim of this article, we will not discuss them in further detail.

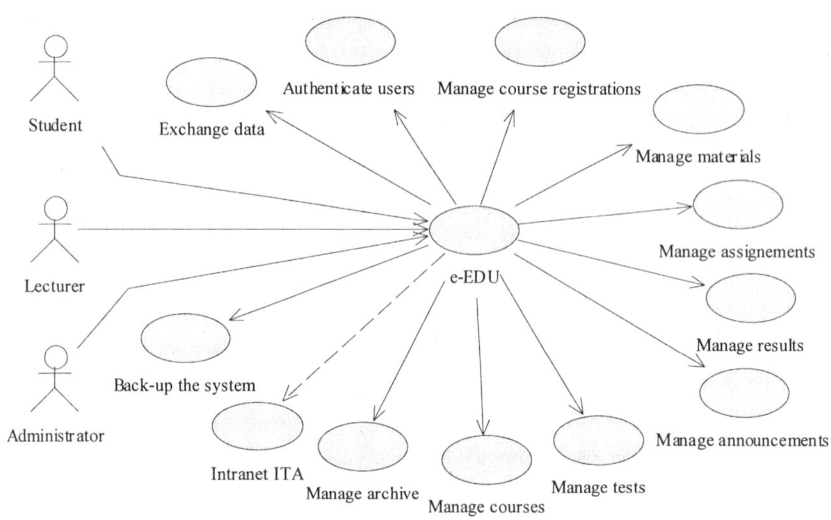

Figure 2. Main functional groups of the e-EDU system

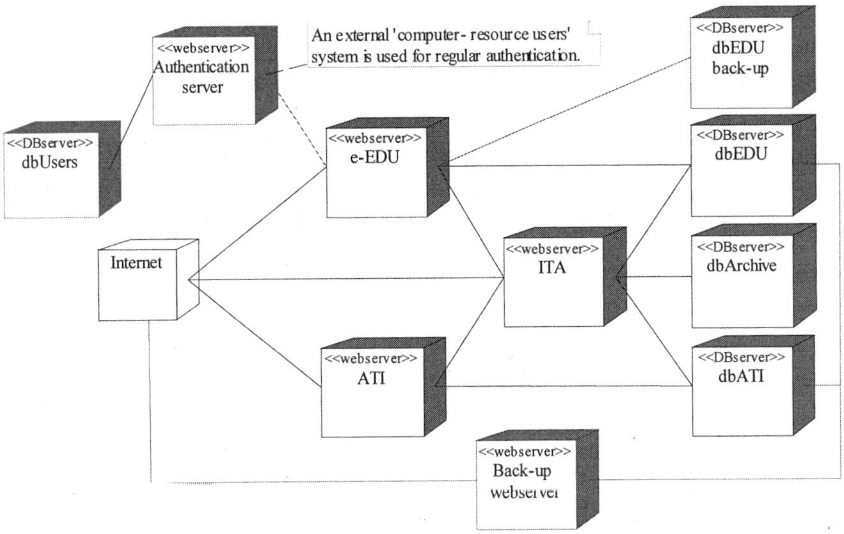

Figure 3. Architecture of the e-EDU system

3.1 The Students' View

The students' view is implemented as a standalone web portal edu.pld.ttu.ee accessible all over the world via the Internet. The system has a built-in language support, so that users can select the language in which they prefer to interact with the system already at the stage of login. The default language is Estonian. To be able to work in the web application, users first need to authenticate themselves, which can be done in three different ways:

- User enters his or her account information for the computer resources at The Computer Centre of TUT and can have an automatic access to the system;
- User enters his or her e-EDU username and password;
- User identifies himself or herself using the Estonian Citizen ID-Card. A Personal Key Infrastructure (PKI) [4] is used for identification.

After a successful authentication, users can access the functionality of the students' view of e-EDU. They can manage their personal data in the system, for instance change their e-mail address, customize the e-EDU application for their preferences (preferred system language, colours, remainder, whether they want to get the announcements also to their e-mail, etc), subscribe to subjects and of course, manage the courses they have registered for.

In order to be able to operate with a certain course, users are required to register for it. During the registration process they have to choose the lectures and the study group in which they will take part of the lessons. This allows the academic personnel to know the actual distribution of students of a particular course. On the other hand, students can freely choose another time, study group or lecturer to take a course with. After a successful registration, users can (Fig. 4.,5.):

- Take their personal assignments and view the state of taken assignments;
- View the results of the assessments;
- View reports about their success compared to other course students;
- Surf the study materials, both, free and password protected materials;
- Take online tests;
- Check-in for a lesson online (computer classes only);

- Activate the remainder function, not to miss the deadlines for the assignments;
- Take part in the subject communication board;
- Read course announcements, descriptions (extended course information cards) and calendars.

Each student gets his or her own personal version of a task. After a successful registration procedure students are able to take their assignments. If there is more than one assignment in a course, lecturers have the possibility to set preconditions to assignments release. For example, in order to take *task 3*, student has to have *task 2* defended and its status set to "OK". This will unable students to perform tasks, which have prerequisites, and the acquiring of needed knowledge in the right order can be assured more accurately. At the same time, it acts as an external motivator – each task has its deadline and preconditions. Moreover, the length of the term is limited (at TUT it is 16 weeks), which leads to restricted time limitations. Each assignment can be viewed as a web page, printed out or saved as a PDF-file. The assignments are presented as personal forms: it has a student's name, task body, variable task part, task deadline, date of task taking and the current status of the assignment among with other information on it. Also, barcodes can be generated for the tasks. For example, students get their personal home tasks (for instance a blank form with student's name, task information and barcode on it), print them out with a barcode on it and submit them to the lecturer. The barcode then makes the procedure of evaluating and results entry easier and more efficient.

The tests system acts like the assignments subsystem. If there are tests put up by the lecturer, and they are made available, students can take them. Tests can be either online web forms for filling in or in a printable format, labelled with student information and barcode.

General course description represents the extended information card of a subject. The course calendar shows the topics for an active study week, highlighting them among the rest of the records. Course materials are represented by their type, whether they are materials given during a lecture, materials needed in tutorials or labs. A search form is also provided.

Each course is provided with a communication board, where students and lecturers can exchange knowledge and discuss problems with each other. It acts more or less like a public forum, with the exception, that each course has its own independent section.

The *e-check-in* is a feature, which is only usable with a direct access to computer resources. It can be used in daily study process as well as with distance learning. For instance, lecturer and students of the distance learning have agreed to be online at the same time to hold an online lecture. In daily study the e-check-in feature allows to check the presence of students in a lecture without sending out a special list with names. Teachers can set the timeframe for e-check-in service or manually open the e-check-in for a period of time, when participants can sign their presence. The system has a built-in feature to allow signing only from certain address range (*i.e.* IP-addresses), if needed. For example, daily study e-check-in can take part only in the computer classes of TUT.

The functionality of the students' view is elaborated taking into account the needs of daily study students as well as the needs of those of distance learning. The central line is to keep everything as simple as possible, providing as much information as needed. In comparison with WebCT and other e-learning systems, e-EDU is more flexible, more automated, user-friendlier and which is most significant, it is in Estonian language. The latter is a key success factor in exploiting such a system, making it as user friendly as possible and at the same time escaping the confusion in users.

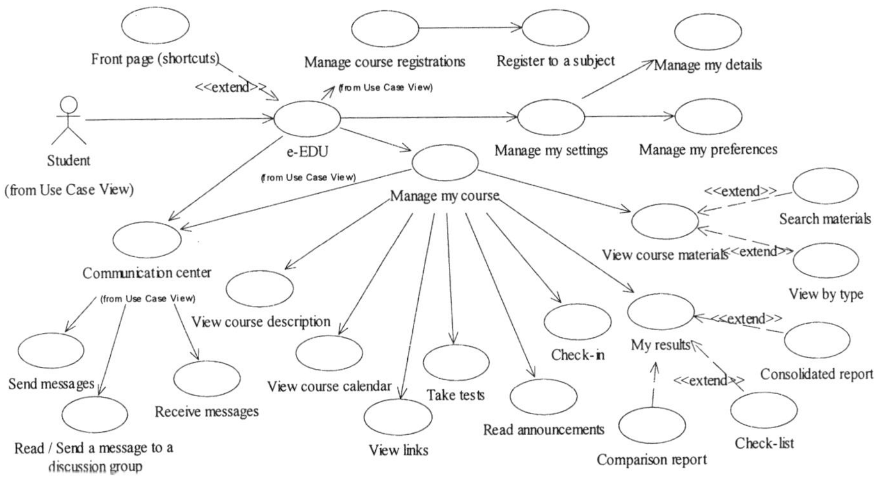

Figure 4. Functionality of the student's view

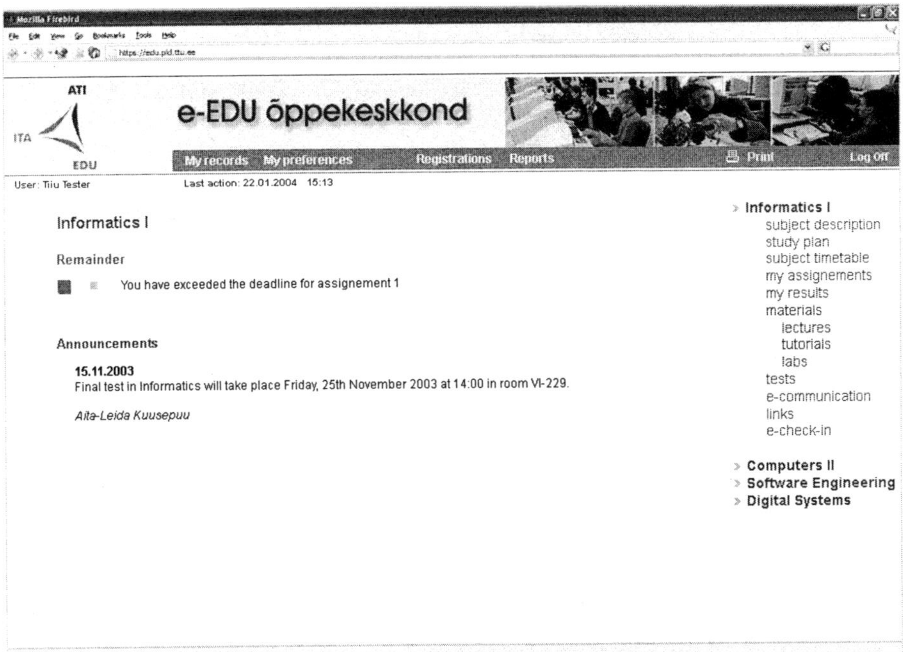

Figure 5. The student's view

3.2 The Teachers' View

The teachers' view is a part of the department's intranet called ITA (Fig. 6.). The section of ITA, which deals with the e-EDU system, has also a secondary interface to work with PDA-s.

Through the system, lecturers can (Fig. 7.):

- Easily see students who attend their courses,
- Upload, delete and reassign course materials,
- Evaluate students' assignments,
- Carry out automatic assessment,
- Send announcements to the course participants,
- Compose and evaluate tests,
- Generate statistical reports,
- Manage examination information in the Examination Centre (Sessikeskus),
- Import and export data in CSV format (for example exchange data with WebCT, Excel, OpenOffice)
- Manage course descriptions, course calendar and schedule.

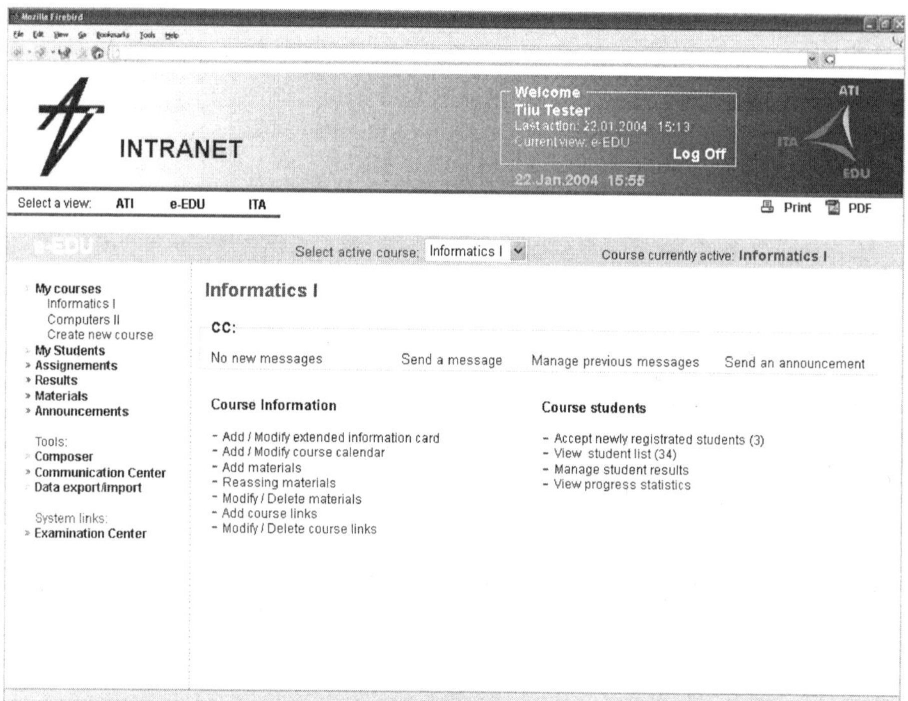

Figure 6. The teacher's view

Although each course has its general description in the central registry of the university, each year minor modifications are being made to the course program and actual content. Therefore, lecturers are ordered to compose extended course content and schedule plans, called extended course information cards. The idea is to provide a description of a course, sufficient for students, so that they are able to make rational subject selections. On the other hand, extended course information cards keep lecturers "on the track" assuring that students attain the knowledge needed. In addition to extended cards, course calendar provides an overview of the course structure as a timeline. There are two different approaches to choose:

- Timeline as academic hours: describes how many academic hours are spent on each topic;

- Timeline as lecture topics: describes what topics will be discussed in a concrete lecture.

The timeline itself shows the logical order of covering the topics during the whole course.

The most significant part of the system is the assignments management. The subsystem uses XML-based documents description technology. Tasks are described as XML-based documents; a coherent style is assigned using XSLT-s. As a result, the assignment management system gives us the following benefits:

- Due to the use of XML and XSLT, tasks have a general coherent outlook, but may have a different style, if needed,
- Tasks can be represented in various forms: HTML, PDF, etc.
- The stationary task body and variable version information are kept separately until a student takes an assignment,
- The distribution system guarantees even allocation of different assignment versions,
- Each student gets his or her own personal version of an assignment, composed from the task body and variable task part, which can be easily stored in an archive,
- Teachers do not have to print out and distribute tasks; everything is organized by the assignment management service and is available in e-EDU.

The exploitation of XML-based document description technology also provides simple interfaces for communication with other systems.

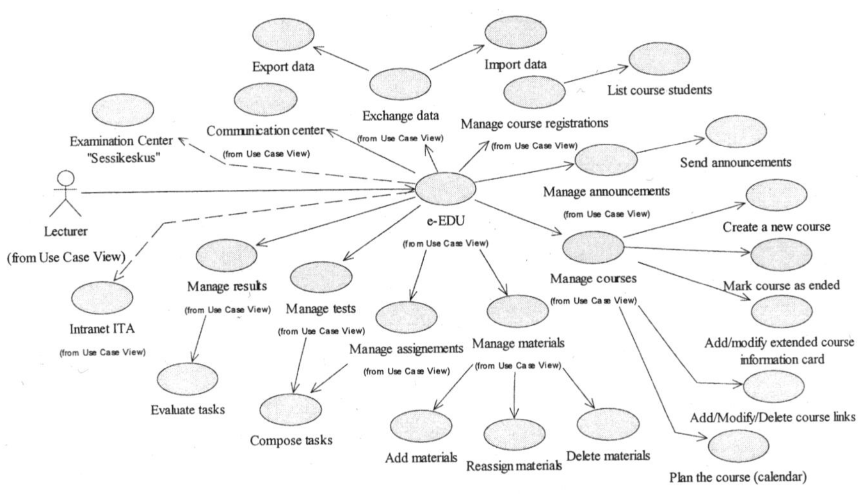

Figure 7. Functionality of the teachers' view

The composition of assignments and tests is implemented also by a web application, where the lecturer has to make selections to provide the system with needed information (i.e. number of task variants) to produce the basic template. After the basic template has been generated, the task information can be entered into the database. By default the tasks are unavailable and lecturer has to give a special order to the system to make the tasks available. The task sub-system generates needed XML structures as well as makes necessary entries into the database.

Results management is common for both, assignments and tests. If a student has taken an assignment or a test, his or her name appears automatically in the task's results list. This way the system assures the integrity of data and implementation of the rules, decrementing the workload of academic personnel. Of course, there is a possibility to add tasks to students or predefined student groups as well.

The default assessment types are:

- Defended and passed;
- In proceeding;
- Cancelled;
- Indefinite.

When a student takes a task, it is in an indefinite state, after the evaluation it may have other states. The final state is either *Defended and Passed* or *Cancelled.* With the cancellation, student has to take the task again. The system assures, that the version does not mach with the cancelled one.

Teachers are able to upload materials in any form they like; it does not matter whether the materials are HTML-files, PDF-files, PowerPoint slide presentations or other types of file. While uploading, they need to specify, whether the materials are for public use or are secured (accessible only through e-EDU). The files can also be classified as lecture, tutorial or lab materials, to provide a better overview of them.

The aim of statistical reports in the teachers' view is to provide a general overview of the progress of students, which is useful and needed information for a lecturer. It enables to adjust the course according to the success of students during the teaching process.

The teachers' view provides a lot of useful tools to make the teaching process easier, simpler and lessen the workload, but it can only provide the functionality built in it, which is rather general and may not correspond to all of the needs of a particular lecturer. However, the system is developed by the actual needs of the department's lecturers and therefore can answer the demands better than any other outsourced software.

3.3 The Administrator's View

The role of an administrator in the e-EDU system is principally to act as the supervisor and technical assistant for the academic personnel as well as for the students. As there should be some kind of superintendence implemented over the two other user groups, the administrator is the only actor, who has the rights to set up new and erase old courses, add, delete and modify data concerned with users, both, students and academic personnel (Fig. 8.). Moreover, the administrator's tasks are not limited with simple course and user management. They can be extended to auditing certain actions taken by the academic personnel. For instance, administrators may have the right to audit and certify the extended course information cards and course calendars created and modified by the academic personnel.

It is evident that with this kind of systems, we are most probably interested in several types of statistics over the applications usage, or even more, there might be a need to get information about the applications usage for a concrete user. Obviously, this sort of information might be delicate and everyone should not have access to such a data.

One year after the course has ended, administrator transfers the course information into archive. After archiving, the data will be available only for administrators, who can excerpt reports, if needed. This approach frees the active server from huge data loads, on the other hand, provides a mechanism, where all the data remains obtainable.

Despite of the technological possibilities and regulations of the department, the role of the administrator should be as minimal as possible, incrementing the independence and responsibility of the academic personnel. Divided responsibility is more or less irresponsibility. As we have shown, the role of an administrator is mainly to function as a superintendent and plays a crucial role in such systems.

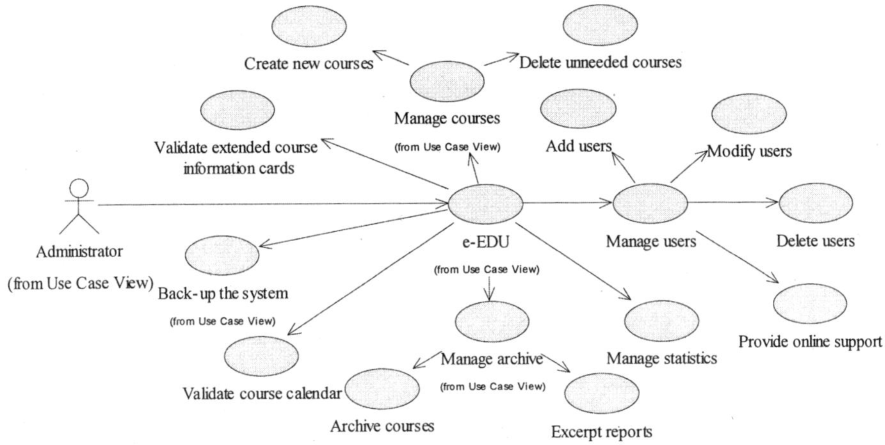

Figure 8. Functions of an administrator

3.4 Overview of the Physical Implementation

The e-EDU system implementation involves many database and web servers as can be seen on Fig. 3. The system is implemented using mainly Apache web servers with PHP modules (currently version 4.x) added [5]. The database servers are implemented using the MySQL database engines (currently versions 3.23.x and 4.0.x) as the most popular open source and the fastest growing database systems in the industry [6]. MySQL has proved to be suitable for such systems, however presently there are some restrictions: there are no triggers, functions or possibilities to run sub-queries. These restrictions do not hold back the development, as there are other possibilities to implement the needed functionality, and hopefully these restrictions will be eliminated by future versions of MySQL.

The main line has been and will be to keep the system as simple as possible, using the available freeware and implementing the functionality mainly by PHP, Javascript and MySQL.

4. Future Work

In the future, further development of the system will be continued, in order to add more functionality, make the system more efficient and more open for integration with other information systems, first and foremost with the department's information systems but also with the university's systems. It has been planned to develop a communication link between the e-EDU system and the study information system (OIS, http://ois.va.ttu.ee) of TUT. Moreover, the e-EDU can be easily integrated with the Career Service Web Portal (CSWP, http://www.ttu.ee/karjaar) of TUT. In the future it would enable the users of e-EDU to access the services offered by CSWP without registering for yet another username and maintaining another password.

In the next version we have planned to introduce the Estonian Digital Signature for the communication between lecturers and students to enable digital signing of documents. It is highly likely that the future also brings M-Services (Mobile Services) to e-EDU, allowing, for instance, students to access their data by just sending a SMS or accessing the portal using the WAP.

5. Conclusions

The rapid IT development has provided us with tools to improve our lives, our possibilities to teach and learn. Web applications are one of the many possible approaches. Because of their open nature, users usually do not need anything more than just a web browser and access to the Internet - web based applications have become very popular, also in the field of education.

The e-EDU system provides a new and localized approach to enliven the daily studies at the university. In addition, it provides means for distance learning. As has been shown, the e-EDU system lessens workload of teachers, providing general means for tasks, materials and results management among with other tools. There is no need for every teacher to manage his or her own mini-systems in Excel, Access or in any other program. Students have a continuous overview of their progress and in case of errors made by a teacher, can quickly announce of it.

There have been many steps towards implementing today's technological means in providing education and e-education. Many of them have failed, whenever the systems have been too complex to manage or just from lack of interest by users. The e-EDU system is making its first steps and until now has been successful.

Acknowledgements

The project has partly been supported by the Estonian Scientific Foundation grant no. 5766. We would like to take this opportunity to thank all those people who have collaborated and supported this project: Margus Kruus, Elmet Orasson.

References

[1] WebCT.com, http://www.webct.com/ (03.12.2003)
[2] LearningSpace, http://www.lotus.com/products/learnspace.nsf/wdocs/homepage/ (03.12.2003)
[3] Robal, T., Viies, V., Kruus, M. The Rational Unified Process with the "4+1" View Model of Software Architecture – a Way for Modelling Web Applications. In.: Proceedings of the Fifth International Baltic Conference, BalticDB&IS 2002, Tallinn, June 3-6, 2002, Vol. 2., pp.119-132
[4] Kalja, A., Vallner, U. Public e-Service Projects in Estonia. In.: Proceedings of the Fifth International Baltic Conference, BalticDB&IS 2002, Tallinn, June 3-6, 2002, Vol. 2., pp. 143-154.
[5] PHP Hypertext Preprocessor, http://www.php.net/ (23.11.2003)
[6] MySQL, http://www.mysql.com/ (23.11.2003)

Author Index